Introduction to Quantum Electronics and Nonlinear Optics

Vitaliy V. Shtykov • Sergey M. Smolskiy

Introduction to Quantum Electronics and Nonlinear Optics

 Springer

Vitaliy V. Shtykov
National Research University
Moscow Power Engineering Institute
Radio Engineering Fundamentals Dept.
Moscow, Russia

Sergey M. Smolskiy
National Research University
Moscow Power Engineering Institute
Radio Signal Generation and Signal Dept.
Moscow, Russia

ISBN 978-3-030-37616-1   ISBN 978-3-030-37614-7   (eBook)
https://doi.org/10.1007/978-3-030-37614-7

This Springer imprint is published by the registered company Springer Nature Switzerland AG
The registered company address is: Gewerbestrasse 11, 6330 Cham, Switzerland

The authors dedicate this book to the 90th anniversary of the National Research University "Moscow Power Engineering Institute," which is the alma mater for both authors and at which the authors have worked during their full scientific and pedagogic lives.

Preface

The end of the 1940s and the beginning of the 1950s were marked by impetuous implementation of fundamental scientific achievements in everyday practice. The results of such implementation provided evidence that electronics development could be the key to application of the very latest ideas in fundamental science. These expectations proved to be totally correct.

It is impossible to present modern electronics and radio physics without semi-conductor devices, quantum generators, and optical–electronic devices. Application of nanotechnologies is the order of the day.

For successful work under these conditions, engineers are needed who are able to conceive ideas that, until recently, were the domain of "pure" theoreticians. When necessary, they should be ready to conduct their own theoretical and experimental investigations of physical phenomena. The new specialty "Radio Physics and Electronics" was introduced in the Department of Radio Engineering at the National Research University "Moscow Power Engineering Institute" (MPEI) in 1955, with deep physical–mathematical training on the initiative of the academician V. A. Kotelnikov to train such experts. This subsequently allowed MPEI graduates to be effectively involved in research on quantum electronics.

This book consists of the lecture materials that have been delivered in various years to these students in the Department of Radio Engineering. The name of this discipline has been changed several times; now it is called "Quantum Electronics and Nonlinear Optics." In different variants of this lecture course, students learn all about the fundamental aspects of linear and nonlinear interactions of an electromagnetic field with a material medium.

The material in this book is based on the general natural scientific and mathematical disciplines that are included in the educational curricula of classical and technical universities, as well as courses for professional disciplines: electrodynamics, circuit and signal theory, oscillation theory, etc.

The authors hope this book will help students in electronics and physics specialties to successfully bridge the gap between disciplines in the natural sciences and engineering.

During the preparation of this book, many experts and university professors provided the authors with invaluable help. We would like to thank our colleagues in the Department of Radio Engineering at MPEI for their enduring attention to our work, valuable discussions, and unwavering support. We are also grateful to our reviewers. Their qualified estimations, advice, and comments helped us at the final stage.

Moscow, Russia   Vitaliy V. Shtykov
                                                                                                      Sergey M. Smolskiy

Acknowledgements

We thank our teachers and our colleagues at the National Research University "Moscow Power Engineering Institute" for detailed discussions of several sections of this book, especially Professors V.A. Kotelnikov, A.M. Nikolaev, N.N. Fedorov, G.D. Lobov, M.V. Kapranov, G.M. Utkin, V.M. Bogachev, V.N. Kuleshov, V. I. Siforov.

We also offer our sincere gratitude to Dr. Sam Harrison, Editor in the Physics Books group at Springer, for his inestimable help at the stage of final manuscript preparation. Authors thank Ms. Pearly Percy Joshuajayakumar, Project Manager, Content Solution, SPi Global, for very accurate corrections of the final manuscript.

Vitaly V. Shtykov
Sergey M. Smolskiy

Introduction

In encyclopedias, quantum radio physics is identified with quantum electronics. The latter is defined as the scientific–technological direction related to investigation of methods of amplification, generation, and conversion of electromagnetic waves in the range of units from megahertz to hundreds of terahertz, using stimulated quantum transition or nonlinear interactions of radiation with a substance.

Quantum electronics was born from radio spectroscopy, which began to develop rapidly at the end of the 1940s. At that time, physicists–researchers had, at their disposal, a rather rich arsenal of ultrahigh-frequency-range devices. Because of this, extensive studies of the absorption spectra of molecules were performed in the millimeter and centimeter wave ranges of electromagnetic waves. It was exactly these investigations that led to creation of the first quantum generator. Now, more than half a century separates us from that moment.

Over time, the area of quantum electronics widened so much that the term "quantum radio physics" already seems more appropriate to us. Perhaps, in the near future, radio and electronics engineers will have to deal with nanoelectronics; then investigation of electromagnetic wave interactions with these structures will become an essential part of quantum radio physics.

Against the background of fast scientific–technological progress, the subject of quantum radio physics as an educational discipline sustains the study of radiation interactions with a substance under conditions where quantum postulates must be used. Therefore, its bases are quantum theory, classical electrodynamics, and oscillation theory, which are sufficient for solution of most applied problems. However, for solution of some problems, classical electrodynamics needs to be reformulated in the quantum setting.

Eq. 1. Historical Information

In 1916, Albert Einstein[1] was the first person to prove the possible existence of induced (forced) radiation and pointed out its coherence. In 1923, Paul Ehrenfest confirmed Einstein's conclusions. In 1927–1933, Paul Dirac[2] created the quantum theory of stimulated radiation. Conditions for stimulated radiation detection and the means for its realization were formulated by Rudolf Ladenburg and Hans Kopfermann (from Germany) in 1928. In 1939, Valentin Fabrikant[3] (from the USSR) was the first person to observe and investigate amplification of light passing through gas discharge plasma. Stimulated radiation in the form of short radio pulses was first discovered by the American physicists Edward Purcell[4] and Robert Pound in 1950.

In 1951, Fabrikant and his colleagues submitted an application to patent a method for amplifying electromagnetic radiation by passing it through a medium with an inverted population of quantum levels. Unfortunately, their application was not published until 1959 and thus was unable to have any favorable effect on research into quantum generator creation.

Quantum electronics came into existence as a scientific–technological direction at the end of 1954. At that time its theoretical fundamentals were laid and the first device—the quantum molecular generator—was created.

---

[1]Albert Einstein (1879–1955) was born in Germany. He was a physicist–theorist and one of the founders of modern physics. In 1921, he received the Nobel Prize in Physics. He had a colossal capacity for work. He authored more than 600 publications on very different themes. He made significant contributions to the creation of quantum mechanics, the development of statistical physics, and cosmology.

[2]Paul Dirac (1901–1984) was a British physicist and one of the founders of quantum mechanics. He received the Nobel Prize in Physics in 1933. He became a member of the Royal Society in 1930. He was awarded the Royal Medal in 1939 and became a foreign member of the US National Academy of Sciences in 1949. He worked as a physics professor at Cambridge University until his retirement in 1968.

[3]Valentin Fabrikant (1907–1991) was a doctor of physics and mathematics sciences, and a professor. He was an academician at the Academy of Pedagogical Sciences and a laureate of the State Award of the USSR. In 1930, on the recommendation of Sergei Vavilov, he began teaching in the Department of Physics at the National Research University "Moscow Power Engineering Institute" (MPEI). He worked as the head of this department from 1943 to 1977. For his pedagogical and research activity he was decorated with four USSR orders and two Sergei Vavilov Gold Medals. In 1992, the Academic Council of MPEI renamed the MPEI Department of Physics after him.

[4]Edward Purcell (1912–1997) was an American physicist, who shared the 1952 Nobel Prize in Physics for his independent discovery (published in 1946) of nuclear magnetic resonance (NMR) in liquids and in solids. NMR has since become widely used to study the molecular structure of pure materials and the composition of mixtures. Further information can be found in the literature on magnetic resonance imaging (MRI).

What hampered the progress of quantum devices earlier? To clarify the situation we remind the reader of some of the principles on which quantum electronics is based.

As we have already mentioned, the phenomenon of stimulated radiation (induced transition) was introduced by Einstein. To describe the thermodynamic equilibrium between a field and atoms, he assumed that an atom located in an induced state could give back its energy in the form of radiation (a quantum) in two ways.

The first way is spontaneous radiation when the atom radiates energy spontaneously. Before the appearance of quantum mechanics, the phenomenon of spontaneous radiation was described classically. The atom was considered as an oscillator with friction whose amplitude decreased over time. All traditional light sources generate light as a result of spontaneous radiation. Therefore, the phenomenon of spontaneous radiation had already been known for a long time by scientists working in the field of optical spectroscopy.

The second way by which an atom can give back its energy is induced transition (stimulated radiation). The phenomenon of stimulated radiation consists in the fact that if the excited atom state interacts with an external field, two indistinguishable quanta with the same frequencies (energies) and wave vectors (impulses) will occur as a result. In 1927, Paul Dirac was the first person to demonstrate this important property of stimulated radiation.

It is absolutely clear that if all of the atoms are in an excited atom state, this particle system will amplify the radiation. We have no doubt that some scientists clearly understood this at the end of the 1930s. Nevertheless, no one demonstrated the possibility of light generator creation. This may be surprising, but optical quantum generators could, in principle, already be created even at that time. However, the discovery made by Fabrikant passed unnoticed. The main reason was the absence of a practical need for such generators. In those years, researchers in radio electronics had only just began to become familiar with decimeter and centimeter waves.

After the end of World War II, when radio spectroscopy began to develop rapidly, the real prerequisites for quantum generator development were established. The scientists working in the field of radio spectroscopy laid the foundations of quantum electronics. This was connected to a series of favorable circumstances, which was not typical for scientists working in the field of optical spectroscopy.

To clarify the sense of difference we will work out the balance equation of the two-level quantum system. Let the population of the lower atom level be n_1, and let the upper one be n_2. Transitions from 2 down to 1 ($2 \rightarrow 1$) are caused by spontaneous and induced actions, whereas transitions from 1 up to 2 ($1 \rightarrow 2$) are caused only by induced actions. In the dynamic equilibrium state, the number of transitions down is equal to the number of transition up during a time unit:

$$n_2(A_{21} + B_{12}u) = n_1 B_{12}u, \qquad \text{(Eq.1)}$$

where u is the spectral density of radiation energy [J/m^{3}Hz], A_{21} is the probability of spontaneous transitions, $B_{12}u$ is the probability of induced transitions, and $B_{12} = B_{21}$ are the Einstein coefficients. Equality (Eq. 1) allows determination of a connection between transition probabilities as follows:

$$A_{21} = B_{12}u(n_1/n_2 - 1).$$

Under conditions of thermal equilibrium, the particle distribution on the levels obeys the Boltzmann law:

$$n_2 = n_1 \exp\left(-\hbar\omega_{21}/k_B T\right);$$

hence,

$$A_{21} = B_{12}u\left(\exp\left(\hbar\omega_{12}/k_B T\right) - 1\right).$$

In the optical band, $\hbar\omega_{12}/k_B T \gg 1$, and in the ultrahigh-frequency range, $\hbar\omega_{12}/k_B T \ll 1$. Thus, at equal energy density in the optical range, spontaneous transitions predominate ($A_{21} > B_{12}u$), and in the ultrahigh-frequency range, induced transitions predominate ($A_{21} < B_{12}u$). That is why in the area of optical spectroscopy we need not take into account stimulated radiation.

In the 1950s, physicists working in the field of microwave spectroscopy began to use molecular beams to increase the resolving capability and sensitivity of radio spectrometers. When the gas pressure decreases, the spectral lines of absorption become narrower but the intensity reduces. It can be increased if a larger number of molecules are situated either in the lower state or in the upper state. It is clear that if molecules are situated in the upper state, this system will amplify radiation.

Everybody knows that each amplifier can be transformed into an oscillator. For this, feedback is necessary. The theory of the usual oscillators in the radiofrequency range is well developed. To describe such oscillators, for instance, the concept of the negative resistance of conductance is introduced. Quantum generators are systems distributed in space; therefore, a medium with negative specific conductance is a similar required element. The self-excitation condition consists in the requirement for loss compensation in the cavity, which, together with an active medium, forms a quantum generator.

In 1953, Professor Joseph Weber from Maryland University published a brief paper[5] in which he showed that if it was possible in any way to invert the population distribution of energy levels, then we could realize an absolutely new method of signal amplification. This was the first such publication to become available, although this problem had already been discussed by different experts in 1951.

---

[5]J. Weber, Trans. Inst. Radio Eng., **PGED-3**, 1 (1953).

In 1954, James Gordon, Charles Townes, and Herbert Zeiger from Columbia University—as well as Nikolay Basov[6] and Alexander Prokhorov[7] from the Physical Institute of the Academy of Sciences of the USSR—were able to provide (independently of each other) amplification and to obtain oscillations on a frequency of 23.867 GHz, using a molecular beam of ammonia. This device was called a "maser" (an acronym of "microwave amplification by stimulated emission of radiation"). For this discovery, Townes, Basov, and Prokhorov shared the Nobel Prize in Physics in 1964.

In 1956, Nicolaas Bloembergen[8] offered an approach to creating inversion based on application of additional radiation (which was called pumping radiation). It permitted users to move on to applications using solid bodies instead of gas in quantum amplifiers and generators. The number of publications in this area of physics began to grow rapidly, and practical devices using the phenomenon of electron paramagnetic resonance were created. We shall discuss this issue below.

In 1958, Arthur Schawlow[9] and Charles Townes published a paper titled *Infrared and Optical Masers*,[10] and in 1960, Theodore Maiman implemented their idea practically with the creation of a pulsed optical quantum generator using a ruby monocrystal. It caused a worldwide sensation, since, on this occasion, the discovery came to light not only in the scientific world but also in the mass media. It is clear now that everything evolved in a natural way: a quantity (a generation frequency) just proceeded to a new quality (a coherent light source).

Approximately 1 year later, Ali Jawan described obtaining continuous generation using gas discharge in a mixture of helium and neon. The numbers of publications, seminars, and conferences on optical quantum generators began to snowball. During a short period, a large number of different optical quantum generators were created.

---

[6]Nikolay Basov (1922–2001) was a Soviet physicist and one of the founders of quantum electronics. In 1964, he was awarded the Nobel Prize in Physics. He devised the idea of semiconductor applications for optical quantum generators in 1959 and developed methods for creation of different types of semiconductor lasers. He completed a series of investigations of the use of powerful pulsed lasers in ruby and neodymium glass, creation of quantum timer standards, and interactions of powerful radiation with a substance.

[7]Alexander Prokhorov (1916–2002) was a Soviet physicist and one of the founders of quantum electronics. In 1964, he was awarded the Nobel Prize in Physics. He was a member of the American Academy of Arts and Sciences. He was appointed as the main editor of the *Great Soviet Encyclopedia* in 1969 and received honorary professorships at Delhi University in 1967 and at Bucharest University in 1971. After being appointed as the deputy head of the Lebedev Physical Institute of the Academy of Sciences of the USSR in 1973, he continued his investigations of laser physics, including application of lasers for multiquantum processing and for thermonuclear synthesis.

[8]Nicolaas Bloembergen (1920–2017) was a Dutch–American physicist and shared the 1981 Nobel Prize in Physics with Arthur Schawlow and Kai Siegbahn.

[9]Arthur Schawlow (1921–1999) was an American physicist. He is best remembered for his work on lasers, for which he shared the 1981 Nobel Prize in Physics with Nicolaas Bloembergen and Kai Siegbahn.

[10]A.L. Schawlow, C.H. Townes, Phys. Rev., **112**, 1940 (1958).

At that time, the name "laser" (an acronym of "light amplification by stimulated emission of radiation") was assigned to them.

The appearance of lasers had a fundamental influence on the development of the engineering sciences and society as a whole. The best scientific forces were involved in research in the field of coherent optics. Investigations were initiated and were directed toward development of absolutely new element bases (photoreceivers, modulators, deflectors, etc.) and new systems (optical radar, communication and navigation systems, etc.). The discovery of lasers is commensurable, in terms of its sense and significance, with the creation of transistors. It is sufficient to state that without quantum generators, it would have been impossible to introduce modern information technologies.

At present, quantum generators operate in the wavelength range of about 200 nm to about 0.1 mm. The output power in the continuous mode reaches hundreds of kilowatts, and that in the pulsed mode reaches several gigawatts.

Eq. 2. Methodical Conception of this Book

Rapid development of quantum electronics began in 1960 and, soon afterward, literature on quantum generators and amplifiers started to appear on the book market. However, the publications that appeared in those first years could not be used as textbooks for engineer training. There were several reasons for this.

The main reason was that in that initial period, the description of laser theory was based on the concepts of spontaneous and induced transitions as the elementary acts of energy exchange between an electromagnetic field and a medium. This permitted users to obtain balance equations for energy level populations in the presence of an electromagnetic field and a pumping source, based on the concept of transition probability in the time unit, starting from simple and visual presentations. In some cases, photons were included in the balance equations as well (e.g., the equations published by Statz and deMars in 1960[11]). However, upon application of the balance equation method, the phase relations were fully lost. That is why such significance is placed on the engineering characteristics of quantum generators; the oscillation frequency, field configuration, mode content, etc. could not be described and calculated.

In the microwave band, the situation was formed in another way. The theory of the maser was fully and naturally based on classical electromagnetic field theory. The medium properties were described using macroscopic observed characteristics (polarization, magnetization, and current density), which can be obtained by means of averaging of microscopic quantities over the particle ensemble. The problem of

---

[11]H. Statz, G. deMars, in *Quantum Electronics*, ed. by C.H. Townes (Columbia University Press, New York, 1960), p. 530.

the quantum generator was reduced to solution of the Maxwell equations. The use of the semiclassical approach appeared to be very fruitful.

Semiclassical approximation forms the methodical basis of this book. It means that to describe the interactions of an electromagnetic field with a substance we use the classical Maxwell equations:

$$\nabla \times \mathbf{H} = \varepsilon_0 \frac{\partial \mathbf{E}}{\partial t} + \frac{\partial \mathbf{P}}{\partial t},$$

$$\nabla \times \mathbf{E} = -\mu_0 \frac{\partial}{\partial t}(\mathbf{H} + \mathbf{M}),$$

and the substance properties are described by vectors of polarization and magnetization, which are connected to fields E and $\mathbf{H}$ through some (in the general case) nonlinear integral–differential operators:

$$\mathbf{P} = \widehat{\mathbf{P}}(\mathbf{E}, \mathbf{H}),$$

$$\mathbf{M} = \widehat{\mathbf{M}}(\mathbf{E}, \mathbf{H}).$$

The specific forms of this relationship are always defined on the basis of some initial postulates; in quantum electronics we have to use quantum postulates. That is why this book contains fundamental theoretical information about the internal structure of a substance and the mathematical apparatus of quantum mechanics (see Chap. 1).

The formalism of the density matrix (see Chap. 1) allows transition from microscopic (quantum) quantities to macroscopic (classical) quantities. As a result, the constitutive equation (laws) appear, describing the reaction of the substance to the electromagnetic field in the form of polarization (see Chap. 2), magnetization (see Chap. 3), and current (see Chap. 4). These equations, together with the Maxwell equations, form the full system of equations, which is sufficient for solution of a wide class of problems on linear and nonlinear interactions of an electromagnetic field with a substance. In particular, quantum generator parameters can be found out with a sufficient degree of accuracy and describe the dynamics of oscillation transients (see Chap. 5). The nonlinear character of the constitutive equations determines the parameters of the steady-state mode of the quantum generator (see Chap. 5), and it also manifests in harmonic generation in laser radiation propagation in various media (see Chap. 6).

Many problems in quantum radio physics in any sense have prototypes in the traditional disciplines of radio engineering profiles: the fundamentals of circuit theory, electromagnetic field theory, oscillation theory, etc. During the discussion, similar prototypes are accompanied by the necessary references. This should make for easier perception of the materials.

Familiarization with any educational discipline cannot be sufficient if the training process is not accompanied by solution of well-designed tasks. Therefore, this book also includes tasks for the reader. These tasks are formulated in the form of some problems for the reader to solve. Solution of these tasks will require attentive study of materials in the appropriate book sections. Some problems require computer applications. For their solution, application of the mathematical package MathCAD, which is popular enough among students, will be sufficient. Use of modern computer software allows the user to apply the task formulation to the real object. We think that most of the tasks included in the book are oriented toward solution at home rather than in classes.

Contents

About the Authors

Vitaly V. Shtykov graduated from the Radio Engineering Faculty of the National Research University "Moscow Power Engineering Institute" (MPEI) in 1963. He received a Ph.D. (Techn.) in 1970. In 1963, he became a teacher in the Department of the Fundamentals of Radio Engineering at MPEI; now he works as a full professor in this department. He delivers various lecture courses in the theoretical fundamentals of radio engineering, radio engineering circuits and signals, and electrodynamics. He has prepared lecture courses in advanced physics, radio physics, functional electronics, acoustic electronics, quantum electronics, and biophysics. He has authored more than 200 scientific and academic publications, and holds 21 patents. His scientific findings are connected with investigations of new physical principles of signal detection and processing. He has examined problems in quantum electronics and nonlinear phenomena in gas discharge plasma and semiconductors, in ferrimagnetic media, and also in the field of acoustic electronics.

Sergey M. Smolskiy was born in 1946. He received a Ph.D. in Engineering in [1974] and a Dr.Sc. in Engineering in 1993. He is a full professor in the Department of Radio Signals Formation and Processing at the National Research University "Moscow Power Engineering Institute." He has worked on theoretical and practical problems in development of transmitting cascades of short-range radar, and his academic experience spans over 40 years. The list of his scientific works and inventions includes over 300 scientific papers, 15 books, and more than 100 technological reports presented at various local and international conferences. He is an active member of the International Academy of Informatization, the International Academy of Electrotechnical Sciences, the International Academy of Sciences of Higher Educational Institutions, and the Institute of Electrical and Electronics Engineers. His scientific work over the last 15 years has been connected to conversion directions in short-range radar systems, radio measuring systems for fuel and energy complexes, radio monitoring systems, etc.

Chapter 1
Excursus on the Atomic–Molecular Theory of Substance

1.1 Introduction: Corpuscular–Wave Dualism

From fundamental physics we know well that a substance consists of molecules and atoms. Depending on the character and intensity of the interaction between atoms and molecules, a substance is in one of the following aggregate states: gas, liquid, solid, or plasma. Quantum electronics deals with all of these states, and we consider them in the order of an increasing degree of interaction between particles. We begin with an excursus on an isolated atom.

Our approach to the subject (at least in this section) does not have a historical character; from a general physics lecture course we already know when researchers arrived at the modern physical picture of the world. We would like to try to systematically state this minimal formulation, which is necessary for understanding the topic of this book.

Thus, it is generally recognized that both a substance and a field have a dual nature. On the one hand, an alternating electromagnetic field in the simplest case takes the form of a plane wave with the following electric vector:

$$\mathbf{E}(\mathbf{r}, t) = \mathrm{Re}\left\{\mathbf{E}_0 \exp\left[j(\omega t - \mathbf{kr})\right]\right\}, \tag{1.1}$$

where $\mathbf{k}$ is the wave vector ($|\mathbf{k}| = k = 2\pi/\lambda = \omega/c$), λ is the wavelength, ω is the frequency, and $c = 3 \cdot 10^8$ m/s is the light speed in a vacuum. At the same time, the electric field can be quantized, and under definite conditions this proves that the particles have impulse $\mathbf{p} = \hbar \cdot \mathbf{k}$ and energy $W = \hbar \cdot \omega$ ($\hbar = h/2\pi$, where $h = 6.62 \cdot 10^{-34}$ J $\cdot$ s is the Plank constant[1]).

---

[1]Max Planck (1858–1947) was a German physicist, the founder of quantum theory, and a foreign honorary member of the Soviet Academy of Sciences. In 1918 he received the Nobel Prize in Physics. The German Physical Society named its highest award the Plank Medal in his honor. He was also a remarkable pianist and a climber.

© Springer Nature Switzerland AG 2020
V. V. Shtykov, S. M. Smolskiy, *Introduction to Quantum Electronics and Nonlinear Optics*, https://doi.org/10.1007/978-3-030-37614-7_1

On the other hand, in accordance with the de Broglie hypothesis, which was brilliantly confirmed, we can attribute some frequency ω and wave vector $\mathbf{k}$ to the substance particle having impulse $\mathbf{p}$ and energy W so that

$$\omega = \frac{W}{\hbar} \tag{1.2}$$

and

$$k = \frac{p}{\hbar} = \frac{2\pi}{\lambda_{Br}}, \tag{1.3}$$

where $\lambda_{Br} = \frac{2\pi\hbar}{mv}$ is the de Broglie wavelength.[2]

Then, in the simplest case, the plain wave

$$\Psi(r, t) = \Psi_0 \exp\left[j\left(\frac{P}{\hbar} r - \frac{W}{\hbar} t\right)\right] \tag{1.4}$$

will correspond to the substance particle.

The sense of wave function $\psi(\mathbf{r}, t)$ consists in the fact that $|\psi(\mathbf{r}, t)|^2 \Delta \mathbf{r} \Delta t = \psi(\mathbf{r}, t) \psi^*(\mathbf{r}, t) \Delta \mathbf{r} \Delta t$ is the probability of detecting the particle in the vicinity $\Delta \mathbf{r} \Delta t$ of the point with coordinates $\mathbf{r}, t$. Thus, the trajectory of particle motion can be interpreted, strictly speaking, only in the probabilistic sense, and the particle itself should be considered as an object belonging to some statistical ensemble.

The substance demonstrates quantum properties in the case when the typical size of the particle interacting with the field is comparable to the de Broglie wavelength. We consider, as an example, hydrogen atom heat motion (its mass $m = 1.7 \cdot 10^{-27}$ kg). For heat motion the average kinetic energy $p^2/2\,m \approx k_B T$ ($k_B = 1.38 \cdot 10^{-23}$ J/k is the Boltzmann constant). At $T \approx 300$ K we have $p \approx 3 \cdot 10^{-24}$ $\frac{J\,s}{m}$, and $\lambda = \frac{h}{p} \approx 2 \cdot 10^{-10}$ m, which is comparable to the hydrogen atom size $r_0 \approx 5 \cdot 10^{-10}$ m. That is why we can conclude that upon description of the field interaction with the substance, the latter should nevertheless be considered from positions of quantum mechanics.

---

[2]Louis de Broglie (1892–1987) was a French physicist and one of the founders of quantum mechanics. In 1929 he received the Nobel Prize in Physics. In 1933 he was elected as a member of the French Academy of Science. He was a founder of the Center of Research on Applied Mathematics at the Poincare Institute for strengthening the connections between physics and applied mathematics.

1.2 Postulates of Quantum Mechanics

The above-mentioned de Broglie hypothesis was an important assumption for the discovery of quantum mechanics, which represents a derivation from axioms and theorems as harmonious as Euclidean[3] geometry. There are several different statements in quantum mechanics. We shall follow the most common of them, based on the Schrodinger[4] equation and Hamiltonian formalism. So, axioms (postulates) of quantum mechanics can be formulated as follows.

1.2.1 The First Postulate

Each state of the physical system is fully described by wave function $\psi(\mathbf{r}_1,\mathbf{r}_2,...,\mathbf{r}_n,t)$, where $\mathbf{r}_i$ are coordinates of particles composing the physical system. From the physical sense of the ψ function the following condition of normalization follows:

$$\int \psi^* \psi d\mathbf{r} = 1.$$

It is not necessary that the spatial coordinates and time should act as arguments of the wave function. However, in any case, these must be, without fail, variables that can be measured simultaneously.[5] As we remember, we cannot exactly measure coordinate $\mathbf{r}$ and impulse $\mathbf{p}$ simultaneously according to the uncertainty principle; therefore, they cannot be considered together as arguments of the ψ function.

1.2.2 The Second Postulate

Some linear operator corresponds to each variable observable in the physical system. This postulate requires a brief comment.

---

[3]Euclid (circa 365 BC–circa 300 BC) was an ancient Greek mathematician who worked in Alexandria. His main work, *Principles* (which includes 15 books), contains descriptions of planimetry, stereometry, and a series of problems related to number theory, algebra, the general theory of ratios, and methods of area and volume determination, including the elements of limits (the exhaustion method).

[4]Erwin Schrödinger (1887–1961) was an Austrian physicist–theorist and one of the founders of quantum theory. In 1933 he received the Nobel Prize in Physics, and in 1937 he was awarded the Max Plank Medal by the German Physical Society.

[5]Later, we discuss the problems of simultaneous measurement of physical variables.

The following variables are observable: coordinate x, impulse component p_x, energy W, electric field $\mathbf{E}$, magnetic field $\mathbf{H}$, electric displacement field $\mathbf{D}$, etc. The operator is a rule in accordance with which one function transforms into another.

Operator $\widehat{A}$ is called linear if it satisfies the superposition principle:

$$\widehat{A}(u_1 + u_2) = \widehat{A}u_1 + \widehat{A}u_2; \quad \widehat{A}(cu) = c \cdot \widehat{A}u,$$

where c is an arbitrary constant. According to this postulate, operator $\widehat{\mathbf{r}}$ corresponds to the observed variable of coordinate $\mathbf{r}$, operator $\widehat{\mathbf{p}}$ to impulse $\mathbf{p}$, operator $\widehat{W}$ to energy W (the Hamiltonian operator $\widehat{H}$ corresponds to the full energy), and so on. All of these operators should be linear, but this postulate does not give us the rule by which the operator can be obtained.

Let us recall some information from operator algebra. Suppose we have the equation

$$\widehat{A}u = A \cdot u, \tag{1.5}$$

where A is some constant. Nonzero solutions to this equation exist for some $A = A_n$ only, and for each A_n we have the identity

$$\widehat{A}u_n = A_n \cdot u_n.$$

The plurality $\{A_n\}$ are called eigenvalues of the operator $\widehat{A}$, and $\{u_n\}$ are their plurality eigenfunctions. The spectrum of eigenvalues and eigenfunctions may be discrete or continuous.

1.2.3 The Third Postulate

The possible results of measurement of physical variable A are the eigenvalues of variable operator $\widehat{A}$.

In contrast to classical physics, quantum mechanics allows one to obtain not only continuous measurement results but also a discrete spectrum of measurement results (for instance, the energy spectrum of the hydrogen atom; see Sect. 1.4.2).

How can we interpret this postulate? Let the atom be in the state described by the ψ function. We measure its energy and obtain the value W_1. As a result of interaction with the measuring instrument, the atom transfers to the state described by the eigenfunction of the ψ_1 function. We return it back to state ψ and repeat the measurement. As a result we obtain value W_2 and the new state ψ_2. The probabilistic nature of the microcosm consists in noncontrollable instrument impact on the object during measurement.

If at repeated measurements of A we obtain the same result A_k each time, we say that variable A is exactly measurable. If this is not so, we shall obtain different results during multiple measurements in a random manner. Of course, by conducting a sufficient number of measurements, we can find out the mean value of A. Nevertheless, we can ask the question of how to determine this average value of <A> without measurement through the wave function.

1.2.4 The Fourth Postulate

If the physical system is in state ψ, the average value of the observable variable A is equal to[6]

$$\langle A \rangle = \int \psi^* \widehat{A} \psi d\mathbf{r}. \tag{1.6}$$

If ψ is the function of the coordinates, we should integrate over all coordinate space. One can write the expression (1.6) via scalar product notation as

$$\langle A \rangle = \left(\psi, \widehat{A}\psi \right),$$

where $\left(\psi, \widehat{A}\psi \right) = \int \psi^* \widehat{A}\psi d\mathbf{r}$. Frequently, in the physics literature, one can see the Dirac notation

$$\langle A \rangle = \left\langle \psi \left| \widehat{A} \right| \psi \right\rangle,$$

where $\langle \psi |$ is a bra vector and $| \psi \rangle$ is a ket vector. The sense of these designations will be explained in the next chapters.

That is why, having the ψ function of the system, we know the average value of any variable observed in it. Hence, the wave function completely describes any physical system. A question arises: if we know $\psi(t_0)$, and the measurement is fulfilled in another time moment t_1, then how can we obtain $\psi(t_1)$?

1.2.5 The Fifth Postulate

The behavior in time of a physical system is described by the Schrödinger equation

---

[6]In (1.6) and subsequently, $d\mathbf{r}$ is the volume element of the selected coordinate system—for instance, dx, dy, dz.

$$jh\frac{\partial \psi}{\partial t} = \hat{H}\psi, \tag{1.7}$$

where $\hat{H}$ is the full energy operator (the Hamiltonian of the system).

We will not try to prove the Schrödinger equation; the Schrödinger equation is simply a hypothesis that allows for calculation of the behavior of microscopic systems.

Equation (1.7) and the other four postulates constitute the formal basis of quantum mechanics, and its conclusions are perfectly confirmed by experimentation. So, it remains for us to submit to any possible contradiction with "common sense."

1.3 Operator Properties of Observable Variables

1.3.1 The Form of Operators of Some Observable Variables

Generally speaking, the form of observable variable operators can be guessed so that theoretical results coincide with experimental data. We can use some leading considerations. If $|\psi(\mathbf{r})|^2$ is the probability density for obtaining a particle in the vicinity of the $\mathbf{r}$ point, then

$$\langle \mathbf{r} \rangle = \int \mathbf{r}|\psi|^2 \, d\mathbf{r}.$$

Comparing this equation with (1.6) we can see that the coordinate operator is the coordinate itself, acting by multiplication:

$$\hat{\mathbf{r}} = \mathbf{r}. \tag{1.8}$$

If one looks at the wave function of the particle with the strictly defined impulse $\mathbf{p}$ (1.4), then it becomes clear that $|\psi|^2 = \text{const}$, i.e., the particle is not localized in space (Fig. 1.1a). The wave packet of the wave group[7] (Fig. 1.1b) consists of a set of plane waves with different wave numbers (or different impulses):

$$\psi(x, t) = \int\limits_{-\infty}^{\infty} C(p_x) \, e^{j\frac{p_x}{\hbar}x} e^{-j\frac{W}{\hbar}t} \, dp_x \tag{1.9}$$

corresponds to the particle localized along direction x, where $C(p_x)$ is the Fourier transformation of the ψ function, whose sense consists in the fact that $|C|^2$ is the probability density for detecting a particle with impulse p_x. We have no right now to

---

[7]The envelope of the wave packet moves in the space with the group velocity $V_{\text{gr}} = \frac{d\omega}{dk} = \frac{1}{\hbar}\frac{dW}{dk}$.

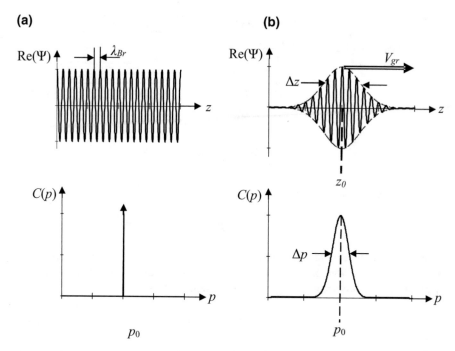

Fig. 1.1 Shape of wave functions in coordinate and impulse spaces for a particle with a definite impulse value (**a**) and a wave group (**b**). The wave packet moves in the space with the group velocity $V_{gr} = \frac{d\omega}{dk} = \frac{1}{\hbar}\frac{dW}{dk}$, which is the analogue of wave numbers (or different impulses) of the classical velocity of the particle. These properties have analogues in signal theory and electrodynamics

assert that (1.9) describes a particle with definite impulse p_x, but we can determine the average value of its impulse without any trouble:

$$\langle p_x \rangle = \int\limits_{-\infty}^{\infty} C^*(p_x)p_x C(p_x)dp_x.$$

On the other hand, because of (1.6),

$$\langle p_x \rangle = \int\limits_{-\infty}^{\infty} \psi^*(x)p_x\psi(x)dx.$$

Taking into consideration that the Fourier transformation of $p_x C(p_x)$ gives $-j\hbar\partial/\partial x$, and comparing both above-mentioned equations, we can draw the conclusion that

Table 1.1 Operators of observable variables in coordinate representation

Observable variable	Designation	Operator
Coordinate	$\mathbf{r}$	$\mathbf{r}$
Impulse	$\mathbf{p}$	$-j\hbar\nabla$
Kinetic energy	$W_{kin} = -\frac{p^2}{2m}$	$-\frac{\hbar^2}{2m}\nabla^2$
Potential energy	$W_{pot}(\mathbf{r})$	$W_{pot}(\mathbf{r})$
Impulse moment	$\mathbf{L} = [\mathbf{r}\,\mathbf{p}]$	$-j\hbar[\mathbf{r}\nabla^2]$

$$p_x = -j\hbar\frac{\partial}{\partial x}$$

and in the three-dimensional case,

$$\widehat{\mathbf{p}} = -j\hbar\nabla.$$

Now you can easily prove by yourself that if in classical physics some observable variable is a function of coordinates and impulses—say, $A = f(\mathbf{r}, \mathbf{p})$—then the quantum operator corresponding to it can be written as follows:

$$\widehat{A} = f(\mathbf{r}, -j\hbar\nabla) \tag{1.10}$$

From (1.10), for example, it follows that the potential energy operator $\widehat{W}_{pot}(\mathbf{r})$ is this energy itself, and the kinetic energy operator is equal to $-\frac{\hbar^2}{2m}\nabla^2$. The forms of some operators are presented in Table 1.1.

1.3.2   Properties of Operator Eigenfunctions of Observable Variables

Since all observable variables should be described by real numbers, all operator eigenvalues should also be real, in accordance with the third postulate.

Operator $\widehat{A}$ is a Hermitian operator if, for any two functions f and g, the following equality takes place:

$$\int f^*\widehat{A}g\,d\mathbf{r} = \int g\left(\widehat{A}f\right)^* d\mathbf{r}. \tag{1.11}$$

We may substitute $f = g = u_n$ in (1.11), where u_n is an eigenfunction of the $\widehat{A}$ operator. As a result, we obtain from (1.11) $A_n = A_n*$, which means the A_n eigenvalue is a real number. It is easy to prove the opposite statement as well: from the reality of all eigenvalues we can prove that the operator is Hermitian. Therefore, all operators of observable variables are Hermitian.

Hermitian operators have two very important properties:

1. Eigenfunctions of Hermitian operators are orthogonal, i.e.,

$$(u_m, u_n) = \int u_m^* u_n dr = 0, \quad m \neq n.$$

Since eigenfunctions are defined up to multiplication by a constant, they can be always normalized so that

$$(u_m, u_n) = \begin{cases} \delta_{mn} & \text{—discrete spectrum} \\ \delta(m-n) & \text{—continuous spectrum} \end{cases}$$

Here, $\delta_{mn} = \begin{cases} 0 & m \neq n \\ 1 & m = n \end{cases}$ is the Kronecker delta and $\delta(x)$ is the Dirac delta function.

Thus, for each observable variable we have an orthonormal system of eigenfunctions.

2. The system of eigenfunctions of the Hermitian operator is a complete system.

All of the mathematical apparatus used for research into electromagnetic waves interacting with a substance is constructed on the above-mentioned properties of eigenfunctions of quantum mechanical operators.

1.3.3   The Condition of Joint Accurate Measurability of Two Variables

In the discussion of the third postulate we stated that variable A is exactly measurable if at each measurement the result is the same—for instance, A_n, i.e.,

$$\langle A \rangle = A_n.$$

From the fourth postulate we see that this is possible in the case when the physical system is in the eigenstate of the $\widehat{A}$ operator. Really,

$$\langle A \rangle = \int u_n^* A u_n d\mathbf{r} = A_n \int u_n^* u_n d\mathbf{r} = A.$$

Can we simultaneously exactly measure variables A and B? Yes, if the wave function of the system is simultaneously the eigenfunction of both $\widehat{A}$ and $\widehat{B}$ operators. In this case, $\widehat{A}$ and $\widehat{B}$ commute, i.e.,

$$\widehat{A}\widehat{B} = \widehat{B}\widehat{A}.$$

Let us prove this statement.

Let $\psi = u_n$ be the eigenfunction for both $\widehat{A}$ and $\widehat{B}$ operators.

$$\widehat{A}\, u_n = A_n u_n, \widehat{B}\, u_n = B_n u_n.$$

We act by operators $\widehat{A}\widehat{B}$ and $\widehat{B}\widehat{A}$ on function u_n.

$$\widehat{A}\widehat{B}\, u_n = A_n B_n u_n, \widehat{B}\widehat{A}\, u_n = A_n B_n u_n.$$

The coincidence of these results demonstrates that the operators commute.

Pairs of operators do not commute; in general, permutation is not allowed for operators and the commutator.

$$\left[\widehat{A}\widehat{B}\right] = \widehat{A}\widehat{B} - \widehat{B}\widehat{A} \neq 0.$$

In this case, the values A and B cannot be simultaneously measured accurately. It can be shown that if the system is in state ψ, the root mean square error for mutual measurements obeys the inequality

$$\Delta A \Delta B \geq 0.5 \cdot \left| \int \psi^* \left[\widehat{A}, \widehat{B}\right] \psi\, d\mathbf{r} \right|. \tag{1.12}$$

Inequality (1.12) is the mathematical formulation of the Heisenberg[8] uncertainty principle. For example, for the coordinate and the impulse, it follows from (1.12) that

$$\Delta x \Delta p_x \geq \frac{\hbar}{2}.$$

Now we know exactly enough to consider the main properties of the atom.

---

[8]Werner Heisenberg (1901–1976) was a German physicist–theorist and one of the creators of quantum mechanics. He was awarded the Nobel Prize in Physics in 1932, the Max Plank Medal by the German Physical Society in 1933, a bronze medal by the National Academy of Sciences of the USA in 1964, and the Nils Bohr International Gold Medal by the Danish Society of Construction, Electrical and Mechanical Engineers in 1970.

1.4 Electron States of Atoms

1.4.1 The Time-Independent Schrödinger Equation

If the atom does not interact with anything, its energy does not depend on time in an explicit form, i.e., the atom Hamiltonian $\widehat{H}$ is not a function of time. In this case we can solve the Schrödinger equation using the variable separation method.

Let us represent the wave function in the form

$$\psi(\mathbf{r}, t) = \psi(\mathbf{r})\varphi(t)$$

and substitute it in (1.7). After transformation we obtain

$$j\hbar \frac{1}{\varphi} \frac{\partial \varphi}{\partial t} = \frac{1}{\psi} \widehat{H} \psi.$$

This equality is possible only if both the right and the left parts are equal to some separation constant W. Then, solving the equation for φ, we find $\varphi(t) = \exp(-j\,Wt/\hbar)$. The coordinate part of the wave function obeys the time-independent Schrödinger equation

$$\widehat{H} \psi = W \psi. \tag{1.13}$$

The eigenfunctions ψ_n of the Hamiltonian $\widehat{H}$ are solutions to this equation. The eigenvalues W_n corresponding to them have the sense of energy of the atom being in states

$$\psi(\mathbf{r}, t) = \psi(\mathbf{r}) \exp\left(-j\frac{Wt}{\hbar}\right). \tag{1.14}$$

These states are called stationary since the energy of the atom in these states does not change in time and it can be measured accurately. Due to the orthogonality of the functions (1.14), any other state of the atom can be represented in the form of a superposition of stationary states:

$$\psi(\mathbf{r}, t) = \sum_n c_n \psi_n(\mathbf{r}) \exp\left(-j\frac{W_n t}{\hbar}\right).$$

Therefore, the top-priority task when studying any conservative physical system, including an isolated atom, is determination of the coordinate parts of wave functions $\psi_n(\mathbf{r})$ describing the atom configuration in stationary states, as well as its energy W_n. For this we must solve the problem of eigenvalues.

1.4.2   Stationary States of the Single-Electron Atom

First, it is necessary to find out the Hamiltonian form of the single-electron atom. The full energy of the atom is formed from the kinetic energy of an electron and the potential energy of the Coulomb interaction of an electron and a nucleus,

$$W = W_{\text{pot}} + W_{\text{kin}} = -\frac{Ze^2}{4\pi\varepsilon_0 r} + \frac{p^2}{2m_e}, \tag{1.15}$$

where m_e and p are the mass and impulse of the electron, respectively; e is its charge; Z is the number of protons in the nucleus; and r is the distance from the nucleus to the electron. Using (1.15), in accordance with Table 1.1, we can write the Hamiltonian as

$$\hat{H} = -\frac{Ze^2}{4\pi\varepsilon_0 r} - \frac{\hbar^2}{2m_e}\nabla^2 \tag{1.16}$$

and then the time-independent Schrödinger equation (1.13) can be written in the form

$$\frac{\hbar^2}{2m_e}\nabla^2\psi + \left(\frac{Ze^2}{4\pi\varepsilon_0 r} + W\right)\psi = 0. \tag{1.17}$$

The solution to this equation is suitable to be looked for in the spherical coordinate system. Representing the required function in the form

$$\psi(\mathbf{r}) = R(r)Y(\varphi, \theta)$$

and having done the standard procedure of variable separation, we can reduce (1.17) to two independent equations for the radial and angular parts of the wave function,

$$\frac{1}{r^2}\frac{\partial}{\partial r}\left(r^2\frac{\partial R}{\partial r}\right) + \left(\frac{Ze^2}{4\pi\varepsilon_0 r} + W\right)\frac{2m_e}{\hbar^2}R - \frac{\lambda}{r^2}R = 0 \tag{1.18}$$

and

$$\frac{1}{\sin^2(\theta)}\frac{\partial^2 Y}{\partial\varphi^2} + \frac{1}{\sin(\theta)}\frac{\partial}{\partial\theta}\left(\sin(\theta)\frac{\partial Y}{\partial\theta}\right) + \lambda Y = 0, \tag{1.19}$$

where λ is the separation constant.

Equation (1.19) is well known from the standard lecture courses in mathematical physics. The following spherical harmonics are its solutions:

$$Y_{lm}(\theta, \varphi) = P_l^{|m|}(\cos \theta) \cdot e^{jm\varphi}, \tag{1.20}$$

where $P_l^{|m|}(x)$ represents the adjoint Legendre function. The solutions in (1.20) exist in the case only where the separation constant $\lambda = l(l + 1)$, where $l = |m|, |m| + 1, |m| + 2 \ldots$; and m in turn is equal to any integer.

The equation in the form of (1.18) is well known to mathematicians. Its solutions forms the orthogonal system of functions

$$R_{nl} = e^{-\rho/2} \rho^l L_{n+l}^{(2l+1)}(\rho), \tag{1.21}$$

where $L_k^{(\alpha)}(\rho) = \frac{d^\alpha}{d\rho^\alpha} L_k(\rho)$, $L_k(\rho)$ is the so-called Laguerre polynomial, $k = n + l$, $\alpha = 2l + 1$, $\rho = \frac{2Zr}{na_0}$, and $a_0 = \frac{4\pi\varepsilon_0\hbar^2}{e^2 m_e} \approx 0.05$ nm is a parameter characterizing the atom size (a_0 is exactly equal to the radius of the first Bohr orbit for hydrogen). Nonzero solutions to Eq. (1.18) in the form of (1.21) exist only for

$$W = W_n = -Z^2 A_0/n^2, \tag{1.22}$$

which are eigenvalues of the Hamiltonian. Here, the constant

$$A_0 = \frac{m_e e^4}{2(4\pi\varepsilon_0\hbar)^2} \approx 13.6 \text{ eV}.$$

Let us sum up all of the above-mentioned facts. The atom energy in the stationary state W_n depends on the main quantum number, which can take on values $n = 1, 2, 3 \ldots$ The stationary state itself is described by the wave function

$$\psi_{nlm}(\varphi, \theta) = C_{nlm} R_{nl}(r) Y_{nlm}(\varphi, \theta)$$

and, besides the main quantum number, is defined also by other quantum numbers: $l = 0, 1, 2 \ldots n - 1$; $m = 0, \pm1, \pm2, \pm3, \ldots \pm l$ (C_{nlm} is a constant of normalization).

The situation where several different states of the particle correspond to the one eigenvalue W_n (a single energy level) is called degeneracytion. For instance, the level W_2 is degenerate on l and m fourfold as for $n = 2$ the values $l = 0, m = 0$ and $l = 1, m = 0, 1, -1$ are possible, respectively.

What is the physical sense of quantum numbers? To answer this question it is necessary to write the operator of the square impulse in the spherical coordinate system:

$$\hat{L}^2 = -\hbar \left[\frac{1}{\sin(\theta)} \frac{\partial^2}{\partial\varphi^2} + \frac{1}{\sin(\theta)} \frac{\partial}{\partial\theta} \left(\sin(\theta) \frac{\partial}{\partial\theta} \right) \right]. \tag{1.23}$$

Comparing this expression with (1.19) we notice that solutions of the form (1.20) are also eigenfunctions of the operator of the square moment of the atom impulse and appropriate eigenvalues are equal to $\hbar^2 l(l + 1)$. It means that being in the stationary state the atom has the impulse moment

$$L = \hbar\sqrt{l(l+1)}, \qquad (1.24)$$

connected to the orbital motion of the electron. For this reason the number l is called the orbital quantum number.

It follows from (1.12) that L^2 is exactly measurable simultaneously with the moment projection on any axis in the space (usually the z axis is used as such an axis). Hence, the (1.20) functions are also eigenfunctions of the operator $\widehat{L}_z$ having in spherical coordinates the form $j\hbar\frac{\partial}{\partial\phi}$. It is easy to make sure that eigenvalues of this operator are equal to $m\hbar$, i.e., in the stationary state the projection of the orbital moment on the z axis is

$$\widehat{L}_z = m\hbar. \qquad (1.25)$$

We know that the mechanical moment of the electron is accompanied by a magnetic moment; therefore, m is called the magnetic quantum number.

As for the main quantum number, it defines the energy of the atom's electron state, in accordance with (1.22). State $n = 1$ is called the main state, since it corresponds to the minimal energy (note the sign in formula (1.22)!). Usually energy W_1 is taken as zero energy and in this case W_∞ represents the energy of atom ionization (for hydrogen, $W_\infty = A_0 = 13.6$ eV).

The set of energy levels W_n is represented graphically in the form of the energy diagram. As we see from Fig. 1.2, in which the energy diagram of the hydrogen atom

Fig. 1.2 Energy diagram of a hydrogen atom. The energy level depends upon one (main) quantum number n. A wave function of the lowest level has spherical symmetry. Higher levels are degenerate: several wave functions, which differ by values of orbital l and azimuth m numbers, correspond to one energy value. The same situation occurs in a resonator with a cubic form

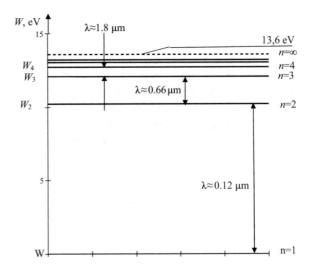

is presented, the electron levels are located nonequidistantly. From fundamental physics we know that electromagnetic waves are absorbed effectively by a substance in the case where the field quantum energy $\hbar\omega$ is equal to the distance between levels. Since typical distances are 1–10 eV, the absorption spectrum is located in the optical range (from ultraviolet at transitions from the W_1 level to infrared at transitions from the level W_3) and has a linear character. The resonance wavelengths of some transitions in the hydrogen atom are shown in Fig. 1.2.

We would like to remind readers that the electron states in the atom corresponding to different values of the quantum number $n = 1$ are designated by lowercase Latin letters: 1s ($n = 1, l = 0$), 2p ($n = 2, l = 1$), 3d ($n = 3, l = 2$), etc.

What does the atom look like in its stationary state? Having known ψ_{nlm} we have information about the probability density for electron detection in the vicinity of a point with the coordinate $\mathbf{r}$. Uncertainty of electron location can be vividly considered as "smearing" of its charge density into some "cloud" whose configuration is defined by function $|\psi_{nlm}|^2$.

Let us examine, as an example, possible stationary distributions of the atom electron density, which has energy W_3. At $n = 3$, three values of the orbital quantum number are allowed. If $l = 0$, then $m = 0$ and the atom orbital moment is absent. In that case, $Y(\theta, \varphi) = $ const, and the electron cloud has a spherical symmetry that corresponds to the so-called s-electron. From Fig. 1.3 we see that for $n = 3$ there are three density maximums along the radial coordinate. If $l = 1$, we deal with the p-electron, for which three different projections of the impulse moment on the z axis ($m = -1, 0, 1$) are possible, and in this case $|\psi|^2$ depends upon the angle θ. The shapes of appropriate electron clouds are also shown in Fig. 1.3. At $l = 2$ (d-electron) the number of possible stationary distributions of electron density increases to five. If the atom is in the main state, the electron can have s-configuration only; the typical radius of the electron cloud for hydrogen is equal to $a_0 \approx 0.05$ nm as it follows from (1.21).

Fig. 1.3 Electron configuration of a hydrogen atom. The *shading* shows the density of the electron cloud. The configuration corresponds to the main quantum number equal to 3. The designations s (sharp), p (principal), d (diffuse), etc. relate to spectroscopic investigations

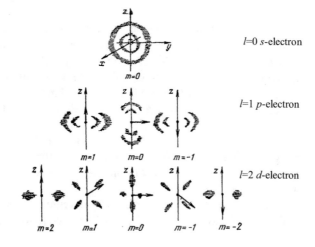

We have considered states of the single-electron atom. For the multielectron atom the situation is much more complicated since its Hamiltonian must contain items describing the potential and kinetic energy of each electron, as well as the energy of their interaction. The equation obtained in this case cannot typically be solved exactly. Therefore, in most cases one can use an approximation in which we neglect the electron interaction and study the motion of each electron separately in the central-symmetric electric field formed by the nucleus and all other electrons (in the so-called matched field). The state of each electron is also characterized by a set of quantum numbers n, l, m, but because of the difference of the field from the Coulomb one, the atom energy in the stationary state depends not only upon the main quantum number but also upon the orbital quantum number (degeneracy of l is absent). The complete description of the atom electron state requires the indication of all electron states, as well as their total orbital moment.

1.4.3 Spin of Electrons and Other Particles

Until now we have not taken into consideration relativistic effects. If we take them into account during the writing of the Hamiltonian (1.16) (as was done by Dirac in 1928) we shall see that the electron and the nucleus have their own mechanical moment—spin S—and corresponding to it the spin magnetic moment. Experimental data testify to the fact that most elementary particles (including electrons, protons, and neutrons) may have two possible spin projections only on the chosen z axis in the space; $S_z = s\hbar$, where $s = \pm 1/2$ is a spin quantum number. We can say that these particles have spin 1/2. By analogy to the orbital moment (see (1.24)) we can define the value of the spin moment as

$$S = \hbar\sqrt{s(s+1)} = \frac{\sqrt{3}}{2}\hbar.$$

It should be taken into account that particle spin is purely a quantum concept and classical analogies (for instance, a top rotating around its axis) are useless here.

If one takes into consideration the spin presence, the electron wave function should depend not only upon coordinate variables but also upon the spin variable $s = \pm 1/2$, i.e., $\psi = \psi(x,y,z,s)$. Since the orbital motion of the electron gives birth to an electric current and to the magnetic field corresponding to the first, the spin magnetic moment should interact with this field. This is the so-called spin–orbital interaction. It leads to the situation in which $\psi(x,y,z,1/2) \neq \psi(x,y,z,-1/2)$, and energies in these states become something different. As a result, energy level splitting arises (in the absence of an external field), i.e., a thin multiple-spectrum structure is present.

The character of atom interaction with an external magnetic field is defined by the total magnetic moment of the electron connected to its total impulse moment:

Table 1.2 Scheme of states of the helium atom

First electron	Second electron	L	S = 0 (singlets)		S = 1 (triplets)	
			J	Symbol	J	Symbol
$1s$	$1s$	0	0	$^1S_0$	1	$^3S_0$
$1s$	$2p$	1	1	$^1P_1$	0,1,2	$^3P_0, ^3P_1, ^3P_2$
$1s$	$3d$	2	2	$^1D_2$	1,2,3	$^3D_1, ^3D_2, ^3D_3$
$1s$	$4f$	3	3	$^1F_3$	2,3,4	$^3F_2, ^3F_3, ^3F_4$

$$\mathbf{J} = \mathbf{L} + \mathbf{S}.$$

The total impulse moment is quantized similarly to the manner of orbital moment quantization. Its value is equal, by analogy to (1.24), to

$$J = \hbar\sqrt{j(j+1)}, \tag{1.26}$$

where $j = |l \pm 1/2|$ is the internal quantum number, and the projection on the z axis takes values, by analogy to (1.25), of

$$J_z = m_j\hbar, \tag{1.27}$$

where $m_j = \pm 1/2, \pm 3/2, \ldots, \pm j$. This is the situation with the single-electron atom.

In the multielectron atom, the total impulse moment of all electrons may be summed in two different ways. In lightweight atoms, the interaction of orbital electron moments L_i and L_k between themselves and their spins S_i and S_k, as a rule, is stronger than the spin–orbital interaction between L_i and S_i. Therefore, when finding J_Σ it is necessary to first find out L_Σ and S_Σ, and then find the sum. This is the case of weak coupling. In heavy atoms, the case of strong coupling frequently happens, i.e., a strong interaction of L_i and S_i arises. Therefore, we first need to find the J_i values of each electron and then sum them to obtain J_Σ. The results for the two types of coupling are naturally different. In the general case we need to use the so-called Hund rules.

Stationary states of a multielectron atom with definite energy can be specified by the quantum numbers L, S, J, such that the orbital moment $\mathbf{L} = \hbar L$, the spin moment $\mathbf{S} = \hbar S$, and, lastly, the total moment $\mathbf{J} = \hbar J$. Atom states as a whole are called terms[9] and are designated by uppercase Latin letters nS ($L = 0$), nP ($L = 1$), nD ($L = 2$), nF ($L = 3$), etc. In Table 1.2 we show possible configurations of the helium atom and term designations.

---

[9]This term was introduced at the end of the nineteenth century by J.R. Rydberg (1854–1919), a Swedish physicist, mathematician, and member of the Swedish Academy of Sciences (1919). A definite energy value—the energy level—corresponds to each term. Rydberg offered the formula to determine the wavelength of any spectral line. In 1980 he introduced the physical constant that would eventually be named after him.

The term designation contains the value $2S + 1$ (multiplet) in the left upper index and the value of the orbital quantum number L in the right lower index. This designation system can be found in the literature on quantum electronics.

1.5   Molecules

Having examined the atom we can proceed to a more complicated physical object: a molecule. Firstly, we should clarify how atoms may be combined to form molecules. For this purpose we focus our attention on the simplest two-atom molecule, hydrogen (H_2).

The ability of atoms to form chemical couplings is characterized by their valence, and the electric interaction between nuclei and electrons forms the basis of this property. The potential energy of this interaction U depends upon the distance R between atom nuclei, and it is formed from the energy of the Coulomb interaction of the nuclei and the energy of the electrons of both atoms $W(R)$:

$$U(R) = \frac{e^2}{4\pi\varepsilon_0 R} + W(R). \tag{1.28}$$

It is clear that a stable molecule is formed in the case where function $U(R)$ has a minimum.

In order to determine $U(R)$ it is necessary to find out the electron energy $W(R)$ as an eigenvalue of the Hamiltonian (compare this with (1.16)):

$$\widehat{H} = -\frac{\hbar^2}{2m_e}\nabla_1^2 - \frac{\hbar^2}{2m_e}\nabla_2^2 - \frac{e^2}{4\pi\varepsilon_0}\left(\frac{1}{r_{11}} + \frac{1}{r_{22}} + \frac{1}{r_{12}} + \frac{1}{r_{21}} - \frac{1}{r}\right). \tag{1.29}$$

The first two items correspond to the kinetic electron energy, and the expression in parentheses represents the energy of the Coulomb interaction of electrons with their "own" nuclei, as well as with the nuclei of the adjacent atoms and with each other. The geometry of the task is shown in Fig. 1.4. Eigenfunctions of the

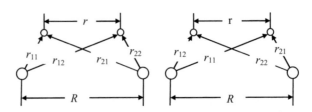

Fig. 1.4 Two indistinguishable models of a hydrogen molecule. The distances between all pairs of interacting particles are shown. An analytical solution to the Schrödinger equation, which takes into consideration all interactions, is impossible

Hamiltonian (1.29) $\psi(r_{11}, r_{22})$ describe the stationary electron states of a molecule with a distance between nuclei equal to R.

Unfortunately, it is impossible to exactly solve the Schrödinger equation with this Hamiltonian. The approximate solution is also sufficiently inconvenient. Therefore, we shall limit ourselves to a short review of the results.

As our zero approximation we would take a solution for the case of a large distance between atoms, when one atom does not practically disturb the electron state of another (we may neglect the last three items in (1.29)). In this case, $\psi = \psi_1(r_{11})\psi_2(r_{22})$, and appropriate energy $W(R) \approx W_1 + W_2$, where $\psi_{1,2}$, $W_{1,2}$ are the wave functions and energy of the undisturbed first and second atoms. Nevertheless, the following fact is principally important in quantum mechanics: if similar particles have a chance to be at the same point in space we cannot distinguish them in the future because the trajectory concept loses its sense as a result of the uncertainty principle. Therefore, at the atom approach, electrons become indistinguishable. This means that with the permutation "yes" the electron index should not change the molecule state, i.e., the wave function will remain unchanged (a symmetric state) or will undergo a change of sign (an asymmetric state). The zero approximation satisfying this condition takes the form

$$\psi_s = \psi_1(r_{11})\psi_2(r_{22}) + \psi_1(r_{12})\psi_2(r_{21})$$

or

$$\psi_a = \psi_1(r_{11})\psi_2(r_{22}) - \psi_1(r_{12})\psi_2(r_{21}). \tag{1.30}$$

The appearance of the second item in (1.30) can be interpreted as a result of electron exchange between atoms in the molecule. Substituting (1.30) into the Schrödinger equation and taking into consideration all items in (1.29) we can find out the correction to the electron energy. It turns out that splitting one electron's state $W_1 + W_2$ into two terms W_s and W_a is larger for the lesser value of R. This phenomenon is reminiscent of the arising of two frequencies of natural oscillations in a system of two similar coupled oscillatory circuits. The characteristic of potential energy (1.28) of both states with respect to R for the main undisturbed atom states is presented in Fig. 1.5. We can see that only a symmetric state can ensure the arising of the stable molecule.

Up to this point we have ignored the electron spin presence in the molecule. Running slightly ahead, we should note that electrons in any physical system can be described only by the antisymmetric complete wave function (1.26). Since in the main state of the hydrogen molecule, the coordinate part of the ψ function is symmetric, its spin part should be antisymmetric, i.e., electron spins have the opposite signs and electrons are in different states in spite of equality of all other quantum numbers. This fact is a demonstration of the fundamental Pauli principle: in one physical system it is impossible to have two electrons in similar states.

If we tried to calculate $U(R)$ while ignoring the electron exchange, we would obtain in the result (presented by the dotted curve in Fig. 1.5) incorrect values of the

Fig. 1.5 Characteristic of the potential energy of a two-atom molecule versus the distance between atoms. Wave functions can be presented approximately in the form of odd and even combinations of wave functions, which depend on the distances r_{11} or r_{22} in Fig. 1.4. W_D energy of molecule dissociation

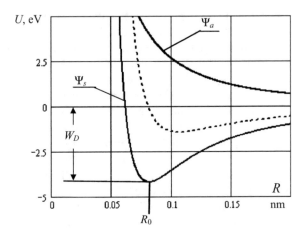

dissociation energy of the molecule W_D and the distance between atoms R_0. This speaks to an important contribution of the interaction arising as a result of electron exchange between atoms: so-called exchange interaction. Its role is especially significant in molecules containing similar or chemically near atoms (for homopolar bonds). In the case of heteropolar bonding in ionic molecules the main role is played by the Coulomb interaction as the electron density distribution in the molecule is sharply nonsymmetric. But in this case the ion repulsion in the molecule cannot be explained without taking attraction into the consideration of the exchange interaction.

1.6 Vibrational and Rotational States of Molecules

Earlier we supposed that atom nuclei are fixed. It is clear that in connection with the complicated molecule structure in comparison with the atom, new degrees of freedom arise in it, corresponding to oscillations of nuclei around the equilibrium state and to molecule rotation as a single whole.

1.6.1 The Vibrational Spectrum

Let us examine vibrational motion by using the example of the hydrogen molecule, where one type of oscillation is possible.

The Hamiltonian describing vibrational energy is formed by the kinetic energy of the motion of both nuclei W_K and their potential energy $U(R)$, obtained in the previous section. We know from theoretical mechanics that the motion of two particles with masses m_1 and m_2 in the field $U(R)$ is equivalent to the motion of one particle with the reduced mass $\mu_m = m_1 m_2/(m_1 + m_2)$. Then,

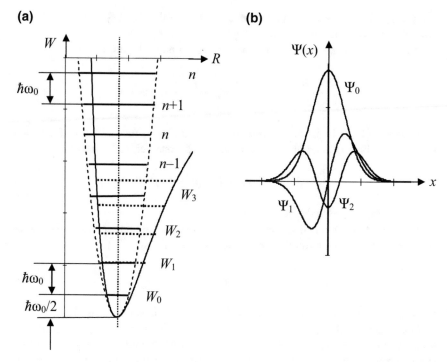

Fig. 1.6 Energy diagram of vibrational levels (**a**) and configuration of wave functions (**b**). Wave functions correspond to parabolic approximation of the potential energy characteristic versus the distance between atoms—the model of the linear harmonic quantum oscillator. Real functions differ from those shown in the figure; the energy levels are nonequidistant and the wave functions are nonsymmetric. Asymmetry is responsible for heat expansion of a substance

$$W_k = -\frac{\hbar^2}{2M_m}\frac{\partial^2}{\partial R^2}.$$

If we limit ourselves to consideration of vibrations with energy much less than the dissociation energy W_D, the potential energy can be expanded into the Taylor series in the vicinity of the equilibrium point R_0 and limited only by the quadratic term (see dotted curve in Fig. 1.6a):

$$U(R) = -W_D + \frac{\mu_m \omega_0^2}{2} x^2, \tag{1.31}$$

where $x = R - R_0$ and ω_0 is a coefficient depending upon the value of the second derivative of $U(R)$ at the point $R = R_0$ (i.e., upon the stiffness of the molecule) and having the sense of the natural frequency of the harmonic oscillator. Perhaps you remember from general physics that the potential energy of an elastic pendulum performing harmonic oscillations with frequency ω_0 is described exactly by

Eq. (1.31). For the hydrogen molecule, $\omega_0 \approx 8.2 \cdot 10^{14}$ rad/s ($f_0 \approx 1.3 \cdot 10^{14}$ Hz), which corresponds to the light wavelength of $\lambda_0 \approx 2.3$ μm.

As a result, the stationary Schrödinger equation takes the form

$$\frac{\hbar^2}{2\mu_m} \frac{\partial^2}{\partial x^2} \psi + \left(W + W_D - \frac{\mu_m \omega_0^2}{2} x^2 \right) \psi = 0. \qquad (1.32)$$

Equation (1.32) is the equation of the quantum harmonic oscillator, for which nonzero and limited solutions exist for definite discrete values of energy:

$$W_v = -W_D + \hbar\omega_0(v + 0.5), \qquad (1.33)$$

where v is the vibrational quantum number.

Appropriate eigenfunctions describing stationary oscillating states of the molecule can be expressed through Hermite polynomials $H_k(x)$:

$$\psi_v(x) = C_v e^{-\xi^2/2} H_v(\xi), \qquad (1.34)$$

where $\xi = x\sqrt{\mu_m \omega_0/\hbar}$. Graphs of these functions are depicted in Fig. 1.6b.

Let us discuss the result obtained. First, as follows from (1.33), the oscillating levels of the molecule are equidistantly located with the interval $\hbar\omega_0$. As will be shown later, under field action, quantum transitions happen between adjacent levels only. Hence, the oscillating system "responds" to the field varying with a frequency close to ω_0 only. Nevertheless, it is true for small v values only, while an oscillator behaves as a harmonic one. With oscillating quantum number growth, the potential function deviation from parabolic law increases and the levels begin to be concentrated (see Fig. 1.6a), merging at $U > 0$ into a continuous spectrum that corresponds to molecule dissociation. Secondly, a quantum oscillator, in contrast to a classical one, cannot be in a state of absolute quiescence. Even for $v = 0$ the oscillation energy is nonzero and is equal to $0.5\hbar\omega_0$, and the wave function is smeared near zero. If it were not so, we would know both the atom coordinate and impulse simultaneously, which contradicts the uncertainty principle. Thirdly, quantum transitions between the vibrational levels of the hydrogen molecule under the action of an electromagnetic field arise at a wavelength of $\lambda_0 = 2.3$ μm, which corresponds to the infrared band. For heavier molecules the value of ω_0 is smaller and the frequencies of the quantum transition can be located in the submillimeter band.

This is the situation in the case of the simplest two-atom molecules. More complicated molecules do not have such a simple equidistant spectrum structure. Let us consider, for example, the linear molecule CO_2. As shown in Fig. 1.7, three types of vibrations are possible in it from the classical point of view: symmetric (a) and nonsymmetric (b) expanding modes, as well as the bending mode (c). Each type of oscillation has its own resonance frequency ω_1, ω_2, ω_3. From the quantum point of view, the transitions between vibrational states of the molecule, which can be characterized by three numbers v_1, v_2, v_3 and are designated by the symbol

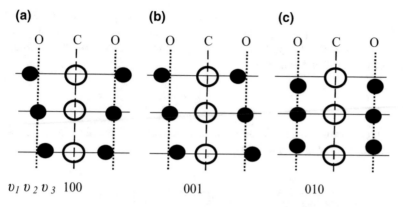

Fig. 1.7 (**a–c**) Three possible types of classical oscillation of a molecule of carbonic gas. In the quantum language, the transition between the main state and the energy level with some set of quantum numbers $v_1\, v_2\, v_3$ corresponds to the oscillation type shown in the figure

"(v_1, v_2, v_3)" (for instance, (000) is the ground state), correspond to these frequencies. The energy in the stationary states of that molecule represents a sum of three states, i.e., instead of (1.33) we have

$$W_{v_1\, v_2\, v_3} = -W_D + \hbar\omega_1(v_1 + 0.5) + \hbar\omega_2(v_2 + 0.5) + \hbar\omega_3(v_3 + 0.5). \quad (1.35)$$

If the molecule has a more complicated spatial structure and/or contains a larger number of atoms, the number of different vibrational types becomes greater still. For instance, besides expanding and bending modes, torsion modes may also exist.

1.6.2 The Rotational Spectrum

Definite energy corresponds to molecule rotation as a single whole. If with translation (headway) motion the kinetic energy is equal to $W_K = p^2/2m$, then with rotational motion $-W_{kin} = L^2/2I$, where I is the molecule inertia moment with respect to some rotational axis passing through its center of mass. It is clear that this value depends upon the choice of the rotational axis. From classical mechanics we know that any rigid body is characterized by the inertia ellipsoid allowing determination of the I value with respect to any axis as the square of the distance along this axis from the mass center to the ellipsoid surface. The three main inertia moments I_x, I_y, I_z correspond to its main axes. If they are not the same, we can talk about an nonsymmetric top. If two of them are the same, the molecule represents a symmetric top. Lastly, if one of the inertia moments is equal to zero and the two others are equal to each other, we have a linear molecule. We shall study this simplest case by considering the H_2 molecule.

In the case of the H_2 molecule the main rotational axis passes through the middle of the line connecting the hydrogen atoms, at an angle of 90° to it. Relatively, for the two-atom molecule, $I = \mu_m r_0^2$. As we see from the expression for W_K, the Hamiltonian of rotational movement is

$$\hat{H} = \hat{L}^2/2I;$$

therefore, because of (1.23) the stationary Schrödinger equation in the spherical coordinate system takes the form

$$\frac{\hbar^2}{2I}\left[\frac{1}{\sin^2(\theta)}\frac{\partial^2\psi}{\partial\varphi^2} + \frac{1}{\sin(\theta)}\frac{\partial}{\partial\theta}\left(\sin(\theta)\frac{\partial\psi}{\partial\theta}\right)\right] + W\psi = 0. \tag{1.36}$$

The same equation was examined when we found out the angular part of the wave function describing the electron state in the atom (see (1.19)). Nonzero solutions to Eq. (1.36) represent spherical harmonics of the (1.20) type; they exist only for

$$W = W_J = \frac{\hbar^2}{2I}J(J+1) = B\hbar J(J+1), \tag{1.37}$$

where $J = 0, 1, 2 \ldots$ represents the rotational quantum number and B is called the rotational constant. Thus, the rotational motion energy can take a discrete series of values, as shown in Fig. 1.8a; the energy levels are clearly located in a nonequidistant manner. As we remember, the electron state, besides the orbital number l, is defined also by the magnetic quantum number m. The rotational state, besides the number J indicating the value of the molecule impulse moment, depends also upon some integer number $M = -J, -J+1, \ldots, J-1$, with J determining the projection of the impulse moment on the axis allocated in the space. Since the energy does not depend upon this number, each rotational energy level of the linear molecule is $(2J+1)$ times degenerated.

In the case of a molecule in the form of a symmetric top we may expect complication of the energy diagram in comparison with the case of a linear molecule. Indeed, the rotational state depends not only upon J and M but also upon the quantum number K, defining the value of projection of the impulse moment on the

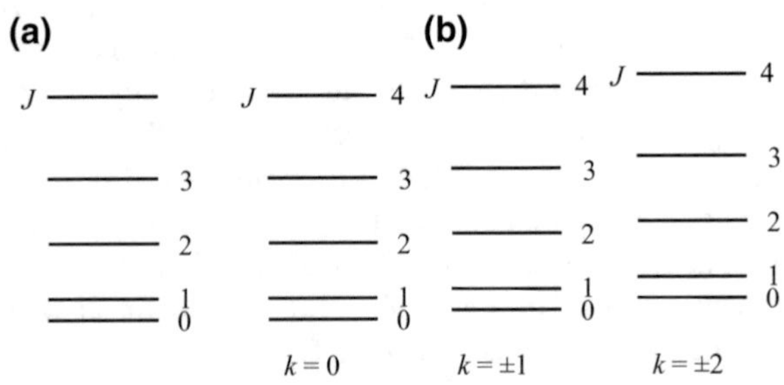

Fig. 1.8 Rotational spectra of a linear molecule (a) and a molecule in the form of a symmetric top (b)

symmetry axis of the molecule. The energy also depends upon this number taking the same values as M; as a result, the energy diagram becomes more complicated (see Fig. 1.8b).

For the H_2 molecule, $R_0 = 0.08$ nm and estimation of the inertia moment results in $B = 1.5 \cdot 10^{12}$ Hz. This corresponds for the quantum transition $J = 0 \rightarrow 1$ to the resonance wavelength of $\lambda_0 = 0.1$ mm. For heavier molecules with larger inertia moments, as we see from (1.37), the transition may happen at lower frequencies in the millimeter and centimeter wavebands.

1.6.3   Total Energy Diagram of the Molecule

Thus, the molecule has the energy of electron, vibrational, and rotational motion; the interval between electron energy levels is 1–10 eV, that between vibrational levels is 0.1–1 eV, and that between rotational levels is hundredths of an electron volt. Therefore, the electron term decomposes into a system of vibrational levels, each of which, in turn, has a thin rotational structure, which is illustrated in Fig. 1.9 by a total energy diagram of the molecule.

Which transitions may take place for the molecule? First, transitions with a change in the electron state may occur. They give birth to the electron oscillation spectrum located in the optical band. Secondly, transitions without changing the electron state lead, as a result, to the vibration (al) – rotation (al) spectrum being

Fig. 1.9 Total energy diagram of a molecule. *1* electron–vibrational transitions, *2* vibrational transitions, *3* purely rotational transitions

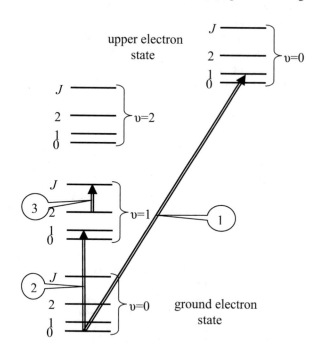

observed in the far-infrared and submillimeter bands. Lastly, purely rotational transitions are responsible for absorption in the ultrahigh-frequency (UHF) band.

1.7   Internal Structure of the Solid Body

Now we can start to study the internal structure of the solid body. Its density is so great that interatomic links take on special significance. These links form up atoms in a definite order, fixing the specific spatial structure of the solid state. Solid bodies ordered in space are known as monocrystals. The internal harmony of monocrystals is demonstrated in the beauty and elegance of their external forms, and in the originality of their properties (mechanical, optical, etc.).

Crystals have attracted the rapt attention of humanity since ancient times, and for a long time they were synonymous with riches and power. In the twentieth century, the situation changed. First, crystals began to find uses in engineering; then, methods were found to grow them artificially on an industrial scale. Without any peculiar difficulties we can, for instance, obtain a ruby. Moreover, crystals began to appear that do not exist in nature—for example, the "fianit"[10] [cubic zirconia], which is used for decoration.

Certainly, there are solids containing a huge number of small crystals (composed of many crystallites of varying size and orientation). Such materials are called polycrystalline materials. Production of such materials does not require specific tricks and has been known for a long time. For instance, all technical metals are polycrystalline. The peculiarity of polycrystalline material is the property of isotropy, and it is connected to averaging of the orientation of the separate crystallites. We pay for simplicity, as a rule, by property degradation. In spite of technological successes, the cost of monocrystallines is still so high that they are used only in exceptional cases. Seemingly, electronics is the most "gluttonous" consumer of monocrystal materials.

Besides mono- and polycrystalline materials, there are amorphous solids. The most well-known example is glass.[11] Such materials are sometimes called solid solutions. The thing is that in amorphous materials, as well as in liquids, only short-range ordering is observed: a more or less ordered location of immediate neighbors. It is well known that liquids in a calm state are extremely uniform. This is true for amorphous solids as well. Glass is transparent only because it does not contain crystallites, the edges of which diffuse light. The amorphous state is unstable, and

---

[10]The name "fianit" comes from the abbreviation "FIAN", which stands for the Physical Institute of the Academy of Sciences (Russia), where it was synthesized for the first time. However, this name is not used beyond the former USSR.

[11]Nowadays, an "amorphous solid" is considered to be the overarching concept and "glass" is a more special case; glass is an amorphous solid that transforms into a liquid upon heating through the glass transition. The glass transition is a reversible transition in amorphous materials from a hard and relatively brittle state into a molten or rubber-like state.

crystallization happens gradually. (For instance, glass can have a tendency to cloud.) The crystallites' edges worsen practically all properties. Therefore, it is quite reasonable for material engineers to aspire to obtaining amorphous materials. This can be achieved by very fast cooling (at a speed of a million degrees per second) of a liquid melt.

Nevertheless, some impressive successes in electronics are related to application of monocrystallines. Therefore, we examine the properties of monocrystallines in detail, especially since, without doing so, we cannot describe the properties of polycrystalline materials as well.

1.8 Band Theory of the Solid Body

A crystal represents an assembly of atoms or molecules ordered in space. This assembly can be considered akin to one huge single molecule. We already know something about molecules. Therefore, if we accept such a model we can formulate some general considerations about the structure of the energy levels of this solid body.

Firstly, some number of electrons will probably be "socialized." Secondly, the energy levels of separate atoms should split on some number of sublevels. Since there are approximately 10^{29} atoms in one cubic meter of a solid body, the number of sublevels will be of the same order. We can suppose that sublevels will form bands of allowed values of energy.

Is it possible to obtain accurate quantitative relations or some numerical estimations? The method that should be used for this is more or less evident. We must solve the Schrödinger equation for all nuclei and electrons forming the solid body. If we assume that the nuclei (as heavier particles) are fixed, we obtain, instead of (1.29),

$$\widehat{H} = -\left[\sum_{n=1}^{N_{el}} \frac{\hbar^2}{2m_2} \nabla_n^2 + \frac{e^2}{4\pi\varepsilon_0} \left(\sum_{l=1}^{N_{nuc}} \frac{1}{r_{nl}} + \sum_{m=n+1}^{N_{el}} \frac{1}{r_{nm}} \right) \right]. \tag{1.38}$$

The wave function will now depend upon three spatial coordinates of each electron. Of course, such a huge equation does not easily admit an analytical solution or a numerical solution. In order to emerge from this deadlock, it is necessary, with guidance from experience and intuition, to simplify the physical model, to eliminate the model's incompleteness, and to reasonably use experimental data.

So, we turn away from solving the accurate Schrödinger equation and concern ourselves with a simpler task: we study the movement of only one electron in the total field of all other electrons and all nuclei. If we knew this field, the solution would be accurate, but the locations of the particles are unknown. Hence, we must use some approximation. The simplest approximation was offered in 1930 by

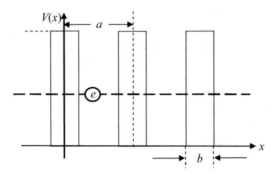

Fig. 1.10 One-dimensional model of the periodic potential in a solid body. For such a potential distribution the wave function is described by trigonometric functions in areas of zero potential and by hyperbolic functions in areas of high potential for $W < V_0$. At area boundaries, the boundary conditions must be fulfilled. The requirement of periodicity of the wave function is an additional condition, which follows from the Bloch theorem

Kronig[12] and Penney[13] (Fig. 1.10). If restricted to the one-dimensional case, the Schrödinger equation for an electron moving in the periodic potential $W_{pot} = V(x)$ takes the form

$$\frac{\hbar^2}{2m_e} \frac{d^2\psi}{dx^2} + [W - V(x)]\psi = 0. \tag{1.39}$$

According to the Bloch[14] theorem, the general solution to the one-electron Schrödinger equation for an electron in a periodic field takes the form

$$\psi(\mathbf{r}) = u_k \exp(j\mathbf{kr}),$$

where $u_k(\mathbf{r})$ is a function with the same spatial periodicity as the crystal lattice.

Equation (1.39) is an excellent example from higher mathematics. Its solution is readily available to anybody who has studied a course of differential equations. For more meticulous students we can say it is necessary to solve (1.39) for two regions using function ψ and its derivative continuity. As a result we obtain the so-called dispersion equation[15]:

---

[12]Ralph Kronig was a German–American physicist (1904–1995). He is noted for the discovery of particle spin and for his theory of x-ray absorption spectroscopy. His theories include the Kronig–Penney model, the Coster–Kronig transition, and the Kramers–Kronig relation.

[13]William Penney (1909–1991) was a British mathematician, professor of mathematical physics at Imperial College, London, and later the Rector of Imperial College.

[14]In mathematics, Eq. (1.39) is known as the Hill equation, the form of solutions to which is defined by the Floquet theorem.

[15]The exact solution has a complicated form. The mentioned equation is obtained if $b \to 0$, $V_0 \to \infty$ at the constant value of bV_0.

Fig. 1.11 Graphic solution to the dispersion equation, obtained for the periodic potential $V(x) =$

$\sum_k b V_0 \delta(x + ka)$. Solutions exist in limited areas. The permitted energy band of energy values corresponds to each such zone. The width of the permitted band grows with an energy increase. The last energy band will not have an upper boundary

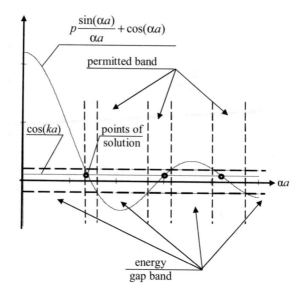

$$\cos(ka) = p\frac{\sin(\alpha a)}{\alpha a} + \cos(\alpha a), \tag{1.40}$$

where $\alpha = \frac{1}{\hbar}\sqrt{2m_e W}$ and $p = \frac{m_e(V_0 - W)}{\hbar^2}ab$.

To find function $W(k)$ we have plotted the right part of (1.40) as a function of αa. The left part of Eq. (1.40) varies from -1 to $+1$. Therefore, the solution to the dispersion equation exists only in regions, shown by shading in Fig. 1.11.

In this case, $k = \alpha$ and $W = \frac{k^2 \hbar^2}{2m_e}$ as for the free electron.

We know nothing about the numerical values of a, b, and V_0; nevertheless, the Kronig–Penney model characterizes the most important regularities. We will see that the energy spectrum has allowed and prohibited bands of energy. With energy growth, the width of the allowed band increases and at the limit the band becomes half-infinite. There is nothing surprising in this. The deep energy levels of separate atoms do not practically interact with each other, while, on the other hand, electrons with energy larger than V_0 practically travel freely over the full crystal.

If in this case $p \rightarrow \infty$, Eq. (1.40) takes the form

$$\sin(\alpha a) = 0.$$

Hence, $W_n = \frac{\pi^2 \hbar^2}{2m_e a^2}n^2$. These values coincide with the discrete energy levels of the potential well with width a. On the other hand, at $p \rightarrow 0$ we have

$$\cos(ka) = \cos(\alpha a).$$

Fig. 1.12 Reduced energy
diagram of two permitted
bands: conductivity (*top*)
and valence (*bottom*). The
direction of convexity
changes from one band to
another. This relates to the
sign of the derivative of the
function in Fig. 1.11 within
the limits of the permitted
bands

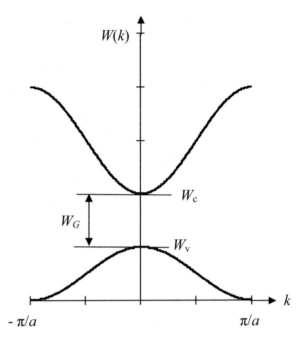

Since $W(k)$ is a periodic function we may be quite limited by the part of the energy
diagram (in the area from $-\frac{\pi}{a}$ to $\frac{\pi}{a}$, which is the first Brillouin zone), as in Fig. 1.12.
The characteristic $W(k)$ is well approximated by the following function:

$$W_N(k) = E_N - 2A_N \cos (ka).$$

This approximation may be generalized in the three-dimensional case:

$$W_N = E_N - 2A_{N,x} \cos k_x a_x - 2A_{N,y} \cos k_y a_y - 2A_{N,z} \cos k_z a_z.$$

Experimental data allow determination of the constants included in this function.

1.9 Motion of the Electron in an Applied Field: The Effective Mass

If the wave function is known, by using (1.6) we may find out any observable
quantity. The accuracy of the answer will depend on the accuracy of the model. Still,
it would not be bad to obtain relations similar to the usual classical ones. This would
allow application of our everyday experience for solution of some problems related
to electron movement inside bands. It turns out that such relations can be obtained
within the limits of the quasiclassical approach. This is possible because within the

limits of one band we may neglect the discrete character of the spectrum of energy eigenvalues.

In Sect. 1.3 we examined the wave group (1.9), which can be assimilated with the particle moving with the group velocity

$$V_{gr} = \frac{d\omega}{dk} = \frac{1}{\hbar}\frac{dW}{dk}.$$

In the general case, the vector of group velocity can be found out through the gradient operation in the k-space with the coordinates k_x, k_y, k_z:

$$\mathbf{V}_{gr} = \frac{1}{\hbar}\nabla_{\mathbf{k}}W.$$

For the one-dimensional model,

$$V_{gr} = \frac{1}{\hbar}2A_N a \sin (ka).$$

The function of the group velocity versus k is shown in Fig. 1.13. For $k = 0$ the electron velocity is equal to zero, and this does not contradict common sense, as impulse $p = \hbar k = 0$. After that, all goes in accordance with the classical scenario: the greater the electron energy, the greater the velocity. But above the middle of the band the velocity begins to decrease, and it vanishes at $|k| = \pi/a$. This phenomenon is purely a quantum one and it influences the perioduic potential of ions. Even more surprising peculiarities of electron behavior are observed in the three-dimensional model.

Let us suppose now that external force $\mathbf{F}$ acts on the electron. We shall find out the energy deviation using the classical definition of the work,

$$\Delta W = FV_{gr}\Delta t.$$

On the other hand,

$$\Delta W = \Delta_{\mathbf{k}}W \cdot \Delta\mathbf{k};$$

hence,

$$\nabla_{\mathbf{k}}W\Delta\mathbf{k} = F\frac{1}{\hbar}\nabla_{\mathbf{k}}W\Delta t.$$

Therefore,

Fig. 1.13 Characteristics of
energy, group velocity, and
effective mass with respect
to the wave vector in the
simplest one-dimensional
model of a solid body. The
effective mass not only
changes in value but also
changes its sign. The
positively charged holes
correspond to electrons with
a negative mass near the
upper end of the permitted
band. The narrower the
permitted band, the heavier
the charge carriers

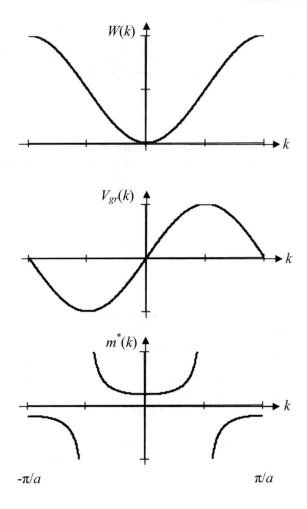

$$\hbar \frac{d\mathbf{k}}{dt} = \mathbf{F}. \tag{1.41}$$

Equation (1.41) is the quantum analogue of the Newton equation. It contains the
Plank constant; therefore, its application relates, as usual, to quantum postulate
adaptation. Let us move forward just a little on the way of analogues and find the
electron acceleration.

If

$$V_{x\,gr} = \frac{1}{\hbar} \frac{\partial W}{\partial k_x},$$

the acceleration is

$$a_x = \frac{dV_{x\,gr}}{dt} = \frac{1}{\hbar}\left(\frac{\partial^2 W}{\partial k_x^2}\frac{\partial k_x}{\partial t} + \frac{\partial^2 W}{\partial k_x \partial k_y}\frac{\partial k_y}{\partial t} + \frac{\partial^2 W}{\partial k_x \partial k_z}\frac{\partial k_z}{\partial t}\right).$$

Using (1.41) we obtain

$$a_x = \frac{1}{\hbar^2}\left(\frac{\partial^2 W}{\partial k_x^2}F_x + \frac{\partial^2 W}{\partial k_x \partial k_y}F_y + \frac{\partial^2 W}{\partial k_x \partial k_z}F_z\right).$$

In the same manner we may obtain another two components of acceleration vector **a**. Coefficients before F_x, F_y, and F_z can be considered as elements of the reverse mass tensor. In the isotropic case the connection between acceleration and force takes the usual form

$$\mathbf{a} = \frac{d\mathbf{V}_{gr}}{dt} = \frac{1}{\hbar^2}\frac{\partial^2 W}{\partial k^2}\mathbf{F} = \frac{\mathbf{F}}{m^*}.$$

A quantity

$$m^* = \left(\frac{1}{\hbar^2}\frac{\partial^2 W}{\partial k^2}\right)^{-1}$$

may be defined as the effective mass of the electron. For a one-dimensional model,

$$m^* = \frac{\hbar^2}{2A_N a^2 \cos(ka)}.$$

The characteristic of $m*$ versus k is shown in Fig. 1.13. We may discover much of interest in this plot. Firstly, the effective mass depends on energy. We can even resign ourselves to this, as it is clear that the crystal ion influences the dynamics of electron movement. Secondly, at the band upper limit the mass becomes negative. This is quite surprising! It has been obtained that electrons move toward the force. It is difficult to devise a classical explanation of this fact. Since, in the end, force of an electromagnetic nature acts on the electron, i.e.,

$$\mathbf{F} = e\mathbf{E} + e[\mathbf{vB}],$$

we can set a few at rest, considering that near the upper limit, particles with a positive charge are moving. This artificial approach has turned out to be so suitable that everybody has begun to use it and a specific name for electrons with a negative mass has even been devised: they are called "holes."

The hole—as an electron absence in the necessary place—is not a material object and hence cannot carry a current. As a matter of fact, nevertheless, the electron itself moves, releasing one place and occupying another one. However, since the hypothesis of the hole (as a real particle with a positive charge) has allowed creation of diodes, transistors, microcircuits, personal computers, and other "delights" of scientific–technological progress, its application is completely justified.

After such surprising things, it will not be strange to us that in anisotropic crystals the field directed along the x axis can move an electron in the direction of the z axis. Therefore, we leave it to readers to investigate such "trifles" by themselves.

1.10 Matrix Formulation of Quantum Mechanics

The Schrödinger equation has helped us to clarify the behavior and character of atoms and molecules isolated from any external influences. Our goal is the analysis of electromagnetic field interactions with these objects. According to thing logic, we would take into consideration the energy of its interaction in the Hamiltonian and find out the new wave function, again solving the Schrödinger equation. Then, using the fourth postulate of quantum mechanics we could calculate any interesting observable quantity—for example, the dipole moment of the molecule. This implies determination of the spatial distribution $\psi(x,y,z)$ for each new external influence, and this is a rather complex problem.

Fortunately, an elegant mathematical approach was constructed a long time ago, allowing essential simplification of analysis of the behavior of quantum objects under external influences. It permits us to avoid solution of the Schrödinger equation in the coordinate space. This approach is based on matrix algebra, and now we shall describe it.

1.10.1 Wave Functions in the Matrix Representation

Since the eigenfunction systems of the observable quantity operators are orthonormal and complete (see Sect. 1.3), any wave function can be presented in the form of a series or an integral over the basis of eigenfunctions.

Let us choose the system of eigenfunctions of the $\hat{A}$ operator as the basis system. Here, for instance, the Hamiltonian of the harmonic oscillator may serve as the $\hat{A}$ operator. Then any wave function can be written in the form of the generalized Fourier series:

$$\psi(\mathbf{r}) = \sum_n c_n u_n(\mathbf{r}), \tag{1.42}$$

where c_n are expansion coefficients:

$$c_n = \int_{-\infty}^{\infty} u_n^*(\mathbf{r})\psi(\mathbf{r})d\mathbf{r}. \qquad (1.43)$$

Equation (1.43) can be rewritten in the form of a scalar product $(c_n = (u_n, \psi))$ or in Dirac designations $(c_n = <u_n|\psi>)$. Expression (1.42) means that some vector column of the expansion coefficients

$$\psi(\mathbf{r}) \leftrightarrow |\psi\rangle = \begin{pmatrix} c_1 \\ c_2 \\ \vdots \\ \vdots \end{pmatrix}$$

corresponds to wave function $\psi(\mathbf{r})$ in the coordinate representation. This is a so-called ket vector. It describes the wave function in representation of the eigenfunctions of the $\widehat{A}$ operator. The vector line (a bra vector)

$$\psi(\mathbf{r}) \leftrightarrow \langle\psi| = \begin{pmatrix} c_1^* & c_{2_1}^* & \cdot\cdot \end{pmatrix}$$

can be brought into correspondence to the conjugate complex wave function.

Since, in the general case, the basis function system is infinite, the bra and ket vectors are infinite dimensional.

What is the physical sense of the coefficients c_n? To answer this question we obtain the average value of the observable quantity A. In accordance with (1.6) and taking into account that $u_n(\mathbf{r})$ are eigenfunctions of the $\widehat{A}$ operator, we have

$$<A> = \int \psi^* \widehat{A} \sum_n c_n u_n(\mathbf{r})d\mathbf{r} = \sum_n A_n c_n \int u_n \psi^* d\mathbf{r} = \sum_n A_n |c_n|^2. \qquad (1.44)$$

Hence, $|c_n|^2$ represents the probability of detection during A measurement of our physical system described by wave function ψ in state $u_n(\mathbf{r})$. Having discussed the third postulate we have already stated that during the measurement process the instrument interacts with the physical system in such a manner that it transfers this system in an uncontrollable way into one of the eigenstates of the observable quantity operator. It is clear that if $\psi = u_k$, then $|\psi\rangle$ represents a column with zeroes in all positions besides the one where the unit is situated.

The convenience of matrix interpretation becomes evident in examination of the scalar product of any wave functions expanded in a series with the same basis:

$$\psi = \sum_n c_n u_n; \varphi = \sum_n b_n u_n$$

Because of the orthogonality of the basis

$$(\psi, \varphi) = \int \psi^* \varphi dr = \int \sum_n \sum_k c_n^* b_k u_n^* u_k dr = \sum_n c_n^* b_n \qquad (1.45)$$

(i.e., the scalar product), which in the coordinate representation is calculated through the integral in three-dimensional space, in the matrix representation it turns out to be the simple product of bra and ket vectors:

$$(\psi, \varphi) = \begin{pmatrix} c_1^* & c_2^* & \cdot & \cdot \end{pmatrix} \begin{pmatrix} b_1 \\ b_2 \\ \vdots \end{pmatrix} = \langle \psi | \varphi \rangle.$$

Now we hope the sense of the Dirac designation is fully clear (brackets means parentheses). By the way, the normalization condition $(\psi, \psi) = \langle \psi | \psi \rangle = 1$ in accordance with (1.42) takes the form

$$\sum |c_n|^2 = 1,$$

which completely corresponds to the physical sense of the coefficients.

1.10.2 Representation of Operators as Matrices

Let some quantum operator be such as

$$\varphi = \widehat{D} \psi. \qquad (1.46)$$

Wave functions in representation of the eigenfunctions of some other operator $\widehat{A}$ are columns. The question arises: what can operator $\widehat{D}$ look like in this representation? It is clear that it should take the form of a matrix with elements d_{ij}. In this case, in accordance with the rules of matrix algebra, Eq. (1.46) takes the form

$$b_n = \sum_k d_{nk} c_k; \qquad (1.47)$$

i.e., the action of the linear integral differential operator reduces to simple operations of addition and multiplication.

How do we find the matrix elements of the $\widehat{D}$ operator? We multiply the right and left parts of (1.46) by the basis function u_i^* and integrate over the coordinates

$$\int u_i^* \sum_n b_n u_n d\mathbf{r} = \int u_i^* \sum_k c_k \widehat{D} u_k d\mathbf{r}.$$

Taking into consideration the basis orthogonality and changing the order of summing and integration, we obtain

$$b_n = \sum_k \left(\int u_n^* \widehat{D} u_k d\mathbf{r} \right) c_k.$$

After comparison of this expression with (1.47), it is easy to make sure that

$$d_{nk} = \int u_n^* \widehat{D} u_k d\mathbf{r} = \left\langle u_n^* \left| \widehat{D} \right| u_k \right\rangle. \tag{1.48}$$

As we know, any operator in quantum mechanics is Hermitian. What does this mean from the point of view of matrix representation? If we are to use (1.48) for determination of the Hermitian property of (1.11), it becomes clear that for a Hermitian operator always

$$d_{nk} = d_{kn}^*. \tag{1.49}$$

If this condition is not fulfilled, such a matrix cannot correspond to an observable quantity.

The search for an average value of observable quantity D in the representation of eigenfunctions of the $\widehat{A}$ operator is reduced also to operations with matrices:

$$< D > = \left\langle \psi \left| \widehat{D} \right| \psi \right\rangle = \sum_k c_k^* \sum_n d_{kn} c_n = \sum_{k,n} c_k^* d_{kn} c_n. \tag{1.50}$$

As we see, the matrix approach allows avoidance of complex integral calculation. Obviously, we must choose as a basis such functions for which operators look like the simplest form.

Let us examine now which form has the $\widehat{A}$ operator in representation of its own eigenfunctions:

$$a_{mn} = \int u_m^* \widehat{A} u_n d\mathbf{r} = A_n \int u_m^* u_n d\mathbf{r} = A_n \delta_{mn}.$$

From this, the conclusion follows that in representation of eigenfunctions the operator is a diagonal matrix, in which diagonal elements are equal to eigenvalues. So, for instance, the Hamiltonian of the harmonic oscillator in representation of eigenfunctions (in so-called energy representation) has a form in accordance with (1.33):

$$\widehat{H} = \hbar\omega_0 \begin{vmatrix} 1/2 & 0 & 0 & \cdots \\ 0 & 3/2 & 0 & \cdots \\ 0 & 0 & 5/2 & \cdots \\ \cdots & \cdots & \cdots & \cdots \end{vmatrix}.$$

According to the Schrödinger equation, in the general case, wave functions depend upon time. This means that in (1.46) the coefficients c_n should be considered as functions of time t. We use matrix representation of operators to obtain the evolution equation for $c_n(t)$. We substitute (1.46) in the Schrödinger equation (1.7):

$$j\hbar \sum_n \frac{dc_n}{dt} u_n = \sum_n c_n \widehat{H} u_n.$$

Calculating the scalar product of both parts of this equation with function u_m we obtain

$$j\hbar \frac{dc_m}{dt} = \sum_n c_n \widehat{H}_{mn}.$$

The Hamilton operator is the Hermitian one; therefore, the evolution equation is

$$-j\hbar \frac{dc_m^*}{dt} = \sum_n c_n^* \widehat{H}_{mn}.$$

Using Dirac designations we may write equations for bra and ket vectors:

$$j\hbar \frac{d|\psi\rangle}{dt} = \widehat{H}|\psi\rangle, \quad -j\hbar \frac{d\langle\psi|}{dt} = \langle\psi|\widehat{H}.$$

In conclusion, it makes sense to state that the formal approach presented here is not something specific, which is typical for quantum mechanics. This specific direction in mathematics was developed a long time ago and successfully—namely, the theory of representation of linear vector spaces and linear operators, with achievements that have found applications in many areas of science, particularly in signal theory and the theory of electromagnetic fields and waves.

1.11 Description of Particle Ensembles by the Density Matrix

It is now time for us to transfer from examining isolated particles—atoms and molecules—to examining systems of many weakly interacting particles. Gases of molecules and atoms, gaseous plasma, gases of electrons and holes in semiconductors and in metals, etc. are examples of such systems. Strictly speaking, the gas particle state should be described by the wave function depending on the coordinates

and spins of all particles constituting this particle ensemble: $\psi(\mathbf{r}_1, s_1, \mathbf{r}_2, s_2, \mathbf{r}_3, s_3, \ldots$
$\mathbf{r}_n, s_n, t)$. But, because in practice $n \sim 10^{10}$ and more, an approach based on the
multiparticle wave function gives us little. It is much more productive to use a
statistical approach and transfer to the wave function for the single particle $\psi(\mathbf{r}, s, t)$,
considering a number of particles being in any one-particle state. The strict procedure of such a transfer is called secondary quantization. However, we shall not delve
deeper into it. Instead of it, we use the formalism of a density matrix, which is
sometimes also called a statistical matrix.

1.11.1   The Density Matrix

If all particles constituting a physical system are described by the same wave
function ψ, we can talk about an ensemble of pure states. Its statistical character is
related to the noncontrollable uncertainty caused by the instrument during any
measurement, and has a purely quantum nature. In many-particle systems the
different particles or their groups can be described by different functions ψ_1, ψ_2,
ψ_3, $\ldots \psi_n$. In this case we can talk about an ensemble of mixed states. For large
numbers of particles our knowledge about a system will certainly be incomplete, and
we can speak only about the fact that the particle is described by wave function ψ_k
with some probability. This probability has no relation to the wave behavior of the
substance and is an example of the "classical" probability peculiar to any large
system.

Now we assume that particles in the ensemble may take an arbitrarily large but
discrete set of states $\{\psi_k\}$. We obtain the average value of some observable quantity
D in this ensemble:

$$\langle D \rangle = \sum pk \langle \psi_k | \hat{D} | \psi_k \rangle. \tag{1.51}$$

The wave function can be expanded into a series of eigenfunctions of some
operator $\hat{A}$ with the discrete spectrum:

$$\psi_k = \sum_n c_{kn} u_n(\mathbf{r}). \tag{1.52}$$

Then, taking into account (1.50), Eq. (1.51) can be transformed into the form

$$< D > = \sum_k p_k \left(\sum_{m,n} c_{km}^* d_{mn} c_{kn} \right) = \sum_{m,n} \left(\sum_k p_k c_{km}^* c_{kn} \right) d_{mn}.$$

Let us introduce the following designation:

$$\rho_{mn} = \sum_k p_k c^*_{km} c_{kn}. \tag{1.53}$$

A matrix with such elements is called the matrix of density $\hat{\rho}$ in representation of eigenfunctions of the $\hat{A}$ operator. Then the equation for <D> takes the form

$$< D > = \sum_n \left(\hat{\rho}\hat{D}\right)_{nn} = \mathrm{Tr}\left(\hat{\rho}\hat{D}\right), \tag{1.54}$$

i.e., the average value of the observable quantity is equal to the sum of diagonal elements of the product of the density matrix and the matrix representing the operator of this quantity. Having known the density matrix we are able to determine any observable quantity. It means that $\hat{\rho}$ is the irrefragable characteristic of the system representing the generalization of ψ-function concepts in the case of mixed ensembles. Let us discuss the physical sense of separate elements of the density matrix. It diagonal element is

$$\rho_{nn} = \sum_n p_k |c_{kn}|^2,$$

where p_k is the probability that the particle is in state ψ_k and $|c_{kn}|^2$ is the probability at measurement of quantity A of detecting the particle described by function ψ_k being in state u_n. So, ρ_{nn} is the probability of detecting the particle in state u_n. As for nondiagonal elements of the density matrix ρ_{nm}, they describe the transitions of the ensemble particles between the proper states u_n and u_m. This is not evident, but we confirm its accuracy in the chapter 2. By the way, the density matrix, like any other operator in quantum mechanics, is Hermitian; therefore,

$$\rho_{nm} = \rho^*_{mn}. \tag{1.55}$$

It is clear that the density matrix, as the wave function, should satisfy the normalization condition, which takes the form:

$$\sum_n \rho_{nn} = \mathrm{Tr}(\hat{\rho}) = 1. \tag{1.56}$$

This is obvious from the physical sense of the diagonal elements.[16]

---

[16]The abbreviation "tr" comes from "trace"; sometimes the abbreviation "sp" (from "spoor") is used.

It remains for us to clarify the last important issue related to the density matrix—namely, its evolution in time. It turns out that the following equation for the density matrix directly follows from the Schrödinger equation:

$$j\hbar \frac{d\widehat{\rho}}{dt} = \left[\widehat{H}, \widehat{\rho}\right], \tag{1.57}$$

where $\left[\widehat{H}, \widehat{\rho}\right] = \widehat{H}\widehat{\rho} - \widehat{\rho}\widehat{H}$ is a commutator.

It should be noted that the system of eigenfunctions of the Hamiltonian of an undisturbed physical system—for instance, an isolated atom—may most often be used as a basis system. If we are interested in the internal motion of atoms or molecules in a gas or impurity atoms in a crystal, then (as we saw in the previous sections) the Hamiltonian spectrum of eigenvalues is discrete, and in this case, all formulas beginning from (1.52), which we obtained above, are valid. And what should we do in a case where we are interested in the headway motion of free particles or the motion of particles in the periodic field?

1.11.2   The Density Matrix in the Case of a Continuous Spectrum of Eigenvalues

It is known that the wave function of the free particle and energy $W = p^2/2m$ of its forward movement depend continuously upon impulse $\mathbf{p}$, which plays the role of the "index" in the system of basis functions. Hence, at the expansion of the single-particle function in such a basis, the summation in the formula (1.52) for a discretely changing index should be replaced by integration over the impulse.

$$\psi_{\mathbf{r}_0\mathbf{p}_0}(\mathbf{r}) = \int C_{\mathbf{r}_0\mathbf{p}_0}(\mathbf{p})u_{\mathbf{p}}(\mathbf{r})d\mathbf{p}, \tag{1.58}$$

which corresponds to the traditional Fourier transformation. If you compare this equation with (1.9), you will see that (1.58) describes a particle localized in the space. The only difference from (1.9) lies in the fact that in (1.58) all time dependence is "concealed" in the expansion coefficient. Complex indices of the wave function in coordinate and impulse representations mean that the particle is localized at the point with the coordinate $\mathbf{r}_0$ and it has the average impulse $\mathbf{p}_0$ (i.e., at $\mathbf{p} = \mathbf{p}_0$, $C_{\mathbf{r}_0\mathbf{p}_0}$ takes the maximal value).

In other words, (1.58) describes the wave function of the particle located at the point $(\mathbf{r}_0, \mathbf{p}_0)$ in the six-dimensional phase space. Here, we must note the reservation that in fact we cannot speak about particle localization at the point, because of the fact that its wave function is "diffused" in the coordinate and the impulse (see Fig. 1.1) owing to the uncertainty principle (1.12). Here, there is a straight analogue of the property of the spectral density of the impulse: a product of the spectrum width and the impulse duration, as we know, is a constant. For this reason, it is expedient to divide all of the volume of the phase space into cells so that its size $\Delta \mathbf{r} \cdot \Delta \mathbf{p}$ will satisfy the uncertainty principle: $\Delta \mathbf{r} \cdot \Delta \mathbf{p} = \hbar^3$. In this case we can speak about particle localization in the cell of the phase space with a center at the point $(\mathbf{r}_0, \mathbf{p}_0)$, as is shown in Fig. 1.14. But the size of this cell, evidently, is very small. For instance, at $\Delta x \approx 10^{-10}$ m (atom size), $\Delta p_x \approx 6 \cdot 10^{-24}$ J/s.

Now we may introduce the purely classical probability that the particle falls into the cell of the phase space with the center at the point $(\mathbf{r}_0, \mathbf{p}_0)$: $P(\mathbf{r}_0, \mathbf{p}_0) \Delta \mathbf{r} \cdot \Delta \mathbf{p}$. The continuous function $P(\mathbf{r}_0, \mathbf{p}_0)$ is the probability density and now replaces for us the discrete probability p_k in (1.51), (1.53). After this, similarly to (1.53), we may write the element of the density matrix in impulse representation as

$$\rho(\mathbf{p}, \mathbf{p}') = \sum_{\mathbf{r}_0} \sum_{\mathbf{p}_0} P(\mathbf{r}_0, \mathbf{p}_0) C^*_{\mathbf{r}_0 \mathbf{p}_0}(\mathbf{p}') C_{\mathbf{r}_0 \mathbf{p}_0}(\mathbf{p}) \Delta \mathbf{r} \Delta \mathbf{p},$$

where summing is provided for coordination of all cells of the phase space. Remembering the small cell sizes, we can transfer from summing to integration:

$$\rho(\mathbf{p}, \mathbf{p}') = \int \int P(\mathbf{r}_0, \mathbf{p}_0) C^*_{\mathbf{r}_0 \mathbf{p}_0}(\mathbf{p}') C_{\mathbf{r}_0 \mathbf{p}_0}(\mathbf{p}) d\mathbf{r}_0 d\mathbf{p}_0. \tag{1.59}$$

Fig. 1.14 A particle localized in the cell of the phase space. Cells of the phase space have a size equal to $\Delta \mathbf{r} \Delta \mathbf{p} = \hbar^3$, which corresponds to the uncertainty principle

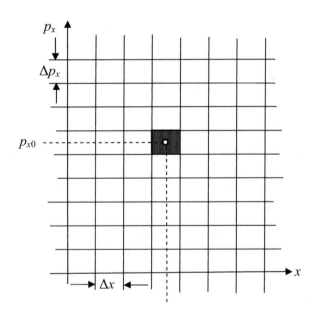

What does the diagonal element of the density matrix represent? Since $\left|C_{\mathbf{r}_0\mathbf{p}_0}(\mathbf{p})\right|^2$ is the density of the probability of detecting the particle with impulse $\mathbf{p}$ in cell $(\mathbf{r}_0, \mathbf{p}_0)$, then $\rho(\mathbf{p}, \mathbf{p})$ represents the probability density for finding the particle with such an impulse in all of the phase space. As we see, we have a full analogy to the case of the discrete spectrum. Hence, formulas (1.54), (1.55), (1.56), and (1.57) remain valid, but the summing must be replaced everywhere with integration over the impulse.

1.12   The Fermi–Dirac and Bose–Einstein Statistics

1.12.1   The Density Matrix in the State of Thermodynamic Equilibrium

In the previous section we stated that the density matrix is most frequently written in representation of eigenfunctions of the undisturbed system Hamiltonian. ρ_{nn} is the probability of finding the particle in the stationary state with energy W_n, and ρ_{nm} describes transitions between energy levels W_n and W_m. Particle transitions from one state to another happen for two reasons.

First, an external force (for instance, an electromagnetic field) can act on the particles. Secondly, particle interaction with any external physical system (a thermostat) always takes place, and this system may be so large that its state is not changed as a result of such interaction. This interaction, as we have already discussed, is rather weak, but it causes relaxation processes, with the result that the particle ensemble (without external disturbance) early or late accomplishes the state of thermodynamic equilibrium. Equilibrium takes place in the case that the average values of the observable quantities in the physical system do not depend upon time. Taking into consideration (1.54) we may say that the equilibrium density matrix ρ^e also does not depend upon time (the index "e" means equilibrium). It turns out that the form of ρ^e can be further specified independently from the nature of the relaxation processes happening in the system.

First, we draw readers' attention to the nondiagonal elements. From condition $\rho^e(t) = \text{const}$ it follows that in view of (1.57), $\left[\hat{H}, \hat{\rho}\right] = 0$. In energy representation the Hamiltonian, as we saw in Sect. 1.7, is a diagonal matrix; thus, the density matrix for thermodynamic equilibrium should also be a diagonal one, i.e.,

$$\rho^e_{mn} = 0, \quad m \neq n. \tag{1.60}$$

Equality to zero of nondiagonal elements means that the number of transitions from state n to state m at equilibrium is equal to the number of reverse transitions.

Now we proceed to the diagonal elements. If we multiply ρ^e_{nn} by the total number of particles in system N, then the time-independent quantity

$$N_n = N\rho^e_{nn} \tag{1.61}$$

will define the number of particles being in state n. This quantity is called the equilibrium state population. The problem of finding a state population distribution in the ensemble at thermodynamic equilibrium is one of the central problems of statistical physics.

But before proceeding to its examination, it is necessary to return once more to the question we previously touched upon when we studied the reasons for molecule formation.

1.12.2   Invisibility of Identical Particles and the Pauli Principle

If we succeeded in some time moment in localizing electrons passing through narrow slots, then in further time moments the wave function would become blurred in the space and would overlap because of electron diffraction in a slot. In the region where wave functions are overlapping it is absolutely impossible to identify this or that electron. For this reason, in quantum mechanics, similar particles (as we have already mentioned) are indistinguishable if they are in one physical system. It follows from this that the system's state should not be changed if any pair of particles are interchanged. This takes place in the case where the complete multiparticle wave function satisfies the condition

$$\psi(q_1, q_2) = \pm\psi(q_2, q_1), \tag{1.62}$$

where $\mathbf{q}$ means the coordinates and the spin of the particle. The "+" sign corresponds to the symmetric wave function, and the "−" sign corresponds to the nonsymmetric one. We can strictly prove that if any system is in a symmetric state, then it will never pass into an antisymmetric state. Thus, only the nature of particles defines by which multiparticle wave function the ensemble of these particles can be described—symmetric or antisymmetric.

First it was discovered and then, with the help of relativistic quantum mechanics, W. Pauli[17] proved that particles with a half-integer spin (electrons, nucleons,

---

[17]Wolfgang Pauli (1900–1958) was an Austrian theoretical physicist and one of the founders of quantum mechanics and relativistic quantum field theory. In 1945 he received the Nobel Prize in Physics, and in 1958 he was awarded the Max Plank Medal by the German Physical Society. He was elected to membership of the Swiss Physical Society, the American Physical Society, and the American Association of Fundamental Sciences, and foreign membership of the Royal Society.

neutrinos) are described by an antisymmetric wave function. They are called Fermi[18] particles (fermions). Particles with an integer spin (photons, hydrogen, mesons) are always in symmetric и states; they are called Bose[19] particles (bosons).

Antisymmetry of the fermion state leads to an extremely important result. Let one particle be in state ψ_1 and another particle in state ψ_2. If the wave functions of these particles are overlapped, their total two-particle wave function is equal, owing to (1.62):

$$\psi = \psi_1(q_1)\psi_2(q_2) - \psi_1(q_2)\psi_2(q_1) = \begin{vmatrix} \psi_1(q_1) & \psi_1(q_2) \\ \psi_2(q_1) & \psi_2(q_2) \end{vmatrix};$$

compare this with (1.30). This result can be generalized to the case of a system with N fermions:

$$\psi = \begin{vmatrix} \psi_1(q_1) & \cdots & \psi_1(q_N) \\ \cdot \cdot & \cdot \cdot & \cdot \cdot \\ \psi_N(q_1) & \cdots & \psi_N(q_N) \end{vmatrix}. \tag{1.63}$$

Let us assume that two particles are in the same state. Hence, two lines of the determinant coincide and $\psi = 0$, which tells us about the unrealizability of such a situation. The Pauli principle follows from that: in a system of interacting fermions, one state can be occupied by only one particle. We have already encountered the action of this principle in the example of the hydrogen molecule.

1.12.3   Equilibrium Distribution of the State Population

Let us return to the issue of the diagonal elements of the density matrix ρ_{ii} related by Eq. (1.61) with populations of the stationary states N_i. There are several different approaches to determination of population distribution at thermodynamic equilibrium. We can start from the principle of detailed equilibrium, equating the transition number from state i into state j with the number of reverse transitions. We may use the so-called canonic Gibbs distribution. It seems to us that the following approach is the most visual.

---

[18]Enrico Fermi (1901–1954) was an Italian physicist and one of the founders of nuclear physics. He received the Nobel Prize in Physics in 1938. He created the first nuclear reactor and in 1942 discovered the chain reaction. The 100th element in the Mendeleev table—fermium—was named in his honor. In the USA, the Institute for Nuclear Studies in Chicago was renamed the Enrico Fermi Institute for Nuclear Studies in 1955, and the Enrico Fermi Award was established in 1956.

[19]Satyendra Bose (1894–1974) was an Indian physicist from Bengal, who specialized in mathematical physics and was one of the founders of quantum statistics. He was elected to membership of the Royal Society in 1958.

It is known from thermodynamics that the entropy of a physical system

$$S = k_B \ln (P_T).$$

Here, $k_B = 1.38 \cdot 10^{-23}$ J/K is the Boltzmann[20] constant; $P_T > 1$ is the thermo-dynamic probability of a macroscopic state equal to a number of possible micro-scopic realizations of the given state of the physical system.

In the equilibrium state, entropy takes the maximal value. This principle lies at the basis of statistical physics. It means that at equilibrium the particle system is in a very chaotic and very uncertain state (out of all possible states).

Starting from this, our task is determination of the number of PT different variants of realization of some particle distribution over states N_1, N_2, ..., N_i and the further search for a distribution that provides the maximum P_T. We should take into consideration that the sum over all states is:

$$\sum_i N_i W_i = W; \quad \sum_i N_i = N, \tag{1.64}$$

where W is the total energy of the system, which is considered fixed. Calculation of P_T should be provided, taking into account that identical particles are indistinguish-able according to the Pauli principle. The procedure for statistical calculation is not very complex, but it requires certain concentration and knowledge of combinatorial analysis; therefore, we do not give it here.

As a result of P_T maximization, and using the fact that $dS = dW/T$, where T is the system temperature, we can obtain the equilibrium distribution of the stationary state population:

$$N_i = \left(e^{\frac{W_i - \mu}{k_B T}} \pm 1 \right)^{-1}. \tag{1.65}$$

Here, the "+" sign relates to fermions (the Fermi–Dirac statistics), and the "−" sign to bosons (the Bose–Einstein statistics). Quantity μ is called the chemical potential, and it can be found from the normalization condition (1.64).

It should be taken into account that some states may have the same energy, i.e., degeneration may happen. If energy level W_i is g_i-times degenerated, its population is

$$N_{W_i} = N_i g_i.$$

---

[20]Ludwig Boltzmann (1844–1906) was an Austrian physicist and one of the founders of statistical physics. His scientific interests covered almost all fields of physics and some areas of mathematics. He authored publications in mathematics, mechanics, hydrodynamics, elasticity theory, electro-magnetic field theory, optics, thermodynamics, and kinetic theory of gases.

Fig. 1.15 Equilibrium distribution of the population of stationary states. The *solid curve* denotes Fermi statistics (fermions) and the *dotted curve* denotes Bose statistics (bosons). Fermions (e.g., electrons) are always distributed in energy more or less uniformly, and bosons (e.g., photons) tend to gather in a state with the same value of energy; therefore, the energy of coherent radiation (the same value of $\hbar\omega$) can be arbitrarily large

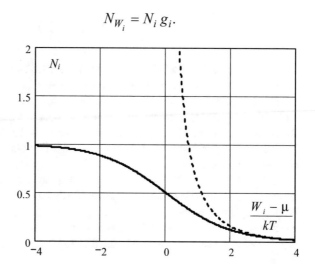

$$N_{W_i} = N_i \, g_i.$$

If summing is fulfilled not over states but over different energy levels, the normalization condition (1.64) can be rewritten in the form

$$\sum_{\text{over levels}} N_{W_i} = N. \tag{1.66}$$

The typical view of the Fermi–Dirac and Bose–Einstein distributions is shown in Fig. 1.15. It is evident that the monotonous decrease of state populations with energy growth is their mutual trait; the lower the system temperature, the more sharp the decrease. The difference lies in the fact that in the case of fermions, in full correspondence with the Pauli principle, $N_i \leq 1$,[21] while for bosons there are no restrictions on the state population.

Let us now discuss the population distribution of energy levels for electrons, oscillation, and rotational spectra. On the one hand, for $T = 300$ K the value $k_B = 0.025$ eV. On the other hand, the typical energy values are 10 eV for electron states and 1 eV for oscillation states. Hence, $\exp\left(\frac{W_i}{k_B T}\right) \gg 1$, and populations of these states at room temperature are described by the distribution (1.65) "tails," which for fermions and bosons exponentially decrease in the same manner. Really, we can neglect the unit in the denominator of (1.65); the expression takes the form:

---

[21] It should be taken into account that N_i and (1.65) have a sense of probability that states with energy W_i are occupied.

$$N_i = C \exp \left(\frac{W_i}{k_B T} \right), \tag{1.67}$$

which corresponds to the Maxwell–Bolzmann distribution. By the way, if at calculation of thermodynamic probability we consider a particle as distinguishable, as a result we obtain exactly the distribution (1.67) instead of (1.65). This testifies to the fact that population distribution happens in accordance with the Maxwell–Boltzmann statistics, when the symmetry property of the multiparticle wave function is inessential—for instance, for localized impurity atoms in a crystal, which are clearly distinguishable.

As for rotational movement, its energy has an order of 0.01 eV. This value is commensurable with the value of $k_B T$. Nevertheless, as we shall see later, owing to the large mass of molecules at real temperatures and concentrations, the molecule gas always is described over forward movement degrees of freedom by the statistic (1.67), i.e., it is nondegenerate. Hence, it is nondegenerate over rotational degrees of freedom.

Thus, the population of electron, vibrational, and rotational states for the most practically important situations obeys the Maxwell–Boltzmann statistic (1.67). The population of appropriate energy levels can be calculated as

$$N_{W_i} = N \frac{g_i \exp \left(-W_i / k_B T \right)}{\sum\limits_{j=1}^{\infty} g_j \exp \left(-W_j / k_B T \right)}, \tag{1.68}$$

where summing in the denominator is provided over all different levels. Obviously, (1.68) satisfies the normalization condition (1.66). In accordance with (1.68), the population of the upper levels grows with the temperature increase, but it does not exceed the population of the lower levels at any temperature.

1.12.4   The Equilibrium Distribution Function for Translational Motion of Fermions

Up to now we have examined the population of discrete energy levels. As for translational motion, its energy in the case of a free particle gas, as we know, changes continuously. In accordance with the results of the previous section, the element of the density matrix (1.55) describing this motion, being multiplied by the number of system particles, gives the average density in the impulse space. In other words, if we take into consideration the distribution function over impulses $f(\mathbf{p})$, then

$$f(\mathbf{p})d\mathbf{p} = N\rho(\mathbf{p}, \mathbf{p})d\mathbf{p} \tag{1.69}$$

represents the average number of particles with impulses, localized in the interval from $\mathbf{p}$ to $\mathbf{p} + d\mathbf{p}$. Let us find the distribution function $f(\mathbf{p})$ in the thermodynamic equilibrium state for the free Fermi particles.

To use the known distribution of the discrete state population (1.65) instead of infinite space, we shall consider the space divided into cubic areas of a large enough L and require that the wave functions of particles in these areas satisfy the periodicity condition. This requirement may seem rather artificial, but at an L much larger than the de Broglie wavelength it does not lead to a noticeable community limitation. However, at the same time, it allows passing from the continuous state spectrum to the discrete spectrum, although the discrete value, as we shall see now, is very small. In this situation, if we know how many discrete states $g(\mathbf{p})d\mathbf{p}$ fit in the impulse interval from $\mathbf{p}$ to $\mathbf{p} + d\mathbf{p}$, we shall be able to find the distribution function, using the following formula:

$$f(\mathbf{p})d\mathbf{p} = N_{W(\mathbf{p})}g(\mathbf{p})d\mathbf{p}, \tag{1.70}$$

where $N_{W(\mathbf{p})}$ is the population of the state with energy $W(\mathbf{p})$. So, our nearest task is the search for $g(\mathbf{p})$, which is called the state density.

The Hamiltonian of the translational motion corresponds to the operator of kinetic energy (see Table 1.1), and the stationary Schrödinger equation takes the form of the Helmholtz equation:

$$\nabla^2 \psi + \frac{2mW}{\hbar^2}\psi = 0. \tag{1.71}$$

Its solution represents plane waves with the wave number $k = \sqrt{2mW}/\hbar = p/\hbar$. The condition of periodicity is satisfied by the following eigenfunctions:

$$\psi_{n_x,n_y,n_z} = C\,\exp\left[j2\pi\left(n_x x + n_y y + n_z z\right)/L\right]; n_{x,y,z} = 0,\,\pm 1,\,\pm 2,\,\ldots,$$

describing free particles with the impulse

$$p_x = 2\pi\hbar n_x/L;\;\; p_y = 2\pi\hbar n_y/L;\;\; p_z = 2\pi\hbar n_z/L.$$

As we see, the discrete value of the impulse projection on any axis is equal to $2\pi\hbar/L$, and it can be made indefinitely small with L growth. To find out the number of states fitted to the interval $d\mathbf{p} = dp_x dp_y dp_z$, it is enough to divide this interval by the discrete value of the impulse space $(2\pi\hbar/L)^3 = (2\pi\hbar)^3/V$, where V is the volume occupied by the N-particle system. Then, in accordance with (1.70),

$$f(\mathbf{p}) = \frac{2V}{(2\pi\hbar)^3}\frac{1}{\exp\left(\frac{W(\mathbf{p})-\mu}{k_BT}\right)+1}. \tag{1.72}$$

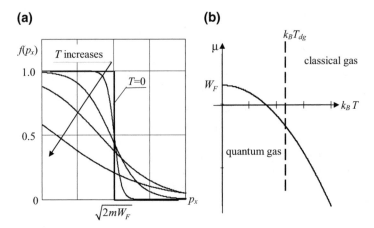

Fig. 1.16 The Fermi–Dirac distribution function (law). (**a**) Distribution of fermions on the impulse for different temperatures. (**b**) Function of chemical potential versus temperature. At a high temperature the distribution function differs a little from the exponent, which corresponds to the Maxwell–Boltzmann classical distribution law. In this case the ensemble of fermions is considered a nondegenerate particle gas. There are so few particles and so many free places that we can forget about the Pauli principle

The "redundant" two in the numerator considers the fact that each of the states (1.71) can be occupied by two fermions with the opposite spin.

Figure 1.16a shows a family of curves describing distribution functions (1.72) for different temperature values. At $T = 0$, all states with impulse $p \leq \sqrt{2mW_F}$ are occupied. Here, W_F is so-called Fermi energy (of a Fermi level), which is equal to the value of the chemical potential for the temperature $T = 0$. Particles with the larger impulse value are absent in the system. But, during temperature growth, the distribution over impulses becomes increasingly diffused and its tail decreases more slowly. Half of the maximal value $f(\mathbf{p})$ always corresponds to the value $p = \sqrt{2m\mu}$ (if $\mu > 0$). It should be noted that the chemical potential itself is not the temperature function (it turns out so as a result of (1.64) normalization) and, as we see from Fig. 1.16b, for a large enough temperature it is even negative. In the literature the extended interpretation of the term "Fermi level" is used very often, in which the chemical potential is always considered as Fermi energy.

It is clear that at a definite temperature, the exponent in the (1.72) denominator essentially exceeds 1 and then (1.72) is transformed into the form

$$f(p) = \frac{N}{(2\pi mk_BT)^{3/2}} \exp\left(-\frac{p^2}{2mk_BT}\right), \qquad (1.73)$$

which corresponds to the Maxwell distribution, which describes a classical nondegenerate gas. The multiplier before the exponent in (1.73) arises as a result of normalization of the distribution function

$$\int f(\mathbf{p})d\mathbf{p} = N.$$

Now we can clarify under which condition Eq. (1.73) is wrong, i.e., the gas is degenerating (when it is impossible to neglect quantum effects), and we need to use the accurate distribution function (1.72). This condition is obvious enough:

$$\exp\left(-\frac{\mu}{k_B T}\right) \leq 1.$$

We transform it into a more visual view, for which we approximately express the chemical potential through the gas parameters under the condition that the gas can still be considered nondegenerate. From comparison (1.72) and (1.73) it follows that

$$\frac{2V}{(2\pi\hbar)^3} \exp\left(\frac{\mu}{k_B T}\right) = \frac{N}{(2\pi m k_B T)^{3/2}}.$$

Then the condition of gas degeneration takes the form

$$T \leq T_{dg} = \frac{h^2 n^{2/3}}{2^{5/3}\pi m k_B}, \tag{1.74}$$

where $n = N/V$ is the particle concentration. Quantity T_{dg} is called the degeneration temperature. From (1.74) it follows that the gas is degenerate in the case that its temperature is low enough and the particle concentration is high enough. If $T \gg T_{dg}$ we may use the Maxwell distribution for description of the thermodynamic equilibrium state.

Which gases of those that really exist are degenerate and which is not? The answer to this question is given by Table 1.3.

In all mentioned examples the conductance electrons in the metal represent the degenerate gas owing to their high concentrations. As for the molecule gas, owing to its large mass, $T_{dg} \leq 10$ K even for the concentration $n \sim 10^{28}$ m^3, which corresponds to the condensed medium. Therefore, the molecule gas can always be considered a nondegenerate one.

Table 1.3 Temperature of degenerate gas

Type of gas	n (m^{-3})	T (K)	T_{dg} (K)	Type of electron gas
Electrons in copper	10^{29}	300	$5 \cdot 10^4$	Degenerate
Electrons in a semiconductor	10^{23}	300	10	Nondegenerate
Electrons in thermonuclear plasma	10^{21}	10^7	0.4	Nondegenerate
Electrons in gas discharge plasma	10^{18}	10^4–10^5	$4 \cdot 10^{-3}$	Nondegenerate

1.13   Problems for Chapter 1

1.1. A bullet with a mass of 10 g flies at a velocity of 1000 m/s. Determine the
de Broglie wavelength. Why might we use equations from classical physics
for description of bullet flight?

1.2. Explain why electron movement in the electron ray tube of a visual display can
be described by classical equations. The electron mass is 9.1×10^{-31} kg and
the velocity is 10^6 m/s.

1.3. Determine the stationary states of the electron in a "potential box" with a width
of 0.2 nm and infinitely high walls.

1.4. An electron is situated in a "potential box" with a wall height of 10 eV and a
width of 0.2 nm. Determine the stationary states.
Assistance: For numerical solution of the disperse equation, you may use a
computer. This task is similar to the task describing electromagnetic waves in
a dielectric plate.

1.5. The wave function in the time moment $t = 0$ takes the form

$$\psi(z, 0) = A \exp\left[-\left(\frac{z}{2\Delta z}\right)^2\right].$$

Determine the constant A and the spectral density of the impulses.
   Assistance: Use the table integral

$$\int\limits_0^\infty \exp\left(-r^2 x^2\right) dx = \frac{\sqrt{\pi}}{2r}$$

   and Fourier transformation tables.

1.6. The wave packet in task 1.5 varies with a change in time. Determine the
dependence of the wave function versus time, taking into consideration that in
the free space,

$$W(p) = \frac{p^2}{2m}.$$

Assistance: Use the results of task 1.5 and formula (1.9) with Fourier transformation tables.

1.7. Calculate the radiated frequencies and wavelengths for the hydrogen atom in
the Balmer series. The lower level of these lines corresponds to $n = 2$ and the
upper level to $m = 3, 4, 5. \ldots$.

1.8. The natural oscillation frequency of the CO molecule is equal to
$\omega_0 = 4.087 \times 10^{14}$ 1/s ($\lambda = 4.61$ μm). Draw an energy diagram of the
oscillation spectrum. Show the absorption lines on the frequency
(wavelength) axis.

1.9. The rotational constant B of the linear CO molecule is equal to ≈ 57.9 GHz. Show the energy spectrum of the rotational levels. Determine the frequencies of the first five allowed transitions.

1.10. The crystal lattice constant of silicon is equal to ≈ 0.55 nm. Determine the zone boundaries by using the one-dimensional Kronig–Penney model.
Assistance: Use a computer for numerical solution of the dispersion equation.

1.11. Find the matrix elements of the dipole moment of an electron situated in a "potential box" with a width of 2 nm. Use the results of task 1.3.

1.12. Prove that the matrix representing the product of operators is equal to the matrix product of these operators.

1.13. Using the results of task 1.12, determine the matrix elements of the dipole moment square for an electron located in a "potential box" with a width of 2 nm. Compare the results with direct calculation.

1.14. The entropy of a system can be written as

$$S = k_B \, Sp(\rho \ln (\rho)).$$

For the basis on which the density matrix is diagonal:

(a) Prove that for the system situated in the pure state the entropy is equal to zero.

(b) Determine the expression for system entropy, which can be in any proper state with equal probability.

(c) Prove that in case (b) the entropy is maximal.

1.15. Determine the particle distribution law on rotational levels of the CO molecule, assuming that molecules obey the Maxwell–Boltzmann statistics.

1.16. Determine the temperature of CO gas if it is known that the populations of levels $J = 5$ and $J = 6$ are equal to 10^{22} and 9×10^{21} m^{-3}, relatively. The rotational constant B of the CO molecule is equal to ≈ 57.9 GHz.

1.17. The density operator for a system in thermodynamic equilibrium with a thermostat takes the form

$$\hat{\rho} = \frac{\exp\left(-\widehat{H}/k_B T\right)}{Sp\left(\exp\left(-\widehat{H}/k_B T\right)\right)}.$$

Prove that if

$$\widehat{H}|u_i\rangle = W|u_i\rangle,$$

the diagonal elements of the density matrix give the Maxwell-Boltzmann distribution (1.67).

Chapter 2
Interaction of Electric Dipoles

2.1 Introduction: Multipole Expansion of Radiation Energy and Medium Interaction

Having presented information about a substance's structure, we are ready now to examine the issues of its interaction with electromagnetic waves. As we have already stated, the semiclassical approach entirely suits us when a substance is considered from the quantum point of view and a field is described by Maxwell equations.

In Chap. 1 it we found out that components of a particle's substance may, in stationary states, have a discrete spectrum of energy and may have a continuous or quasicontinuous energy spectrum. In this and later chapters we shall consider what happens with a substance under the action of a field if the particles have a discrete energy spectrum.

In the ensemble of weakly interacting particles, which have a discrete energy spectrum, wave functions are localized in a space with a size of about 10^{-10} m. The wavelength that is interesting for quantum electronics exceeds 10^{-7} m. Hence, the radiation field can be considered as a plane wave within the limits of an atom or a molecule.

The electric field of the wave

$$\mathbf{E}(\mathbf{r}, t) = \mathrm{Re}\left\{\mathbf{E}\exp\left[j(\omega t - \mathbf{kr})\right]\right\}$$

can be expanded into the Taylor series near the coordinate of the particle:

$$\mathbf{E}(\mathbf{R} + \mathbf{r}) = \mathbf{E}(\mathbf{R}) + (\mathbf{r}\nabla)\mathbf{E}(\mathbf{R}) + \dots .$$

If we are to be limited by the first item in the series, the task becomes like an electrostatic task. Hence, the interaction energy can be determined by considering the particle as an elementary electric dipole. For this reason, this approximation is called the electric dipole approximation. The kinetic interaction energy

© Springer Nature Switzerland AG 2020
V. V. Shtykov, S. M. Smolskiy, *Introduction to Quantum Electronics and Nonlinear Optics*, https://doi.org/10.1007/978-3-030-37614-7_2

$$W_{\text{int}} = -\mathbf{dE}$$

will serve as the initial point for determination of the interaction operator.

The second expansion item can be presented in the form

$$(\mathbf{r}\nabla)\mathbf{E}(\mathbf{R}) = -\mathbf{r} \times \mathbf{rot}\mathbf{E} + \mathbf{Q}(\mathbf{r}, \mathbf{E}, \nabla) = \mathbf{r}\frac{\partial \mathbf{B}}{\partial t} + \mathbf{Q}.$$

The first item here defines the magnetic dipole interaction with energy

$$W_{\text{int}} = -\mathbf{mB},$$

and the second item is the so-called quadrupole interaction. We can obtain items of higher order in the same manner. However, in practice, it is usually sufficient to use electric dipole or magnetic dipole approximation.

As we remember, energy levels are located equidistantly. For this reason, the field interacts, as a rule, in a resonance manner with only one pair of levels. The influence of other levels is insignificant. Therefore, in the future we will be limited by consideration of hypothetical two-level particles, which represents a very convenient model for description of substance resonance reaction in a field.

2.2 Equation for the Density Matrix of a Two-Level System

Now we examine two allocated energy levels 1 and 2, which are considered nondegenerate for simplicity (Fig. 2.1). Let energy W_1 correspond to stationary state $u_1(\mathbf{r})$, and energy W_2 to state $u_2(r)$. The $u_{1,2}(\mathbf{r})$ functions are eigenfunctions of the Hamiltonian describing the isolated particle—a so-called undisturbed Hamiltonian $\widehat{H}_0$.

As a matter of fact, in the "gas," two-level particles interact weakly with each other, colliding from time to time and thereby changing the energy and the impulse. Moreover, a gas of two-level particles may interact weakly with a thermostat, in the role of which other gas molecules, the walls of the vessel, and atoms in the crystal lattice may act. Definite energy and the appropriate Hamiltonian $\widehat{H}_{\text{rel}}$ are peculiar to

Fig. 2.1 Energy diagram of a two-level system. W_1 and W_2 are eigenvalues, and $u_1(\mathbf{r})$ and $u_1(\mathbf{r})$ are eigenfunctions of the undisturbed Hamiltonian $\widehat{H}_0$

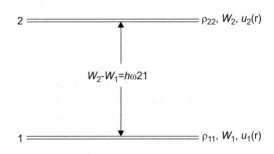

all of these weak interactions. The index "rel" means "relaxation" because exactly these interactions lead to gradual return of the gas to the state of thermodynamic equilibrium if this gas has been taken out of that state in some manner.

Finally, if particles are under the action of an electromagnetic field, the energy of particle interaction with the field, which corresponds to the interaction Hamiltonian $\widehat{H}_{int}$, should be added to its energy. Thus, the total Hamiltonian of the two-level particle gas in the electromagnetic field takes the view

$$\widehat{H} = \widehat{H}_0 + \widehat{H}_{rel} + \widehat{H}_{int}. \tag{2.1}$$

Our goal is to describe the substance reaction in the field. The reaction represents either the substance polarization if we are speaking about the electric field action or the substance magnetization if we are interested in the magnetic field action. Both of these macroscopic observable quantities can be obtained by averaging over the volume of the microscopic observable quantities—electric or magnetic moments of the particle—and we shall be able to determine these last quantities, knowing the density matrix of the two-level particle gas. Therefore, our next task is equation writing for the density matrix of the two-level particles.

According to (1.57), we obtain the commutator $\left[\widehat{H}_0, \widehat{\rho}\right]$. In the energy representation (see Sect. 1.11),

$$\widehat{H}_0 = \begin{pmatrix} W_1 & 0 \\ 0 & W_2 \end{pmatrix}. \tag{2.2}$$

Taking (2.2) into consideration, we find that

$$\left[\widehat{H}_0, \widehat{\rho}\right] = \begin{pmatrix} \rho_{11}W_1 & \rho_{12}W_1 \\ \rho_{21}W_2 & \rho_{22}W_2 \end{pmatrix} - \begin{pmatrix} \rho_{11}W_1 & \rho_{12}W_2 \\ \rho_{21}W_1 & \rho_{22}W_2 \end{pmatrix}$$
$$= \begin{pmatrix} 0 & -\rho_{12}\hbar\omega_{21} \\ \rho_{21}\hbar\omega_{12} & 0 \end{pmatrix}, \tag{2.3}$$

where $\omega_{21} = (W_2 - W_1)/\hbar$ is the resonance frequency of the transition between levels 1 and 2.

Now it is time to determine the commutator $\left[\widehat{H}_{rel}, \widehat{\rho}\right]$. But the problem is that we do not know the explicit view of the $\widehat{H}_{rel}$ operator, and we will hardly be sure of finding it by starting from the physics of relaxation processes: they are too complicated and often large in number. We may use the empirical approach, taking into account the fact that relaxation processes are described by exponential functions. Moreover, we know that when relaxation finishes, nondiagonal elements of the density matrix $\rho_{12}, \rho_{21} = 0$ and its diagonal elements are equal to their equilibrium values $\rho_{11} = \rho_{11}^e$ and $\rho_{22} = \rho_{22}^e$, which correspond to the Maxwell–Boltzmann statistics (see Sect. 1.12). Starting from these considerations and also from (2.3), we may write evolution equations (1.57) for elements of the density matrix of the two-level system in the absence of an electromagnetic field:

$$\frac{d\rho_{11}}{dt} = -\frac{\rho_{11} - \rho^e{}_{11}}{T_1};$$

$$\frac{d\rho_{22}}{dt} = -\frac{\rho_{22} - \rho^e{}_{22}}{T_1};$$

$$\frac{d\rho_{12}}{dt} = j\omega_{21}\rho_{12} - \frac{\rho_{12}}{T_2};$$

$$\frac{d\rho_{21}}{dt} = -j\omega_{21}\rho_{21} - \frac{\rho_{21}}{T_2}.$$

(2.4)

From Eq. (2.4) it clearly follows that at $t \to \infty$, ρ_{11} and ρ_{22} tend exponentially to $\rho^e{}_{11}$ and $\rho^e{}_{22}$, and ρ_{12} and ρ_{21} tend to zero, exponentially performing damping oscillation with the frequency ω_{21}. The relaxation speed for ρ_{11} and ρ_{22} is defined by time T_1, and for ρ_{12} and ρ_{21} it is defined by time T_2. Why are these relaxation time constants different? To answer this question, we need to remember which physical sense the elements of the density matrix have.

The diagonal elements ρ_{11} and ρ_{22} are proportional to the populations of energy levels 1 and 2; hence, they describe the energy of the two-level system. Therefore, its variation at relaxation should be related to the energy exchange between particles and the thermostat. A similar exchange happens in interaction of particles with other gas molecules or with atoms of the lattice, in inelastic collisions between particles, and in spontaneous emission. The typical frequency of the energy exchange process is the sum of the frequencies of all mentioned components and, without doubt, depends upon the physical nature of the two-level particle gas. Therefore,

$$\frac{1}{T_1} = \frac{1}{\tau_{rad}} + \nu_{ne} + \nu_{thr},$$

(2.5)

where τ_{rad} is the radiation lifetime of the particle at the upper level, ν_{ne} is the frequency of inelastic collisions of particles among them, and ν_{thr} is the frequency characterizing the speed of energy exchange with a thermostat.

The quantity T_1 is called the time of dipole–lattice, spin–lattice, or longitudinal relaxation. The accepted terminology will be more understandable in Chap. 3.

Let us pass on to consideration of the nondiagonal elements ρ_{12} and ρ_{21}. They describe particle transitions from level 2 to level 1 and vice versa, which corresponds to particle oscillations with frequency ω_{21} from the classical point of view. That is why ρ_{12} and ρ_{21} are described by oscillating functions of the type exp $[(\pm j\omega_{12} - 1/T_2)t]$. The oscillation amplitude of ρ_{12} and ρ_{21} will be large in the case where all oscillations are provided synchronously. However, the above-mentioned three types of relaxation processes, besides energy exchange, disturb the particle synchronism. As a result, the oscillation amplitudes of ρ_{12} and ρ_{21} decrease. Moreover, the purely elastic particle collisions, which do not change the

particle energy, lead also to losses of phase coherency in their oscillations. Therefore, the typical time of oscillation damping is determined as

$$\frac{1}{T_2} = \frac{1}{T_1} + \nu_{el}, \tag{2.6}$$

where ν_{el} is the frequency of elastic collisions between particles. The quantity T_2 is
called the time of dipole–dipole, spin–spin, or transverse relaxation. It is clear that
always

$$T_2 \leq T_1. \tag{2.7}$$

Of course, the relaxation times and relaxation items introduced by us in (2.4)
describe the actual running processes only approximately, i.e., on the basis of a
model. But similar approximation is used very often in applied calculations. Determination of $\widehat{H}_{rel}$ and calculation of $T_{1,\,2}$ are tasks for "pure physicists." We shall use
either their results or (more probably) experimental data or semiempirical
estimations.

As the conclusion of this section, we show, for illustration, some numerical data.
In quantum electronics the ruby is used very often, representing a crystal of Al_2O_3
with Cr^{3+} ions as the impurity, which are the active particles. Time T_1 for them is
defined mainly by the radiation lifetime at the upper level, and for transition with a
resonance wavelength of $\lambda_{21} \cong 0.69$ μm it is $T_1 \cong 3 \cdot 10^{-3}$ s. Time T_2 is defined
mainly by the frequency of elastic dipole–dipole interactions between the impurity
ions, and at $T = 300$ K it is a small value: $T_2 \cong 7 \cdot 10^{-12}$ s. As the second example,
we examine oscillation–rotation transitions in CO_2 gas with a resonance wavelength
of $\lambda_{21} = 10.6$ μm. The radiation lifetime is very large ($\tau_{rad} = 3$ s); therefore, T_1 is
determined mainly by the frequency of inelastic collisions, and at a pressure of 1 mm
of a mercury column and a temperature of $T = 300$ K it is $T_1 \cong 10^{-3}$ s. The
frequency of elastic collisions can be estimated on the basis of gas–kinetic
presentations:

$$\nu_{el} = n\widetilde{v}S_{eff},$$

where n is the molecule concentration, $\widetilde{v}$ is the average speed of heat motion, and S_{eff}
is the effective section of collisions. If we take $S_{eff} \sim 10^{-17}$ m^2 and take into account
that $n = \frac{P}{k_B T}$, $\widetilde{v} = (8k_B T/\pi M)^{1/2}$, where M is the molecule mass, then $\nu_{el} \cong 10^7$ s^{-1}.
If the pressure in the vessel is small and the free pass length is great enough, we must
take into consideration also elastic collisions of molecules with the vessel walls.
Their frequency is

$$\nu_{wall} \approx \frac{\widetilde{v}}{r},$$

where r is the radius of a long cylindrical vessel and for $r \cong 1$ cm, $\nu_{wall} \cong 4 \cdot 10^4$ s^{-1}. In total we have $T_2 \cong \nu_{el}^{-1} \cong 10^{-7}$ s.

2.3 Electric Dipole Transitions

Now we add Eq. (2.4) by items describing the interaction of particles with the electromagnetic field in electric dipole approximation. Let us find the disturbance Hamiltonian. Using the classical definition of the dipole moment, we write the appropriate quantum operator as

$$\widehat{\mathbf{d}} = q\widehat{\mathbf{r}}. \tag{2.8}$$

Using (2.8), we can write the interaction operator as

$$\widehat{H}_{int} = -\widehat{\mathbf{d}}\mathbf{E}. \tag{2.9}$$

Equation (2.9) constitutes the essence of the so-called electric dipole approximation in analysis of particle interaction with the field. Do not forget that it is valid for the small enough size of particles compared with the radiation wavelength.

To calculate the commutator $\left[H_{rel}, \widehat{\rho}\right]$, it is necessary to find the form of $\widehat{H}_{rel}$ in the energy representation. For this we should calculate the matrix elements of the dipole moment operator, which are in accordance with (1.48):

$$d_{ij} = \int u_i^*(r)\widehat{d}u_j(r)dr = -q\int u_i^*(r)ru_j(r)dr. \tag{2.10}$$

Knowing the wave functions describing stationary states of particles, we can in principle determine the values of the matrix elements. We would do it for a single-electron atom and for the simplest two-atom molecules. In more complicated cases, the calculations become very labor intensive, and in practice one can use spectroscopy measurement data to determine d_{ij}.

Let us consider the diagonal element

$$d_{ii} = -q\int u_i^*(\mathbf{r})\widehat{\mathbf{r}}u_i(\mathbf{r})^2 d\mathbf{r} = -q\int \mathbf{r}|u_i(\mathbf{r})|^2 \, d\mathbf{r}. \tag{2.11}$$

If the particle has a symmetry center, the wave functions $u_i(\mathbf{r})$ are either even or odd functions. This means that the integrand in (2.11) is always an odd function; hence, the diagonal elements $d_{ii} = 0$. The value of d_{ii} has a sense of the average value of the dipole moment of the particle being in the i-th state. Thus, the permanent dipole moment of the central-symmetric atoms is absent. It does not mean that such particles do not interact with the field. As will be clear in the future, the degree of interaction leading to stimulated transitions $i \leftrightarrow j$ depends upon the value of the nondiagonal element of the dipole moment operator d_{ij}.

From (2.10) it follows that for a symmetric particle, $d_{ij} \neq 0$ only in the case that u_i and u_j have the opposite symmetry. Analyzing in which case $d_{ij} \neq 0$, we can determine the following selection rules for electric dipole transitions:

1. For electron levels, $\Delta n = 0, \pm 1$; $\Delta l = \pm 1$.
2. For oscillation levels of the harmonic oscillator, $\Delta v = \pm 1$; in the case of the anharmonic oscillator, Δv is arbitrary.
3. For rotation levels, $\Delta J = \pm 1$.

If the selection rules are not fulfilled, interaction with the field still takes place, but it will be at the expense of moments of higher order. The intensity of such interactions is essentially weaker than that of electric dipole ones.

Let us move on, at last, to writing equations for elements of the density matrix. We shall assume for simplicity that our particles are symmetric and transitions between levels 1 and 2 are allowable. Then, in accordance with (2.9) and taking into account hermicity,

$$\widehat{H}_{\text{rel}} = - \begin{pmatrix} 0 & d \\ d^* & 0 \end{pmatrix} E,$$

where $d \equiv d_{12} = d_{21}^*$. Hence,

$$\left[\widehat{I}, \widehat{\rho}\right] = \left[\begin{pmatrix} \rho_{12} d^* & \rho_{11} d \\ \rho_{22} d^* & \rho_{21} d \end{pmatrix} - \begin{pmatrix} \rho_{21} d & \rho_{22} d \\ \rho_{11} d^* & \rho_{12} d^* \end{pmatrix} \right] E. \tag{2.12}$$

Substituting (2.12) and (2.4), we obtain the following system of equations:

$$\begin{cases} \dfrac{d\rho_{11}}{dt} = \dfrac{1}{j\hbar}(\rho_{12} d^* - \rho_{21} d)E - \dfrac{\rho_{11} - \rho^e_{11}}{T_1}; \\[2mm] \dfrac{d\rho_{22}}{dt} = \dfrac{1}{j\hbar}(\rho_{12} d^* - \rho_{21} d)E - \dfrac{\rho_{22} - \rho^e_{22}}{T_1}; \\[2mm] \dfrac{d\rho_{12}}{dt} = \dfrac{1}{j\hbar}(\rho_{11} - \rho_{22})dE - \dfrac{\rho_{12}}{T_2} + j\omega_{12}\rho_{12}; \\[2mm] \dfrac{d\rho_{21}}{dt} = \dfrac{1}{j\hbar}(\rho_{11} - \rho_{22})d^* E - \dfrac{\rho_{21}}{T_2} - j\omega_{12}\rho_{21}. \end{cases} \tag{2.13}$$

Equation system (2.13) is written in a scalar form. In reality, atoms and molecules have spherical symmetry only in exceptional cases. Therefore, (2.13) should be considered the equation system for one of the coordinate components of field E that interacts with component d.

Specifying the initial conditions and solving this system of linear equations, we would find $\widehat{\rho}(t)$ and, hence, the time evolution of any observable physical quantity. We take an interest in the specific quantity—polarization that arises in the "gas" of two-level particles under the action of the electric field. Therefore, it would be fully reasonable, using (2.13), to obtain an equation for the substance polarization.

2.4 Polarization and Population Difference for Electric Dipole Interactions

Substance polarization is a sum of electric dipole moments of substance particles in the unit of a volume or, more precisely, the sum of observable values of dipole moments, each of which is connected, in accordance with (1.54), to elements of the density matrix and the dipole moment operator:

$$< d >= \mathrm{tr}\left\{ \begin{pmatrix} \rho_{11} & \rho_{12} \\ \rho_{21} & \rho_{22} \end{pmatrix} \begin{pmatrix} 0 & d \\ d^* & 0 \end{pmatrix} \right\} = \rho_{12}d^* + \rho_{12}d. \tag{2.14}$$

We should bear in mind that the matrix element of dipole moment operator $\widehat{\mathbf{d}}$ is a vector whose orientation in the space is defined by the orientation of the particle itself, and the value depends only upon the particle nature. Therefore, determination of polarization is reduced to averaging of dipole moments over the particle orientations:

$$\mathbf{P} = \frac{1}{V} \sum_{i=1}^{N} < \mathbf{d} >_i = n\{< \mathbf{d} >\}, \tag{2.15}$$

which is designated in the formula (2.15) by braces. Here, N is the particle's number in volume V and n is the particle concentration.

If you look attentively at (2.13), you will notice the expression $\rho_{11} - \rho_{22}$ there. Multiplying it by the particle concentration, we obtain

$$n_{12} = n(\rho_{11} - \rho_{22}), \tag{2.16}$$

which is the difference in the populations of energy levels 1 and 2 in the volume unit. This is quite an important characteristic of a substance, directly connected to energy accumulated in the substance. By the way, in the thermodynamic equilibrium state, $n_{12} = n_{12}^e$ is always more than zero and for $t \to \infty$ the quantity is always $n_{12}^e \to 0$ independently of the initial state.

Furthermore, in (2.13) there is also the item $\rho_{12}d^* - \rho_{21}d$, which has no specific physical sense. We shall use it in the form of some auxiliary quantity:

$$\xi = (\rho_{12}d^* - \rho_{21}d)n. \tag{2.17}$$

Now we will find the equations for the polarization and population difference of the two-level particle gas. First of all, we multiply scalarwise the third equation in system (2.13) by $n\mathbf{d}*$ and the fourth equation by $n\mathbf{d}$. Adding and subtracting them— taking into account (2.14), (2.16), and (2.17)—we obtain

$$\frac{d(n < \mathbf{d} >)}{dt} = j\omega_{12}\xi - \frac{n < \mathbf{d} >}{T_2};$$
(2.18)

$$\frac{d\xi}{dt} = \frac{1}{j\hbar}n_{12}[\mathbf{d}^*(\mathbf{d}, \mathbf{E}) + \mathbf{d}(\mathbf{d}^*, \mathbf{E})] - \frac{\xi}{T_2} + j\omega_{12}n < \mathbf{d} > .$$
(2.19)

In the last equation, the expression in square brackets can be presented as a product of some symmetric tensor and the electric field vector:

$$[\mathbf{d}^*(\mathbf{d}, \mathbf{E}) + \mathbf{d}(\mathbf{d}^*, \mathbf{E})] = 2\,\mathrm{Re} \left(\begin{vmatrix} |d_x|^2 & d_x^*d_y & d_x^*d_z \\ d_y^*d_x & |d_y|^2 & d_y^*d_z \\ d_z^*d_x & d_z^*d_y & |d_z|^2 \end{vmatrix} \right) E = 2\hat{\mathbf{d}}^2 \mathbf{E}.$$
(2.20)

Exclude quantity ξ from consideration. For this we express it through the quantities that are interesting for us, with the help of (2.18):

$$\xi = \frac{1}{j\omega_{21}} \left[\frac{d(n < \mathbf{d} >)}{dt} + \frac{n < \mathbf{d} >}{T_2} \right].$$
(2.21)

Substituting (2.21) in (2.19) and taking (2.20) into account, we obtain the rather long equation

$$\frac{1}{j\omega_{21}} \frac{d^2(n < \mathbf{d} >)}{dt^2} = \frac{2}{j\hbar}n_{12}\hat{\mathbf{d}}^2 \mathbf{E}$$
$$- \frac{2}{j\omega_{21}T_2} \frac{d(n < \mathbf{d} >)}{dt} + n < \mathbf{d} > \left[j\omega_{21} - 1/\left(j\omega_{21}T_2^2 \right) \right]$$
(2.22)

in which only the observable quantities are included—the dipole moment and the population difference.

For real systems, $1/T_2 \ll \omega_{21}$. Therefore, we can neglect the last item in square brackets in (2.22).

Now we multiply the first and second equations in system (2.13) by n and subtract one from the other. We find the equation

$$\frac{dn_{12}}{dt} = \frac{2}{j\hbar}\xi E - \frac{n_{12} - n_{12}^e}{T_1},$$

which takes the following form, accounting for (2.21):

$$\frac{dn_{12}}{dt} = \frac{2}{\hbar\omega_{21}} \left[\frac{d(n <\mathbf{d}>)}{dt} + \frac{n <\mathbf{d}>}{T_2} \right] \mathbf{E} - \frac{n_{12} - n_{12}^e}{T_1}. \quad (2.23)$$

This equation can also be slightly simplified to neglect the second item in the square brackets. In doing that, we shall not make a big mistake since $<\mathbf{d}>$, because of (2.14), is defined by nondiagonal elements of the density matrix, which quickly oscillate in time with a frequency of about ω_{21}. Therefore, a time derivative of this quantity is much larger than the second item.

It remains for us now to average over the particle orientation each item included in (2.22) and (2.23). The population difference does not depend upon the particle orientation (generally speaking, it is not always true); averaging of $n<\mathbf{d}>$ in accordance with (2.15) gives us the polarization, but averaging of tensor $\widehat{\mathbf{d}}^2$ represents a very nontrivial problem. If we are dealing with a gas of chaotically oriented particles, it represents an isotropic medium. It is clear that in this case, $\{d_x\} = \{d_y\} = \{d_z\} = 0$ and tensor $\widehat{\mathbf{d}}^2$ becomes diagonal. Moreover, $\{|d_x|^2\} = \{|d_y|^2\} = \{|d_z|^2\} = d^2/3$, where $d = |d|$ is the modulus of the matrix element of the dipole moment operator. This is a result you can easily obtain yourself if you remember something from the theory of random variables. Thus, tensor $\widehat{\mathbf{d}}^2$ transforms into the scalar $d^2/3$.

The situation will be more complicated if we are dealing with ordered–oriented particles, as in the case of crystal bodies, as not only may diagonal elements of the tensor not be equal to each other, but also nondiagonal elements may be nonzero after averaging. In the general case, tensor $\widehat{\mathbf{d}}^2$ has the same symmetry elements as the crystal lattice of a substance. This issue is discussed in detail in Sect. 2.8.

In the future we shall limit ourselves to the case of an isotropic substance, although probably sometimes we will take into consideration the anisotropy if it is important principally. As a result of (2.22) and (2.23), averaging over the orientations, we finally obtain:

$$\frac{d^2\mathbf{P}}{dt^2} + \frac{2}{T_2}\frac{d\mathbf{P}}{dt} + \omega_{12}{}^2\mathbf{P} = \frac{2\omega_{21}d^2}{3\hbar} n_{12}\mathbf{E}; \quad (2.24)$$

$$\frac{dn_{12}}{dt} + \frac{n_{12} - n_{12}^e}{T_1} = -\frac{2}{\hbar\omega_{21}}\frac{d\mathbf{P}}{dt}\mathbf{E}. \quad (2.25)$$

These equations deserve a detailed discussion. Indeed, they represent nothing other than constitutive equations of the medium. The system of Eqs. (2.24) and (2.25), together with the Maxwell equations, give an exhaustive description of electromagnetic wave propagation in a homogeneous medium consisting of two-level particles.[1]

If there are not exterior charges and currents in the medium, then

---

[1]In the case of medium spatial nonuniformity it is necessary to pass from equations for the density matrix to macroscopic parameters depending upon the coordinates. This is quite a nontrivial problem. Therefore, the medium is considered homogeneous everywhere in the future.

$$\mathbf{rot\,H} = \frac{\partial \mathbf{D}}{\partial t}; \quad \mathbf{rotE} = -\frac{\partial \mathbf{B}}{\partial t};$$
$$\mathrm{div}\,\mathbf{D} = 0; \quad \mathrm{div}\,\mathbf{B} = 0,$$

the magnetic induction is $B = \mu_0 H$ and the electric induction is

$$\mathbf{D} = \varepsilon_0\mathbf{E} + \mathbf{P}_{\mathrm{total}} = \varepsilon_0\varepsilon'\mathbf{E} + \mathbf{P},$$

where $\mathbf{P}_{\mathrm{total}} = \mathbf{P} + \mathbf{P}_{\mathrm{add}}$ is the polarization arising not only at the expense of the resonance transitions $1 \leftrightarrow 2$ but also as a result of all other transitions. Although the contribution of each nonresonance transition to the polarization is small, the large number of them can make a noticeable contribution in the induction, which is reflected by some "background" dielectric permeability $\varepsilon' \neq 1$, which should be taken into consideration especially in the case of condensed media.

Let us examine the equation for polarization. If an electric field is absent, then (2.24) represents the equation for the proper oscillations of the harmonic oscillator with losses (Fig. 2.2, left panel). The frequency of natural oscillation is equal to ω_{21}, and it is defined by the difference in the stationary state energy of the particles. The damping time for the oscillations is equal to T_2 and is determined by the interaction process between them and with a thermostat. The field acts on the oscillator as the external stimulating force. Thus, the obtained result is fully in agreement with the classical model of the medium as a set of elastic dipoles. But quantum theory introduces one new and very significant element: the degree of the field impact on a substance depends not upon the total number of oscillators but upon the population difference of the levels. This is supported by the view of the right part of (2.24). If $n_{12} = 0$, a substance, as it were, does not "feel" a field.

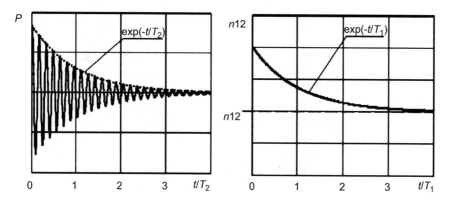

Fig. 2.2 Relaxation of the polarization and the population difference of a two-level system to a state of thermal equilibrium. The polarization behaves in the same way as the current in an oscillating circuit in the theory of electric circuits. The population difference corresponds to the process of energy exchange with a thermostat, which is described by a differential equation of the first order

Let us now examine the equation for population difference. The derivative dn_{12}/dt characterizes the speed of the variation in the energy accumulated in the substance. By what is this variation caused? According to Eq. (2.25), it is caused firstly by energy exchange as a result of relaxation processes (inelastic collisions, spontaneous emission, interaction with a thermostat) and secondly by the work of the electric field on the variation of the substance polarization. The power spent on polarization of the volume unit is equal to $E \cdot dP/dt$. Thus, (2.25) represents the equation for the power balance in the medium and is a consequence of the energy conservation law. If $E = 0$, at $t \to \infty$, because of relaxation processes, we obtain $n_{12} \to n_{12}^e$ (Fig. 2.2, right panel). The equilibrium population difference n_{12}^e is determined by the Boltzmann distribution and, as we have repeatedly emphasized, it is always a positive quantity. Nevertheless, the nonequilibrium population difference can be negative also. In this case it is said that there is a population inversion in the medium. We will understand later how it can be obtained.

At last, we have come to the final issue we would like to discuss here. We consider the electric field here as the external influence and polarization as the medium response to this influence. Is this response a linear one with respect to the external influence? Equations (2.24) and (2.25) testify to the fact that it is not linear, because of the presence of cross items in the right parts of these equations. Thus, a gas of two-level particles is principally a nonlinear medium, although, under some conditions, it can be approximately considered as a linear medium.

2.5 Linear Interaction of an Electromagnetic Field with a Substance

So then, we have a system of differential equations ((2.24) and (2.25)) describing the interaction of an electromagnetic field with a substance in electric dipole approximation. As mentioned above, this interaction has a nonlinear character. The strict solution of system (2.24) and (2.25) represents a complicated enough problem. The outlet, as you have probably guessed, consists in application of approximate linear equations instead of (2.24) and (2.25). It is clear that these equations can be obtained from the initial ones, assuming that the influence is weak (in comparison with something we shall discuss later). What will remain from (2.24) and (2.25) at E tending to zero?

If the intensity of electric field E is small, polarization P will have a similar order of smallness. Therefore, we may consider the right part as zero because it is the quantity of second order of smallness. As a result, (2.25) takes the form

$$\frac{dn_{12}}{dt} + \frac{n_{12} - n_{12}^e}{T_1} = 0. \tag{2.26}$$

The stationary solution to (2.26) is $n_{12} = n_{12}^e$. Substituting it in (2.25), we obtain equations describing the local linear connection of the electromagnetic field with the two-level system:

$$\frac{d^2\mathbf{P}}{dt^2} + \frac{2}{T_2}\frac{d\mathbf{P}}{dt} + \omega_{21}^2\mathbf{P} = \frac{2\omega_{21}d^2}{3\hbar}n_{12}^e\mathbf{E}. \tag{2.27}$$

Thus, in linear approximation we neglect an influence of the external field on particle distribution over energy levels, i.e., on the value of the internal energy of the system. Hence, linear approximation assumes an intensive energy exchange between the system and a thermostat. It means that the time of longitudinal relaxation T_1 is small.

We can solve (2.27) if

$$\mathbf{E}(t) = \mathbf{E}_m \cos(\omega t + \varphi).$$

In this case, polarization also takes the form of harmonic oscillation:

$$\mathbf{P}(t) = \mathbf{P}_m \cos(\omega t + \psi).$$

Now we express instantaneous values through complex amplitudes:

$$\mathbf{E}(t) = \mathrm{Re}\left(\mathbf{E}_m e^{j\varphi} e^{j\omega t}\right) = \mathrm{Re}\left(\dot{\mathbf{E}} e^{j\omega t}\right);$$
$$\mathbf{P}(t) = \mathrm{Re}\left(\mathbf{P}_m e^{j\psi} e^{j\omega t}\right) = \mathrm{Re}\left(\dot{\mathbf{P}} e^{j\omega t}\right);$$

instead of (2.27) we obtain

$$\left(-\omega^2 + j\frac{2}{T_2}\omega + \omega_{21}^2\right)\dot{\mathbf{P}} = \frac{2\omega_{12}d^2}{3\hbar}n_{12}^e\dot{\mathbf{E}}.$$

The solution to this equation is

$$\dot{\mathbf{P}} = \frac{2\omega_{12}d^2}{3\hbar}n_{12}^e\frac{\dot{\mathbf{E}}}{\omega_{21}^2 - \omega^2 + j\omega 2/T_2}. \tag{2.28}$$

The linear expression (2.28) can be presented in the form

$$\dot{\mathbf{P}} = \dot{\chi}(\omega)\dot{\mathbf{E}},$$

and one may determine the complex dielectric susceptibility of the two-level quantum system as

$$\dot\chi(\omega) = \frac{2\omega_{21}d^2n^e_{12}}{3\hbar(\omega_{21}{}^2 - \omega^2 + 2j\omega/T_2)} = -j\frac{\omega_{21}d^2n^e_{12}T_2}{3\hbar\omega}\frac{1}{1+j\xi}, \tag{2.29}$$

where $\xi = (\omega^2 - \omega_{21}{}^2)T_2/2\omega$ is the generalized frequency offset.

The solution in the form (2.29) is typical for any system manifesting resonance properties. You have already encountered the mechanical pendulum and the resonance electric circuit. Now there is another resonance system in front of you. Let us simplify (2.29), assuming (not without good reason) that $\omega \gg T_2{}^{-1}$. In this case we may replace the generalized offset with the approximate expression

$$\xi = \frac{(\omega - \omega_{12})(\omega + \omega_{12})}{2\omega}T_2 \approx (\omega - \omega_{12})T_2 = \Delta\omega T_2 = \frac{\Delta\omega}{\omega_{12}}\omega_{12}T_2 = 2Q_L\frac{\Delta\omega}{\omega_{12}},$$

where $Q_L = \frac{1}{2}\omega_{21}T_2$ is the Q factor of the spectral line. Here, you easily recognize well-known equations for the general offset of the usual resonance circuit. Frequency ω in the denominator of (2.29) can be replaced by ω_{21}. In this case, (2.29) takes the following form:

$$\chi(\omega) = -j\frac{d^2n^e_{12}T_2}{3\hbar}\frac{1}{1+j\xi}. \tag{2.30}$$

The complex dielectric susceptibility contains real $\chi'(\omega) = \chi_0\frac{-\xi}{1+\xi^2}$ and imaginary $\chi''(\omega) = -\chi_0\frac{1}{1+\xi^2}$ parts. Here, $\chi_0 = \frac{d^2n^e_{12}T_2}{3\hbar}$. The real part makes a contribution to the dielectric permeability of the medium, and the imaginary part describes energy absorption by the two-level system.

The complex amplitude of electric induction is

$$\dot{\mathbf{D}} = \varepsilon_0\dot{\mathbf{E}} + \dot{\mathbf{P}}_{total} = \varepsilon_0\varepsilon'\dot{\mathbf{E}} + \dot\chi\dot{\mathbf{E}} = (\varepsilon'\varepsilon_0 + \dot\chi)\dot{\mathbf{E}}$$
$$= [(\varepsilon'\varepsilon_0 + \chi') + j\chi'']\dot{\mathbf{E}} = \dot\varepsilon_a\dot{\mathbf{E}}$$

Here, $\dot\varepsilon_a$ is the complex dielectric permeability. For media with specific conductivity σ,

$$\dot\varepsilon_a = \varepsilon_a - j\sigma/\omega.$$

Thus, the total dielectric permeability is

$$\dot\varepsilon_a = \varepsilon'\varepsilon_0 + \chi_0/(1+\xi^2) - j\chi_0\xi/(1+\xi^2)$$

and the equivalent specific conductivity is

$$\sigma = \omega \chi'' \approx \chi_0 \omega_{12}/(1+\xi^2) = \sigma_0/(1+\xi^2). \tag{2.31}$$

Using the term "equivalent" we emphasize by this that in the two-level system there are not free charge carriers and the conduction current cannot flow. The appearance of σ relates only to phase relations between polarization **P** and field E. For brevity in the future we will not distinguish between "true" and "equivalent" conductivity.

If the specific conductivity is known, we may determine the power that is absorbed in the unit of a substance volume:

$$P_{\text{abs}} = \sigma \frac{|\dot{\mathbf{E}}|^2}{2} \approx \frac{\sigma_0 |\dot{\mathbf{E}}|^2}{2(1+\xi^2)}.$$

The characteristics of χ' and σ versus the frequency are shown in Fig. 2.3. The value of $\Delta\omega = 2/T_2$ is accepted as the absorption line width. The absorption line Q factor and the generalized offset can be conveniently expressed through $\Delta\omega$:

$$Q_{\text{line}} = \omega_{21}/\Delta\omega; \quad \xi \approx 2\frac{\omega - \omega_{21}}{\Delta\omega}.$$

Poynting vector flow **I** of the plane wave changes according to the law

$$\mathbf{I}(z) = \mathbf{I}(0) \exp\left(-2\alpha z\right) = \mathbf{I}(0) \exp\left(-\alpha_p z\right).$$

Here, α is the constant of field damping and $\alpha_p = 2\alpha$ is the constant of power damping. According to this, in media with small linear losses the last is equal:

$$\alpha = \frac{\sigma}{2}\sqrt{\frac{\mu_a}{\varepsilon_a}}.$$

Substituting (2.31) here we obtain

Fig. 2.3 Normalized function of the conductivity (*solid curve*) and the real part of the dielectric susceptibility (*dotted curve*) versus the generalized offset. In circuit theory, the real and imaginary parts of the complex impedance of a parallel circuit behave in the same manner

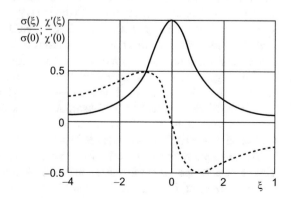

(a)   **(b)**

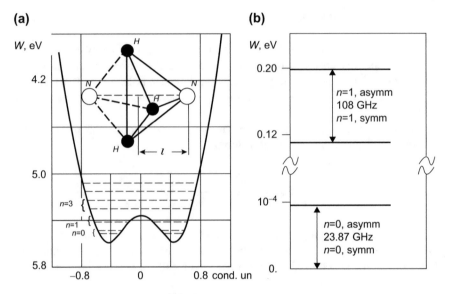

Fig. 2.4 Structure of the ammonia molecule (**a**) and its energy diagram (**b**). Two energy minimums correspond to two possible positions of the nitrogen atoms. The lower part of the energy diagram can be considered as a pair of coupled harmonic oscillators. Coupling leads to fission (removal of degeneration) of levels, which are accepted as inversion levels

$$\alpha = \frac{\sigma_0}{2(1 + \xi^2)} \sqrt{\frac{\mu_a}{\varepsilon_a}}. \tag{2.32}$$

Let us evaluate the orders of quantities concerning the susceptibility of the two-level system. As an example, we consider interaction of electromagnetic waves in the ultrahigh-frequency (UHF) range with ammonia molecules.

Experimental research studies of absorption of waves in the UHF range in ammonia vapors at a pressure of about 10^2 Pa (about 1 mm of a mercury column) allow detection of a strong absorption line at a frequency of 23.866 GHz. The resonance interaction is connected to the motion of the nitrogen atom with respect to the plane containing three atoms of hydrogen (Fig. 2.4a). The lower pair of energy levels (Fig. 2.4b) corresponds to the 23.866 GHz frequency. The wave functions have different parity, and this transition is allowed. The complete spectrum has a fine structure of rotational states of molecules. Ammonia has been the subject of numerous investigations, which have resulted in frequency measurements of more than half a hundred lined of a fine structure near 23.866 GHz.

It is possible to calculate the absorption index at power α_p according to the measurement results. The characteristic of α_p versus the frequency for ammonia at a pressure near 110 Pa is presented in Fig. 2.5.

The absorption line width on level 0.5 according to Fig. 2.5 is about 50 MHz. This value corresponds to the relaxation time $T_2 \cong 6.5 \cdot 10^{-9}$ s. The concentration of NH_3 molecules at $T = 300$ K and $P = 110$ Pa is $\cong 2.7 \cdot 10^{22}$ m^{-3}. According to

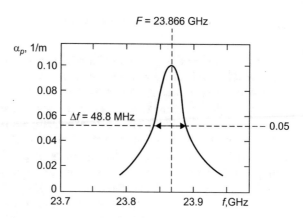

Fig. 2.5 Experimental characteristic of the damping constant versus the frequency of electromagnetic waves in ammonia at a pressure of 110 Pa. The spectral line width is near 50 MHz. This corresponds to a Q factor near 5000. In the molecular generator, the pressure is decreased so much that the Q factor reaches about 10^7

Table 2.1 Spectroscopic characteristics of some linear molecules

Molecule	Transition frequency (GHz)	Dipole moment (debyes)	Line width for 1 mm of a mercury column (MHz)	Absorption (1/m)
ClF	30.38	0.88	10.0	0.036
CH$_3$CN	35.69	3.9	25.0	0.068
HDO	50.24	1.8	4.0	0.16
HDO	80.58	1.8	4.0	0.5
HCN	172.7	3.0	25.0	10.0
HCN	365.9	3.0	25.0	37.0

estimation for the Boltzmann distribution, $n_1 \cong 1.7 \cdot 10^{21}$ m^{-3}, and the population difference $n_{12}^e \cong 6.6 \cdot 10^{18}$ m^{-3}. The value of $\sigma = 2.65 \cdot 10^{-4}$ Sim/m corresponds to the absorption index $\alpha_\Pi = 0.1$ m^{-1}. Using this value, we can determine the dipole moment of the transition $d = \sqrt{\frac{3\sigma\hbar}{n_{12}^e T_2 \omega_{12}}} \approx 3.6 \cdot 10^{-30}$ C $\cdot$ m.

In publications, authors often give the dipole moment value not in C $\cdot$ m but in debyes. These two units are connected with each other by the following relation:

$$1 \text{ debye} = (1/3) \cdot 10^{-29} \text{ C} \cdot \text{m}.$$

So, the dipole moment of the transition for ammonia is about 1.1 debyes, which corresponds well to the values obtained by the different approaches.

The results of spectroscopic investigations of some molecules are presented in Table 2.1.

The section of experimental physics dealing with investigations of radio wave interaction with a substance in the frequency range from 10^3 to 10^{12} Hz is known as radio spectroscopy.

2.6 The Spectral Absorption Line Shape

Investigation of the absorption spectra of atoms and molecules is a high-capacity instrument for study of their internal structures. It is important to know not only the frequency values but also the absorption line shape.

In the simplest case, the frequency characteristics of absorption are defined by the frequency-dependent multiplier in formulas (2.30), (2.31), and (2.32), which corresponds to the so-called Lorentz line shape. Such a curve type is often encountered during study of the frequency characteristics of physical systems—for instance, during study of the resonance phenomenon in high-Q electric circuits. Hendrik Lorentz[2] initially obtained this line shape in a slightly different form as the statistical distribution of the radiation frequencies of collided molecules in a gas, when the frequency of collisions between molecules is small in comparison with the radiation frequency of the undisturbed system. These processes increase the absorption line width. Similar widening is called uniform (homogeneous).

Since the line shape represents a separate interest, one can often consider and discuss only the type of function that describes absorption. A convention has been adopted in spectroscopy of writing these functions in such a manner that the area under the curve is equal to 1. The function describing the Lorentz absorption line shape and satisfying this condition takes the form

$$g_L(\omega) = \frac{1}{\pi} \cdot \frac{T_2}{1 + (\omega - \omega_{12})^2 T_2^2} = \frac{1}{\pi} \cdot \frac{T_2}{1 + \xi^2}.$$

This is shown in Fig. 2.6. As we have already mentioned, Lorentz line width $\Delta\omega_L$ is the frequency difference between points located on both sides of the peak, in which $g_L(\omega)$ is equal to half of the maximal value. This value can be measured easily, and it is suitable to use it for description of the line shape:

$$g_L(\omega) = \frac{2}{\pi \Delta\omega_L} \cdot \frac{1}{1 + \xi^2}.$$

Now we may write (2.31) and (2.32) as

$$\chi''(\omega) = -\chi_0 \frac{\pi g_L(\omega)}{T_2}; \quad \sigma(\omega) = \sigma_0 \frac{\pi g_L(0)}{T_2}; \quad \alpha(\omega) = \alpha_0 \frac{\pi g_L(\omega)}{T_2}.$$

---

[2]Hendrik Lorentz (1853–1928) was a Dutch physicist. He created the classical electron theory, with the help of which he explained many electric and optical phenomena, including the Zeeman effect. He developed the electrodynamics of moving media, introduced the transformation that bears his name, and came close to creating a theory of relativity. In 1902 he received the Nobel Prize in Physics.

Fig. 2.6 Normalized function of the spectral line shape versus the generalized offset for uniform widening. Such a line is called a Lorentz line

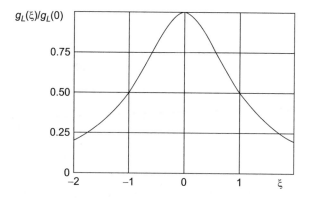

The term "homogeneous" is used for Lorentz widening, for the following reasons. Derivation of the polarization equation (2.24), from which the Lorentz function (2.30) is determined, is based on the evolution equation of the density matrix (2.13). In Eq. (2.13), transition frequency ω_{21} and relaxation time T_2 are the determining parameters. The polarization equation was obtained by means of summing over the atom number in the volume unit (see 2.15). As a result, the frequency characteristics of the total polarization coincide with the frequency characteristics of the separate atom polarization.

Hence, the characteristics of the total macroscopic polarization will be determined simply by the sum of the uniform series of contributions of separate atoms, each of which has the same transition frequency ω_{21} and line width $\Delta\omega_L$. Therefore, in summarizing, we can say that the frequency characteristics of the absorption line presented earlier correspond to the case of the uniformly widened Lorentz line. The uniformly widened line arises every time when widening is caused by relaxation processes acting in the same manner on all atoms that have the same frequency of transition. Interaction with lattice oscillations, collisions, and interaction between atoms may play roles in relaxation processes.

Under definite conditions, another type of line widening is observed, which is known as nonuniform widening. Such widening arises if the line width of the atom or molecule set increases owing to the fact that the atoms or molecules have different ω_{21} frequencies.

As an important example, we can examine Doppler line widening in a gas. In this case, different molecules have different frequencies of transition due to the Doppler frequency shift because of molecule motion. Other sources of nonuniform line widening include other mechanisms leading to displacement of the natural frequency of transition for reasons that vary from one atom to another. Variations in the frequencies of transition often have a Gaussian[3] statistical distribution, and the line at nonuniform widening will have a Gaussian form.

---

[3]Johann Gauss (1777–1855) was a German mathematician and physical scientist. From childhood he demonstrated ability in mathematics. At 10 years of age he found a solution to the problem of

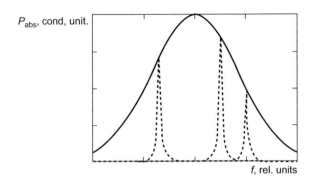

P_{abs}, cond, unit.

f, rel. units

Fig. 2.7 Spectral line formation for nonuniform widening. The line consists of a continuous set of standard Lorentz lines. Each such partial line corresponds to its own jump frequency. The diversity of jump frequencies is caused by the action of spatially nonuniform disturbances. The line shape depends on the distribution law of jump frequencies

In the case of nonuniform line widening, the separate atoms usually have uniformly widened lines, which are essentially narrower than the full line of the set of atoms as a whole. Such narrow uniformly widened lines, each of which corresponds to the absorption line of atoms with the same ω_{21} frequency, are called spin packets (see Chap. 4). The total shape of the uniformly widened line is defined in this case by the ensemble of spin packets, and the ω_{21} frequency region defines the total line width, as shown in Fig. 2.7.

If the field is applied to such an atom ensemble, the polarization is defined by the sum of independent contributions of separate spin packets. The equation for susceptibility in this case can be obtained by integration of a series of overlapped distributions from each spin packet.[4] There is an example of such a calculation below, which can be used for the case of Doppler widening in a gas. However, the calculation results may be used for another mechanism of nonuniform line widening. It relates to the probabilistic character of widening mechanisms and their multiplicity. A central limiting theorem is known in mathematics according to which the sum of a large number of random variables obeying any distribution law has a normal distribution.

---

summing the numbers from 1 to 100. The natural connection between theoretical and applied mathematics is typical of his creative work. Gauss achieved his worldwide reputation after developing a method for calculation of an elliptical planet orbit according to observations. His discoveries had an essential influence on the development of electricity and the theory of magnetism. His name appears on the map of the Moon.

[4]It should be taken into consideration that in order to introduce the local medium parameters, it is necessary to fulfill the integration over the volume, which is much less than a wavelength. The volume integral can be replaced by the integral over frequencies, if there are enough numbers of different types of particles in the volume. Moreover, we should take into account cross-relaxation between particles with different frequencies. The particle movement may cause so-called spatial dispersion—dependence of polarization at the specified point upon fields at another points. So, the real picture is much more complicated.

In the case of Doppler widening in a gas, the molecules in each spin packet have different ω_{21} transition frequencies because of the motion of the molecules. For example, if a molecule moves with velocity v toward an electromagnetic wave, the (nonrelativistic) interaction frequency is

$$\omega_{21} = \tilde{\omega}_{21}\left(1 + \frac{v}{c}\right).$$

Here, $\tilde{\omega}_{21}$ is the transition frequency of the motionless molecule. The effect of the frequency offset (the Doppler[5] effect) gives birth to a line consisting of the set of uniformly widened Lorentz lines, whose maximums correspond to different ω_{21} frequencies.

To determine the molecule distribution over the frequency of transition, it is necessary to use the distribution function for the forward degrees of freedom. According to (1.73), for a classical nondegenerate gas,

$$f(p) = \frac{N}{(2\pi m k_B T)^{\frac{3}{2}}} \exp\left(-\frac{p^2}{2mk_B T}\right).$$

Hence, the molecule number dn in the volume unit, whose velocity is in the interval from v_x to $(v_x + dv_x)$, is equal:

$$dn = n\left(\frac{m}{2\pi m k_B T}\right)^{\frac{1}{2}} \exp\left(-\frac{mv_x{}^2}{2k_B T}\right)dv_x.$$

Here, n is the total number of molecules in the volume unit. Now, taking into consideration the frequency variance as a result of the Doppler effect, the population difference n_{12}^e in (2.30) should be replaced as

$$dn = n_{12}^e F_G(\omega_{21})d\omega_{21},$$

where $F_G(\omega_{21}) = \frac{0,939}{\Delta\omega_G} \exp\left(-\ln(2)\left[2\frac{\omega_{21}-\tilde{\omega}_{21}}{\Delta\omega_G}\right]^2\right)$ is the function describing the particle distribution over the ω_{21} frequencies of transition. The value of $\Delta\omega_G$ defines the width of the Gaussian line on the level equal to half of the maximal value. The function $F_G(\omega_{21})$ is normalized in such a manner that the area under the curve is equal to 1.

---

[5]Christian Doppler (1803–1853) was an Austrian mathematician, physicist, and astronomer. He formulated the Doppler principle in acoustics and optics, which is responsible for the Doppler effect. He published works on light aberration, color theory, etc. He became a member of the Vienna Academy of Science in 1848.

Now we can determine the susceptibilities for nonuniform Doppler widening of the line. When electric field $\mathbf{E}$ is applied to the gas (in an isotropic case), any specified uniformly widened spin packet makes a contribution to the polarization on frequency ω; therefore,

$$\dot{P} = \int_{-\infty}^{\infty} \dot{\chi}(\omega, \omega_{21})\dot{E}d\omega_{21} = \dot{E} \int_{-\infty}^{\infty} \dot{\chi}(\omega, \omega_{21})d\omega_{21} = \dot{\chi}(\omega)\dot{E}. \qquad (2.33)$$

Here, $\dot{\chi}(\omega, \omega_{21}) = -j\frac{\chi_0}{1+j\xi(\omega,\omega_{21})}F_G(\omega_{21})$, and the generalized offset should be considered as the function of transition frequency ω_{21}.

The characteristics of real and imaginary parts of the dielectric susceptibility for the nonuniformly widened spectral line versus the generalized offset $\xi = 2Q_L\frac{\omega-\widetilde{\omega}_{21}}{\omega_{21}}$ are shown in Fig. 2.8. It can be seen that nonuniform widening increases the absorption width by almost one order.

Usually, the line width at uniform widening $2/T_2$ corresponding to the Lorentz spin packet is much less than total width $\Delta\omega_G$ of the Gaussian line. In this condition, calculation of (2.33) is essentially simplified because we may suppose that the spin packet width is vanishingly small. Therefore, for example, at calculation of the imaginary part of susceptibility

$$\chi''(\omega) = -\int_{-\infty}^{\infty} \chi_0 \frac{\pi g_L(\omega, \omega_{21})}{T_2}F_G(\omega_{21})d\omega_{21},$$

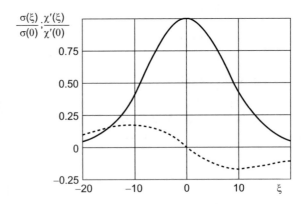

Fig. 2.8 Normalized functions of the real part of the dielectric susceptibility (*dotted curve*) and the specific conductance (*solid curve*) for a nonuniformly widened line versus the generalized offset. The figure shows the result of numerical integration. Each of the presented curves is defined by mutual action of separate uniformly widened spectral lines. The mechanism of nonuniform widening increases the absorption line by almost one order

the slower function $F_G(\omega_{21})$ can be taken out of the integral symbol:

$$\chi''(\omega) = -\pi \frac{\chi_0}{T_2} F_G(\omega) \int\limits_{-\infty}^{\infty} g_L(\omega, \omega_{21}) d\omega_{21}.$$

Taking into consideration the normalization condition of function $g(\omega)$, we obtain

$$\chi''(\omega) = -\pi \frac{\chi_0}{T_2} F_G(\omega) = -\pi \frac{\chi_0}{T_2} g_G(\omega). \tag{2.34}$$

Note that in the above-mentioned conditions, the absorption line follows the type of the distribution law. For a line with Doppler widening,

$$g_G(\xi) = \frac{0,939}{\Delta\omega_G} \exp\left(-\ln(2)\xi^2\right),$$

where $\xi = 2 \cdot \frac{\omega - \tilde{\omega}_{21}}{\Delta\omega_G}$ is the generalized offset.

The shape of the Gaussian line is shown in Fig. 2.9. At a similar width, the Gaussian line decreases faster than the Lorentz one.

From Eq. (2.34) we can see that in the case when spin packets are essentially narrower than the total line, the form of absorption line writing differs from the uniformly widened case only in that Lorentz function $g_L(\omega)$ is replaced by Gaussian function $g_G(\omega)$.

Fig. 2.9 Comparison of Lorentz (*dotted curve*) and Gaussian (*solid curve*) absorption line shapes. At a similar width on level 0.5 the Lorentz line changes more slowly

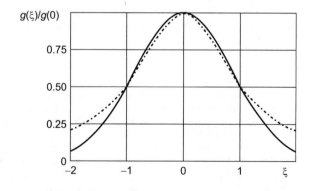

2.7   Kramers–Kronig Relations

From the theory of electric circuits we should know that the real and imaginary parts
of the complex impedance of two-ports are related by the integral Hilbert transformation.[6] The connection between the polarization vector and the vector of electric
field intensity coincides in a form with equations of circuits. Therefore, we can hope
that the real and imaginary parts of the complex dielectric susceptibility connect with
each other in the same manner.

Really, the function of complex variable p,

$$\dot{\chi}(p) = \frac{2\omega_{12}d^2 n_{21}^e}{3\hbar(p^2 + 2p/T_2 + \omega_{21}^2)},$$

can be attributed to so-called fractional rational functions. It has two complex
conjugate poles, which lie in the left half plane. It is known that fractional rational
functions are analytical ones. According to the theory of complex variable functions,
the real and imaginary parts of such complex functions are connected by the Hilbert
transformation. For dielectric susceptibility, this looks like

$$\chi'(\omega) = \frac{1}{\pi} \int\limits_{-\infty}^{\infty} \frac{\chi''(\tilde{\omega})}{\tilde{\omega} - \omega} d\tilde{\omega}; \quad \chi''(\omega) = \frac{1}{\pi} \int\limits_{-\infty}^{\infty} \frac{\chi'(\tilde{\omega})}{\omega - \tilde{\omega}} d\tilde{\omega}. \qquad (2.35)$$

The procedure for derivation of these equations is the same as used earlier. In
conformity with complex dielectric susceptibility, Eq. (2.35) is known as the
Kramers[7]–Kronig relations.

Let us make sure that nonuniformly widened lines also satisfy the relations in
Eq. (2.35). In addition to (2.34), we find under the same conditions the real part of
dielectric susceptibility. For this it is necessary to calculate the following integral:

---

[6]David Hilbert (1862–1943) was a German mathematician and had a great influence on the
development of various sections of mathematics. An assurance in the unlimited force of the
human mind and conviction of the unity of mathematical science and the unity of mathematics
and natural sciences were typical of his creative work. He summarized Euclidean geometry in
infinite-dimension space. His contribution to integral equation theory formed one of the fundamentals of modern functional analysis and led to the coining of the term "Hilbert space."

[7]Hendrik Kramers (1894–1952) was a Dutch physicist, a public figure, and a member of the
Netherlands Royal Academy of Sciences. He made a contribution to mathematical formalism of
quantum mechanics. In 1926 he developed a method for solution of the single-dimension
Schrödinger equation within the limits of so-called quasiclassical approximation, which permitted
determination of a correspondence with the Bohr–Zommerfeld quantization rules in the old
quantum theory. In 1927, independently of Ralph Kronig, he obtained the dispersion relations of
classical electromagnetics, coupling the real and imaginary parts of polarization or the refraction
index (Kramers–Kronig relations).

$$\chi'(\omega) = -\int\limits_{\infty}^{\infty} \chi_0 \frac{\xi(\omega, \omega_{21})}{1 + \xi(\omega, \omega_{21})^2} F_G(\omega_{21}) d\omega_{21}.$$

Since $\xi(\omega, \omega_{21}) = (\omega - \omega_{21})T_2$, at $T_2 \to \infty$ we have

$$\chi'(\omega) = -\int\limits_{-\infty}^{\infty} \frac{\chi_0}{T_2} \frac{1}{\omega - \omega_{21}} F_G(\omega_{21}) d\omega_{21}.$$

Taking (2.34) into consideration we obtain

$$\chi'(\omega) = \frac{1}{\pi} \int\limits_{-\infty}^{\infty} \frac{\chi''(\omega_{21})}{\omega - \omega_{21}} d\omega_{21}.$$

This equation represents the Hilbert transformation. In a similar manner we can also obtain the second integral equation.

2.8 Anisotropy of the Medium

Equation system (2.24) and (2.25) is obtained with the assumption of medium isotropy. In the general case we must use (2.20) in Eq. (2.24). It will allow extension of the linear susceptibility concept on the anisotropic medium. Since anisotropic media play an important role in practical applications, it is expedient to examine the adaptability of the susceptibility concept to them and to discuss some significant consequences of this issue.

In the anisotropic case, the linear equation for polarization components takes the form

$$\frac{d^2 P_i}{dt^2} + \frac{2}{T_2} \frac{dP_i}{dt} + \omega_{21}^2 P_i = \frac{2\omega_{21}}{\hbar} n_{12}^e (d^2)_{ij} E_j, \qquad (2.36)$$

where i,j,k, are, for instance, x,y,z; and $(d^2)_{ij}$ is a matrix element of dipole moment squares (see (2.20)). In (2.36) it is implied that summing is provided over the repeated indices. Writing variables in the complex form, we obtain from Eq. (2.36)

$$\dot{P}_i = \frac{2\omega_{12}}{\hbar} n_{12}^e \frac{1}{\omega_{12}^2 - \omega^2 + j2\omega/T_2} (d^2)_{ij} \dot{E}_j.$$

From this linear equation we may pass to the vector form

$$\dot{\mathbf{P}} = \dot{\chi}(\omega)\dot{\mathbf{E}}.$$

Susceptibility $\dot{\chi}(\omega)$ connects two physical quantities, each of which should not depend upon a random choice of the coordinate system. Therefore, susceptibility should also not (in some sense) depend upon the choice of the coordinate system. It is possible if $\dot{\chi}(\omega)$ is transformed in a definite manner in changing the coordinate system—namely, to keep the required correlation between physical quantities. Tensors have such a property. In particular, since $\dot{\chi}(\omega)$ connects components of two vectors, it is required to have two coordinate indices. Therefore, from the mathematical point of view, this object is the tensor of the second rank.[8]

The second-rank tensor is specified by nine numbers. In the rectangular Cartesian coordinate system, the susceptibility $\dot{\chi}(\omega)$ components can be written in the table form:

$$\dot{\chi}(\omega) = \begin{vmatrix} \dot{\chi}_{xx} & \dot{\chi}_{xy} & \dot{\chi}_{xz} \\ \dot{\chi}_{yx} & \dot{\chi}_{yy} & \dot{\chi}_{yz} \\ \dot{\chi}_{zx} & \dot{\chi}_{zy} & \dot{\chi}_{zz} \end{vmatrix}.$$

Tensor $\dot{\chi}(\omega)$ is the symmetric tensor (see (2.20)). Therefore, only six elements out of nine are independent.

It should be taken into consideration that connection of polarization and the electric field through the material susceptibility tensor bears a general character. It can be proved using thermodynamic equilibrium of the interaction process of the field and the medium. It is possible, of course, to accept the mathematical statement that in the general case, two vectors are interconnected through the second-rank tensor. They usually act in traditional lecture courses in electrodynamics and optics.

The transform properties of the susceptibility tensor represent a special interest in different transformations of the coordinate system. These properties completely reflect the symmetry properties of the medium, which, in turn, are important for determination of the view of the second-rank tensor of linear susceptibility $\dot{\chi}(\omega)$.

---

[8]A number of coordinate indices define the tensor rank; the vector is the first-rank tensor and the scalar is the zero-rank tensor. The second-rank tensor characterizes the most general connection of two vectors when they differ not only in length but also in direction. Higher-rank tensors characterize the connection between tensor quantities of the arbitrary rank.

2.9 Influence of Crystal Symmetry on the View of Material Tensors

The tensor—and hence the physical properties defined by this tensor—are different for various crystals. So, it is a priori clear that the physical properties of the anisotropic crystal and the isotropic medium are principally different.

The internal crystal structure causes the strictly definite anisotropy of its physical properties. This is a consequence of the crystal symmetry. We can show that all crystals existing in nature can be divided according to crystal symmetry into a finite number of symmetry classes or point groups of crystal symmetry. One point group of crystal symmetry, generally speaking, can unify crystals that have various chemical natures but a series of similar physical properties.

The question about the influence of crystal symmetry on the properties is only part of the question about the role of symmetry in nature. The general principle regarding the influence of symmetry on all physical phenomena (without exceptions) was formulated by Curie.[9] In conformity with crystal properties, this principle was examined by Neumann[10] in 1885. The modern formulation of this principle states that the symmetry elements of any physical property of a crystal must include the symmetry elements of the point group of the crystal.

Crystallography is the science that deals with study of the spatial properties of ordered structures—crystals. Let us briefly discuss the main position of crystallography.

2.10 Crystal Symmetry

First of all, we should take into account that crystals represent periodic structures ordered in space—crystal lattices. The joint structure of a crystalline substance can be obtained by transfer (translation) of some elementary cell in the direction of some three vectors (translation vectors) by steps of a fixed length (the lattice period). Therefore, all crystals have the symmetry of translation.

In description of spatial properties we can distinguish the symmetry elements—auxiliary images (points, straight lines, planes), with the help of which the figure (or space) symmetry is detected—as well as the symmetry operations. Simple symmetry operations—reflection and rotation—are described by three symmetry elements: an axis of rotation, a plane, and a center.

---

[9]Pierre Curie (1858–1904) was a French physicist and a pioneer in crystallography (in which he investigated the symmetry of crystals), magnetism, piezoelectricity (which he discovered in 1880), and radioactivity (which he discovered in 1898). He discovered polonium and radium, establishing the complex composition of radium radiation and glass and porcelain coloration under the action of this radiation. In 1903 he received the Nobel Prize in Physics for his research on radioactivity.

[10]Franz Neumann (1798–1895) was a German mineralogist, physicist, and mathematician.

Fig. 2.10 Symmetry axes
in a plain equilateral
triangle; *2* and *3* designate
the axes of the second and
third orders, respectively. At
rotation around any of these
axes by 180° or 120°, the
figure is transformed into the
initial view

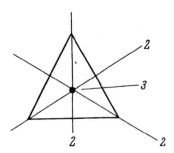

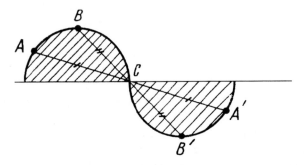

Fig. 2.11 Symmetry of the inversion center type. The figure is transformed into the initial view
with changing of the sign of all point coordinates. The transform matrix is equal to -1. Therefore,
the effects described by the third-rank tensor in media with an inversion center are absent

The rotational axis of symmetry is the straight line at which a figure coincides
with itself at rotation by the angle α around the line. Axes of symmetry differ by the
symmetry order, designated as $n = 360/\alpha$, where $n = 1, 2, 3, \ldots$.

A plain equilateral triangle (Fig. 2.10) has one axis of the third order and three
axes of the second order. A cylinder has one axis ∞, which coincides with its
longitudinal axis and an infinite number of axes **2** perpendicular to it.

The symmetry plane, designated as **m**, divides the figure into two mirror-like parts
located as a subject and its mirror image.

The symmetry center, or the center of inversion, designated as **C**, is a singular
point characterized by the fact that any straight line drawn through it meets, at equal
distances from it, (relatively) similar points of the figure (Fig. 2.11).

Introduction of symmetry elements is not a unique method of symmetry operation
description. So, formally, the symmetry of the figure in Fig. 2.11 can be described by
the inverse axis, which contains both elements of figure symmetry: the rotational
axis and the symmetry center **C**. Such an axis is called an inversion axis of the first
order and is designated as $\bar{1}$. By analogy we may introduce inversion axes of any
order: $\bar{2}, \bar{3}, \bar{4}$, etc.

The symmetry of a material figure is described by a set of symmetry elements. For
instance, a cube has several axes of different orders, planes, and a symmetry center;
not all of the symmetry elements are independent. Indeed, the intersection of two
symmetry planes under angle α gives a symmetry axis of about $n = 180/\alpha$. The

intersection of three symmetry planes in the cube gives the inversion center, etc. Thus, symmetry elements may be added and give birth to new elements.

The full set of symmetry elements of any material figure is called the symmetry group (view). To indicate the symmetry group, it is not necessary to write all symmetry elements. It is quite enough to indicate only those that give birth to all others. Such a way of determining the symmetry group is called topographic nomenclature.

There are many ways to designate nomenclatures. In the future we shall use the international system. Description of nomenclature in this system contains a figure indicating the order of the symmetry axis (including inversion axes) and the letter "**m**" designating the symmetry plane.

Parallelism of the symmetry axis and plane **m** is not designated by a special symbol; parallelism is designated by a slash symbol. For example, the description "**2m**" shows that the axis of the second order is parallel to plane **m**; the designation "**6/m**" indicates that the axis of the sixth order is perpendicular to plane **m**.

The existence of an infinite number of point groups is possible theoretically; however, in crystals, not all symmetry elements are real. It has been proved that in crystals the rotational axes **1**, **2**, **3**, **4**, and **6** and inversion axes corresponding to them can exist, but axes 5, 7, etc., which are not compatible with the translation symmetry, are impossible. If we take into account that there are still the symmetry plane *m* and the symmetry center **C**, it turns out that there are 32 different full sets of symmetry-generating elements, or 32 point groups of crystals. The discovery of the 32 point groups of symmetry is connected with the name of the Russian academician Axel Gadolin,[11] who published his discovery in 1867.

In crystallography, all symmetry groups of crystals are divided into six large classes (systems). The complete list of all 32 point groups of symmetry is given in Appendix 4. The designation system for symmetry groups is presented there, together with examples of crystals.

2.11   The Form of the Linear Dielectric Susceptibility Tensor

Let us demonstrate the application of the Neumann principle concerning the linear susceptibility tensor.

To consider the most general cases, we must study the transformation properties of the second-rank tensor for different rotations of the coordinate system. We shall understand as rotations of the coordinate system the transfer from one aggregate of

---

[11]Axel Gadolin (1828–1892) was a Russian scientist in the field of artillery weaponry, metal processing, minerology, and crystallography. He was an artillery general and the author of a lecture course on metal technology. In 1868 he was awarded the Lomonosov Prize for his derivation (in 1867) of the 32 point groups of crystal symmetry.

mutually perpendicular axes (designated x_1, x_2, x_3 for convenience) to another (designated x'_1, x'_2, x'_3) without changing the origin. Transformation from the old (non-dotted) aggregate of the coordinate axes into the new (dotted) one obeys the equation of coordinate transformation:

$$x'_i = a_{ij} x_j, \qquad (2.37)$$

where a_{ij} are the tangents of angles between the new and old axes. On the repeated indices, as usual, we imply summation. The same can be written in the matrix designation:

$$\mathbf{x}' = \mathbf{A}\mathbf{x}.$$

Here, $\mathbf{A}$ represents the matrix of coordinate transformation.

Coordinate transformation is transformation of components of the radius vector. The components of any other vector—for example, the polarization vector or electric field vector—are transformed in the same manner as the radius vector.

Now we examine the transformation of the second-rank tensor connecting two vectors. In particular, we shall clarify the transformation properties of the susceptibility tensor of the second rank. In this case,

$$\dot{P}_i = \dot{\chi}_{ij} \dot{E}_j.$$

Polarization in the new coordinate system is determined with the help of (2.37). Taking into account the previous expression, we can write

$$\dot{P}'_k = a_{ki} \dot{\chi}_{ij} \dot{E}_j.$$

On the other hand, vector $\mathbf{E}$ is transformed according to the law

$$E'_j = a_{lj} \dot{E}_l,$$

therefore,

$$\dot{P}'_k = a_{ki} \dot{\chi}_{ij} a_{lj} \dot{E}_l.$$

Finally, we obtain

$$\dot{P}'_i = \dot{\chi}'_{ij} \dot{E}'_j,$$

where $\dot{\chi}'_{ij}$ is the susceptibility in the new coordinate system, which is connected with the susceptibility in the old coordinate system, as

$$\dot{\chi}'_{kl} = a_{ki}\dot{\chi}_{ij}a_{lj}.$$

Transformation of tensor $\dot{\chi}'_{ij}$ can be written in the matrix form:

$$\dot{\chi}' = \mathbf{A}\dot{\chi}\mathbf{A}^T = \mathbf{A}\dot{\chi}\mathbf{A}^{-1}. \tag{2.38}$$

Here, $\dot{\chi}$ and $\mathbf{A}$ are matrices from nine elements. $\mathbf{A}^T$ is a matrix transposed with respect to matrix $\mathbf{A}$; i.e., its lines are replaced by columns. $\mathbf{A}^{-1}$ is the inverse matrix satisfying the condition $\mathbf{A}\mathbf{A}^{-1} = \mathbf{A}^{-1}\mathbf{A} = 1$, where $\mathbf{1}$ is the unitary matrix. In orthogonal transformations (i.e., transformations from one Cartesian coordinate system into another), the inverse and transposed matrices are identical to each other. Transformation of type (2.38) is called similarity transformation.

Now we can consider the connection between properties of susceptibility tensor transformation at rotations of the coordinate system and the physical properties of the medium. First, the susceptibility tensor elements are measured in the specified coordinate system. After that, the sample position is changed—for instance, by rotation on angle $\pi/2$—and the measurements are repeated. If the susceptibility tensor does not change, it indicates that the tensor in the old and new coordinate systems is the same; hence,

$$\dot{\chi}' = \dot{\chi}. \tag{2.39}$$

In this case it is said that the susceptibility has the symmetry element specified by the rotation operator. Using (2.39) and (2.38), we can write the new condition:

$$\dot{\chi} = \mathbf{A}\dot{\chi}\mathbf{A}^{-1}.$$

Acting by operator $\mathbf{A}$ from the right, we obtain

$$\dot{\chi}\mathbf{A} = \mathbf{A}\dot{\chi}\mathbf{A}^{-1}\mathbf{A} = \mathbf{A}\dot{\chi}.$$

Hence, if $\mathbf{A}$ corresponds to symmetry transformation, the commutator is

$$[\mathbf{A}, \dot{\chi}] = 0. \tag{2.40}$$

The obtained equation determines how the properties of medium symmetry limit the shape of $\dot{\chi}(\omega)$. This can be best illustrated in a specific example.

As a simple example, we consider a crystal with the symmetry axis of the fourth order. Symmetry operator $\mathbf{A}$ is determined from Eq. (2.37) of coordinate transformation and can be written as

$$\mathbf{A} = \begin{vmatrix} 0 & 1 & 0 \\ -1 & 0 & 0 \\ 0 & 0 & 1 \end{vmatrix}.$$

Using condition (2.40), we obtain

$$
\begin{vmatrix} 0 & 1 & 0 \\ -1 & 0 & 0 \\ 0 & 0 & 1 \end{vmatrix}
\begin{vmatrix} \dot{\chi}_{xx} & \dot{\chi}_{xy} & \dot{\chi}_{xz} \\ \dot{\chi}_{yx} & \dot{\chi}_{yy} & \dot{\chi}_{yz} \\ \dot{\chi}_{zx} & \dot{\chi}_{zy} & \dot{\chi}_{zz} \end{vmatrix}
=
\begin{vmatrix} \dot{\chi}_{xx} & \dot{\chi}_{xy} & \dot{\chi}_{xz} \\ \dot{\chi}_{yx} & \dot{\chi}_{yy} & \dot{\chi}_{yz} \\ \dot{\chi}_{zx} & \dot{\chi}_{zy} & \dot{\chi}_{zz} \end{vmatrix}
\begin{vmatrix} 0 & 1 & 0 \\ -1 & 0 & 0 \\ 0 & 0 & 1 \end{vmatrix}.
$$

This system has nine equations, of which only six are independent. Solving it, we find the conditions that are imposed upon elements of the material tensor. For instance, the first equation, beginning from the top left, takes the form

$$
\dot{\chi}_{xy} = -\dot{\chi}_{xy}.
$$

This means that $\dot{\chi}_{xy} = 0$. Similar relations can be extended to all nondiagonal elements of the tensor. From the second equation,

$$
-\dot{\chi}_{yy} = -\dot{\chi}_{xx},
$$

we can draw the conclusion that the first two diagonal elements are equal. Fulfilling the calculations in such a manner until all elements are determined, we can find the tensor view:

$$
\dot{\chi} = \begin{vmatrix} \dot{\chi}_{xx} & 0 & 0 \\ 0 & \dot{\chi}_{xx} & 0 \\ 0 & 0 & \dot{\chi}_{zz} \end{vmatrix}.
$$

The considered example illustrates the interconnection between the medium symmetry properties and the view of the susceptibility tensor. In Appendix 5 we give the view of tensor $\dot{\chi}$ for all classes of crystals.

The peculiarities of optical wave propagation in crystals were well studied long ago. Much of use can be picked up in lecture courses on engineering electrodynamics, in contrast to optics, where much less attention is given to problems of wave propagation in anisotropic media.

2.12 The Saturation Effect

As was mentioned above, the system of material equations of the two-level system is nonlinear. Chapter 6 of this book is devoted to detailed consideration of nonlinear effects. However, at this stage, for estimation of the area in which the linear approximation is valid, it is necessary to find out the solutions to Eqs. (2.24) and (2.25), which at least roughly describe the nonlinear properties of the medium. We can do this in the following manner.

Let us exclude from examination the transients and try to find out the equilibrium state solution to $E(t) = E_m \cos(\omega t)$. Because of nonlinearity, $P(t)$ and $n_{12}(t)$ will contain harmonics of frequency ω. We take into account that the task makes sense only in the case when $\omega \approx \omega_{21} \gg T_2^{-1}$. In this case the variable part of n_{12} is small in comparison with the constant. Therefore, in the right part of (2.24), we may keep only the constant component of population difference n_0. The solution to Eq. (2.24) will take the form

$$\dot{P} = \frac{2\omega_{12}}{3\hbar} d^2 n_0 \frac{\dot{E}}{\omega_{21}^2 - \omega^2 + 2j\omega T_2}.$$

We can obtain unknown quantity n_0 by solving (2.25). Since polarization changes according to the harmonic law with frequency ω, the component of the right part of (2.25) averaged over the period at determination of the power is

$$\left(\frac{d\mathbf{P}}{dt} \mathbf{E} \right)_{st} = \frac{1}{2} \omega \mathrm{Im} \left(\dot{\mathbf{P}} \dot{\mathbf{E}}^* \right).$$

Then, instead of (2.25), for determination of n_0 we have

$$\frac{n_{12} - n_0}{T_1} = -\frac{1}{\hbar^2} \frac{2\omega}{3} d^2 n_0 \frac{|\dot{E}|^2 2\omega/T_2}{(\omega_{21}^2 - \omega^2)^2 + 4\omega/T_2^2},$$

from where

$$n_0 = \frac{n_{12}^e}{1 + \frac{|\dot{E}|^2}{(1+\xi^2) E_{sat}^2}}, \tag{2.41}$$

where

$$E_{sat}^2 = 3\hbar^2 / (d^2 T_1 T_2).$$

Thus, with the influence of an external alternate electric field on the two-level system, it transfers to the equilibrium state with the new (different from n_{12}^e) value of the population difference. According to (2.41), at $|\dot{E}|^2$ tending to infinity, the level populations are equalized. This relates to the fact that the energy coming from a field exceeds the energy collected by a thermostat. As a result, the system transfers to the equilibrium condition with the acting field. Since the probabilities of the stimulated transitions up and down are equal, n_2 becomes equal to n_1. Using (2.41), we can write

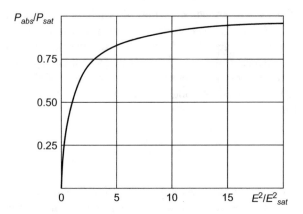

Fig. 2.12 Function of the power absorbed by a two-level system versus the intensity of an electromagnetic field, taking into consideration the saturation effect. The essence of the saturation effect consists in the fact that with the influence of powerful enough radiation, the natural processes of relaxation to an equilibrium with a thermostat give way to processes of relaxation to an equilibrium with the more powerful radiation source

$$\chi'(\omega) = \chi_0 \frac{\xi}{1 + \xi^2 + \left|\dot{E}\right|^2 / E_{sat}^2} ; \quad \sigma(\omega) = \frac{\sigma_0}{1 + \xi^2 + \left|\dot{E}\right|^2 / E_{sat}^2} .$$

The power absorbed in the medium volume unit is equal:

$$P_{abs} = \frac{\sigma_0}{2} \frac{\left|\dot{E}\right|^2}{1 + \xi^2 + \left|\dot{E}\right|^2 / E_{sat}^2} . \qquad (2.42)$$

Figure 2.12 shows the characteristic P_{abs} versus $\left|\dot{E}\right|^2$ for $\xi = 0$. At $\left|\dot{E}\right|^2 \to \infty$ the absorbed power tends to the limit equal to

$$P_{sat} = \frac{\sigma_0 E_{sat}^2}{2} = \frac{\hbar \omega_{21} n_{12}^e}{2 T_1} .$$

This circumstance serves as an occasion for us to call this phenomenon the saturation effect. The saturation effect is an extremely important feature in the process of field interaction with the medium. Let us discuss some consequences of this phenomenon.

First of all, we determine the boundaries of the area of linear interaction of the medium and the field. It is clear that they can be found from the condition

$$\left|\dot{E}\right|^2 \ll E_{sat}^2 .$$

If we are speaking about the field of the plane electromagnetic wave, the condition on the value of $|\dot{\mathbf{E}}|^2$ can be replaced by the condition on the power flow density

$$I \ll I_{\text{sat}},$$

where

$$I_{\text{sat}} = \frac{3}{2} \frac{\hbar^2}{d^2 T_1 T_2} \sqrt{\frac{\varepsilon_a}{\mu_a}}. \tag{2.43}$$

Let us estimate the values of E_{sat} and I_{sat} for ammonia vapors. At a pressure of about 5 Pa (~0.04 mm of a mercury column), $T_1 \cong 0.3$ μs and $T_2 \cong 0.15$ μs. Substituting the numerical values of the parameters, we obtain $E_{\text{sat}} \cong 228$ V/m and $I_{\text{sat}} = 6.7$ mW/cm^2. The power density at several milliwatts per square centimeter is quite accessible, and the saturation effect is easily observed. In gases with pressure growth, T_1 and T_2 decrease. Therefore, the saturation parameters grow approximately proportionally to the pressure square, and at a pressure of about 100 Pa (about 1 mm of a mercury column), the phenomenon becomes hardly noticeable for the available power densities.

The saturation effect distorts the spectral line shape. For the uniformly widened line, instead of (2.31) we have

$$g_L(\omega) = \frac{T_2}{\pi} \cdot \frac{\sqrt{1 + I/I_{\text{sat}}}}{1 + \xi^2 + I/I_{\text{sat}}}.$$

Saturation smooths the spectral line and increases its width.

The saturation effect changes the shape of the nonuniformly widened line in an absolutely different manner. Radiation effectively interacts with spin packets with frequencies close to its frequency ω. As a result, the saturation effect changes the distribution law of the population difference on the transition frequency. In the spectral line shape, the dip appears near frequency ω. This phenomenon is often called the effect of the "hole burning through."

In Fig. 2.13 the spectral line shape is shown with the action of powerful enough radiation. The dip width has a value of about $\Delta\omega_L$. Experimentally, the dip can be discovered during scanning by weak radiation of a medium under the action of powerful radiation. Such a mode of spectral analysis allows study of the properties of the particles of which the nonuniformly widened line consists. The "hole burning through" effect is a rather important phenomenon in gas discharge lasers.

Fig. 2.13 Influence of the saturation effect on a line with nonuniform widening. A similar function can be observed during measurement of the absorption of weak radiation by a medium situated under the action of powerful radiation with frequency ω_{rad}. For measurement convenience, the weak and strong waves will be orthogonal in the space. Physicists like image-bearing expressions and call this phenomenon the effect of the "hole burning through"

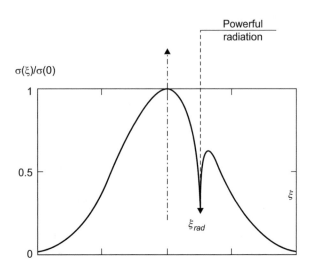

2.13   Problems for Chapter 2

2.1. Prove that damping exponent α caused by transitions between rotation levels is approximately proportional to the cube of the transition frequency.

Assistance: Use the results of task 1.9 and take into consideration the selection rules for rotation transitions.

2.2. Show that damping exponent α caused by transitions between rotation levels of gas molecules does not depend upon pressure. Reveal the formula for α containing the width of absorption band Δf_1 at a pressure of 1 mm of a mercury column (133.32 Pa).

Assistance: Take into account that the relaxation time $T_2 \sim p^{-1}$ and $p = nk_BT$.

2.3. Show that the damping in the gas is approximately inversely proportional to the absolute temperature at a power of 2.5.

Assistance: Consider that the collision frequency is proportional to the average velocity of the particle heat motion.

2.4. Calculate the damping exponent for a gas of ClF molecules on the resonance frequency of the transition $J = 0 \leftrightarrow J = 1$ for room temperature $T = 300$ K. Rotation constant $B \approx 15$ GHz, dipole moment $d \approx 0.9$ debyes, and the width of the absorption band at a pressure of 1 mm of a mercury column is 10 MHz. Compare the result with the data in Table 2.1.

2.5. Using linear approximation, find the impulse characteristics for a uniformly widened two-level quantum system. Calculate the damping time to level 0.1 for ammonia molecules at a frequency of 23.88 GHz for $T_1 \cong 0.3$ μs, $T_2 \cong 0.15$ μs, and $d = 3.47 \cdot 10^{-30}$ C × m.

Instruction: Use Laplace transformation tables.

2.6. Using linear approximation, find the impulse characteristics of the two-level quantum system under conditions of nonuniform widening, assuming that $\Delta f_G \gg \Delta f_L$.

Instruction: Use Laplace transformation tables.

2.7. Assuming that the electric field and the polarization are narrow-band processes, obtain, on the basis of (2.24) and (2.25), a system of abbreviated equations for slowly changing complex amplitudes.

Instruction: Assume that

$$\mathbf{P}(t) = \mathrm{Re}\left\{\dot{\mathbf{P}}(t)\exp\left(j\omega_{21}t\right)\right\};$$

$$\mathbf{E}(t) = \mathrm{Re}\left\{\dot{\mathbf{E}}(t)\exp\left(j\omega_{21}t\right)\right\}.$$

2.8. Using the result obtained in task 2.4, show that for $T_1, T_2 \to \infty$, the population difference changes according to the law

$$n_{12}(t) = n_{12}(0)\cos\left(\frac{|d\,|\,E}{\sqrt{3}\hbar}t\right).$$

Note: Such a behavior of the population difference is used in the impulse method of inversion of the two-level quantum system. Inversion achieves its maximum at $\frac{|d|E}{\sqrt{3}\hbar}t = \pi$.

2.9. On the basis of (2.33), fulfil the calculation of the normalized frequency function of the complex transition susceptibility with the Gaussian absorption line for CO_2. Assume that the central transition frequency $f_{21} = 30$ THz, the width of Doppler widening $\Delta f_G = 50$ MHz, and the relaxation time $T_2 = 5 \cdot 10^{-8}$ s.

Instruction: Use a computer for the calculations.

2.10. Investigate the influence of nonuniform widening on the shape of the frequency function of the complex susceptibility. Use the data presented in task 2.6 as the initial values.

Instruction: Use a computer for the calculations. You may, for instance, simply change Δf_G, but it is better to obtain the preliminary equation depending upon $\Delta f_G / \Delta f_L$ and the generalized offset ξ.

2.11. Reveal the formula for determination of the absorption band, taking into account the saturation effect of the quantum transition. Determine the value of the power flow at which the width increases twice. Calculate this value for ammonia molecules at a frequency of 23.88 GHz for $T_1 \cong 0.3$ μs, $T_2 \cong 0.15$ μs, $d = 3.47 \cdot 10^{-30}$ C × m, and $I_{\mathrm{rad}} \ll I_{\mathrm{sat}}$.

2.12. Reveal the formula and calculate the distribution of the population difference on the transition frequencies f_{21} under conditions of intensive radiation action on the specified frequency f_{rad}. Use the data presented in task 2.6 as the initial values.

Instruction: Use a computer for the calculations.

2.13. Repeat the solution of task 2.9 under the condition that widening is caused by the Doppler effect, and in the medium there is radiation in the form of a standing wave, with a specified frequency f_{rad} that does not coincide with the central frequency of the absorption line.

Chapter 3
Magnetic Dipole Interaction

3.1 Introduction: The Magnetic Dipole Moments of Particles

In the previous chapter we assumed that particle sizes are much smaller than the radiation wavelength. This gave us the possibilities of considering the field within the limits of the particle as being independent of the coordinates and analyzing the influence of a single electric field on a substance. In this chapter, that assumption about particle smallness remains in force and we examine the impact of a single magnetic field on particles.

What is the reason for substance particle interaction with a magnetic field? We consider the classical model of the one-electron atom (Fig. 3.1a). The electron rotating at distance r around the nucleus with velocity $\mathbf{v}$ has the mechanical moment $\mathbf{L} = m_e \mathbf{v} r$. This movement gives birth also to an electric current with a value of $i = -\frac{e\mathbf{v}}{2\pi r}$. It is known that a loop with a current with area S has a magnetic moment with a value of $m = iS = -\frac{eVr}{2} = -\frac{e}{2m_e} L$. Thus, the orbital movement of the electron in the atom gives birth to the orbital magnetic moment:

$$\mathbf{m}_L = -\frac{e}{2m}\mathbf{L}.$$

In a similar manner we can introduce into quantum mechanics the operator of the orbital magnetic moment:

$$\widehat{\mathbf{m}}_L = -\frac{e}{2m_e}\widehat{\mathbf{L}}. \tag{3.1}$$

But, besides the orbital magnetic moment, there is the spin magnetic moment for electrons $\mathbf{S}$, which corresponds to the spin magnetic moment described by the operator:

© Springer Nature Switzerland AG 2020
V. V. Shtykov, S. M. Smolskiy, *Introduction to Quantum Electronics and Nonlinear Optics*, https://doi.org/10.1007/978-3-030-37614-7_3

Fig. 3.1 Magnetic
moments of the electron (**a**)
and the nucleus (**b**). The
electron rotation along the
orbit has orbital and inherent
spin magnetic moments.
The nuclear magnetic
moment is equal to the sum
of the moments of all
protons and neutrons. Its
total number should be odd

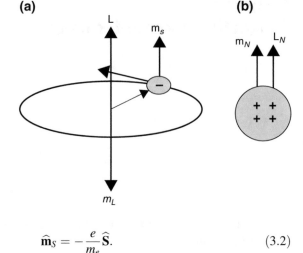

$$\widehat{\mathbf{m}}_S = -\frac{e}{m_e}\widehat{\mathbf{S}}. \tag{3.2}$$

As a result, the external magnetic field interacts with the total magnetic moment,
whose operator is

$$\widehat{\mathbf{m}} = \widehat{\mathbf{m}}_S + \widehat{\mathbf{m}}_L = -\frac{e}{2m_e}g\left(\widehat{\mathbf{L}} + \widehat{\mathbf{S}}\right) = -\frac{e}{2m_e}g\widehat{\mathbf{J}}. \tag{3.3}$$

Here, g is the g factor, or spectroscopic splitting factor. For electron spin,
$g = 2.0023 \approx 2.00$. If the orbital movement prevails, $g = 1$. For complex atoms,
g may be greater than 2.

Comparing (3.1) and (3.2), it is easy to notice that the spin magnetic moment is
twice the orbital moment. This fact was confirmed by classical experiments
performed by Einstein and de Haas,[1] and an explanation can be given on the basis
of relativistic quantum mechanics alone.[2]

Magnetic phenomena connected with orbital and spin moments are called
paramagnetism.

---

[1]Wander Johannes de Haas (1878–1960) was a Dutch physicist and mathematician. He is best
known for the Shubnikov–de Haas effect, the de Haas–van Alphen effect, and the Einstein–de Haas
effect.

[2]In the popular literature and in some textbooks, the classical explanation for magnetic phenomena
in a substance is given. However, strictly according to classical physics, in the heat equilibrium state
after accurate averaging, the total magnetic moment vanishes. The thing is that the magnetic field
changes the trajectory of the charge movement but does not change the system energy as occurs in
an electric field. Hence, the result of magnetic field action on the substance is impossible to register.
All classical explanations in the explicit or implicit form use quantum postulates—for instance, the
presence of stationary orbits.

In the case of a multi-electron atom, Eq. (3.3) remains in force, and as $\mathbf{J}$ we should take the total summary mechanical moment of all electrons, $\mathbf{J}_\Sigma = \mathbf{L} + \mathbf{S}$, to which, in the stationary energy state, the magnetic moment corresponds:

$$\mathbf{m} = -\gamma \mathbf{J}_\Sigma = -\gamma \hbar \mathbf{J} = -g\mu_B \mathbf{J}.$$

Here, $\mathbf{J}_\Sigma = \hbar \mathbf{J}$ and $\mu_B = \frac{e\hbar}{2\,m_e} = 0.927 \cdot 10^{-23} \frac{J}{Tl}$ is the Bohr[3] magneton. We shall use the Greek letter γ for designation of the gyromagnetic ratio (also sometimes known as the magnetogyric ratio in other disciplines). Its numerical value, at $g = 2$, is $\gamma = 2\pi \cdot 28 \cdot 10^9 \frac{rad/s}{T}$. In engineering practice the gyromagnetic constant is usually measured in hertz per Tesla and a value of $\gamma = 28 \frac{GHz}{T}$ is given.

It should be noted that completely occupied electron layers do not make contributions to $\mathbf{J}_\Sigma$. Such a situation, for example, is typical for rare gases. The orbital and spin moments of the included electrons completely compensate for each other. Thus, paramagnetism is connected with the presence of unoccupied electron layers.

If the total moment $\mathbf{J}_\Sigma = 0$, appearance of the induced magnetic moment is possible only in connection with the Lentz law. Under the action of a magnetic field, the electron orbits are changed and so the magnetic flow will remain unchanged. Therefore, the vector of the induced magnetic moment is directed against the opposite vector of the external field. This phenomenon is called diamagnetism. Diamagnetism is also peculiar to conduction electrons, which move inside energy zones. However, determination of their contribution to diamagnetism represents a problem, which is complicated by the peculiarities of energy zone construction. The relative diamagnetic susceptibility is less than zero and does not exceed the modulo 10^{-5}. That is why diamagnets hold no special interest for us.

The atom nucleus also has its own mechanical moment $\mathbf{I}_N$. Hence, it has a nuclear magnetic moment (Fig. 3.1b), whose operator is

$$\widehat{\mathbf{m}}_N = \frac{e}{2M_P} g_N \widehat{\mathbf{I}}_N = \gamma_N \widehat{\mathbf{I}}_N, \tag{3.4}$$

where M_P is the proton mass, g_N is the nuclear g factor, and $\mu_N = \frac{e\hbar}{2M_p}$ is a nuclear magneton. Since the proton mass is triple the electron mass, the nuclear magnetic moment is three orders less than the electron magnetic moment. If we do not take into consideration so-called ultrafine fission, which arises at the expense of interaction of the nuclear and electron moments, we may consider the behavior of magnetic particles without drawing attention to its true nature.

---

[3]Niels Bohr (1885–1962) was a Danish physicist and one of the founders of modern physics. He discovered atom theory and made an essential contribution to the theory of the atom nucleus. In 1922 he received the Nobel Prize in Physics. From 1920 until his death, he worked as the director of the Institute of Theoretical Physics in Copenhagen, which was founded by him and now bears his name. The Bohr Institute has become one of the most significant scientific centers worldwide.

3.2 Magnetic Dipole Transitions

In magnetic dipole approximation it is assumed that the atom is equivalent to a small loop with a current. Then the energy of the interaction of the atom with a magnetic field can be determined as the product of its magnetic moment and the magnetic induction. Therefore, the disturbance Hamiltonian in the magnetic dipole approximation takes the form

$$\widehat{H}_B = -\widehat{m}\mathbf{B}, \tag{3.5}$$

which is absolutely similar to (2.9). But we shall further consider the physical situation, which differs from the one studied in Chap. 2. Instead of dealing with particles that have two possible stationary states u_1 and u_2, we examine particles that are in some undisturbed state u_k all the time, with energy W_k. We place these particles in a permanent magnetic field directed, say, along the z axis and with induction B_{0z}. They will obtain additional energy, and the Hamiltonian of interaction with the permanent magnetic field (3.5) takes into account (3.3):

$$\widehat{H}_B = -\widehat{m}_z B_{0z} = \gamma \widehat{J}_z B_{0z}. \tag{3.6}$$

In the stationary state the atom has projection of the impulse moment to the z axis equal to $J_z = m_J \hbar$, where $m_J = J, J - 1, \ldots, -J$. Values of J_z corresponding to different m_J values represent eigenvalues of the operator $\widehat{J}_z$. This means that the energy of our particles in stationary states that differ by the value of the magnetic quantum number is equal in the field presence to $W_k + \gamma \hbar m_J B_{0z}$. In other words, energy level W_k in a permanent magnetic field is split into as many levels as the number of allowed values of m_j, i.e., $2J + 1$. For example, if $J = 1/2$, splitting into two levels takes place, and if $J = 3/2$, under the action of the field, four energy levels arise, as shown in Fig. 3.2. The value of splitting $\gamma \hbar B_{0z}$ is proportional to the magnetic field intensity. This phenomenon is called the Zeeman[4] effect.

In the future, for simplicity, we shall examine particles for which $J = 1/2$. The hydrogen atom in the main state and atoms with a single valence s electron (for instance, silver) are particles of such a type. In a permanent magnetic field their energy level is split into two levels; therefore, we can consider them as two-level particles.

---

[4]Pieter Zeeman (1865–1943) was a Dutch physicist who shared the 1902 Nobel Prize in Physics with Hendrik Lorentz for his discovery of the Zeeman effect.

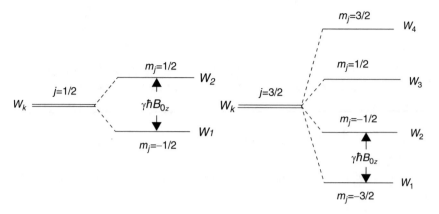

Fig. 3.2 Splitting of the initial energy level into magnetic sublevels in a permanent magnetic field, known as the Zeeman effect. The total number of sublevels depends on the number of allowed values of m_j, which is equal to $2j + 1$. Interaction of the orbital and spin moments leads to level splitting in a zero magnetic field. The fine structure of the energy spectra of complex atoms can be explained by this. Interaction of the nuclear and electron magnetic moments gives birth to ultrafine splitting

3.3 Pauli Matrices

Now our task is the search for the disturbance Hamiltonian in a presentation in which the undisturbed Hamiltonian of the paramagnetic particle gas is described by the diagonal matrix.

We shall assume that the particles are at $B_{0z} = 0$ in the stationary state W_k and located in a permanent magnetic field $B_{0z} \neq 0$ as an undisturbed physical system. Then the undisturbed Hamiltonian at $J = 1/2$ in representation of eigenfunctions will take the form[5]

$$\widehat{H}_0 = \begin{pmatrix} W_k - \dfrac{\gamma\hbar}{2}B_{0z} & 0 \\ 0 & W_k + \dfrac{\gamma\hbar}{2}B_{0z} \end{pmatrix}. \tag{3.7}$$

Comparing this equation with (3.6), we can draw a conclusion that in the same presentation,

---

[5]In the literature, according to the existing tradition, the upper level is numbered 1 and the lower level is numbered 2. Here and subsequently, we use the numeration from the previous chapters, in which the lower level is numbered 1 and the upper level is numbered 2. This does not change the final result.

$$\widehat{J}_z = \frac{\hbar}{2} \begin{pmatrix} -1 & 0 \\ 0 & 1 \end{pmatrix}. \tag{3.8}$$

To write the Hamiltonian of disturbance, we must clarify the form of operators $\widehat{J}_x$ and $\widehat{J}_y$. Calculating commutators, we may show that the total impulse moment obeys the relations

$$\left[\widehat{J}_x, \widehat{J}_y\right] = j\hbar\widehat{J}_z, \quad \left[\widehat{J}_y, \widehat{J}_z\right] = j\hbar\widehat{J}_x, \quad \left[\widehat{J}_z, \widehat{J}_x\right] = j\hbar\widehat{J}_y.$$

We can, without effort, make sure that the operators describing so-called Pauli matrices:

$$\widehat{J}_x = \frac{\hbar}{2} \begin{pmatrix} 0 & 1 \\ 1 & 0 \end{pmatrix}, \quad \widehat{J}_y = \frac{\hbar}{2} \begin{pmatrix} 0 & j \\ -j & 0 \end{pmatrix} = j\frac{\hbar}{2} \begin{pmatrix} 0 & 1 \\ -1 & 0 \end{pmatrix} \tag{3.9}$$

satisfy all of these relations.

Now we are able to write the disturbance Hamiltonian. Let us assume that an alternate magnetic field is directed along the normal to the permanent magnetic field, i.e., $\mathbf{B} = B_x\mathbf{1}_x + B_y\mathbf{1}_y$. Then (3.5) takes the form

$$\widehat{H}_B = \frac{\gamma\hbar}{2} \begin{pmatrix} 0 & B_x + jB_y \\ B_x - jB_y & 0 \end{pmatrix} = \frac{\gamma\hbar}{2} \begin{pmatrix} 0 & B_+ \\ B_- & 0 \end{pmatrix}, \tag{3.10}$$

which is similar to the one the disturbance Hamiltonian (2.12) had in the electric dipole approximation. The disturbing alternate field causes the particles to perform quantum transitions between splitting levels in the permanent magnetic field. By analogy with the electric dipole jumps we should expect that the interaction of particles with the alternate magnetic field will be maximal in the case that its frequency is close to the resonance frequency $\omega_{21} = (W_2 - W_1)/\hbar = \gamma B_{0z} = \omega_H$. Frequency ω_H is called the frequency of the Larmor[6] precession.

3.4 Bloch Equations

We are interested in the reaction of a paramagnetic particle system to an alternate magnetic field. Such a reaction is the substance magnetization. Therefore, our goal is to obtain the equation that describes the behavior of this physical quantity.

---

[6]Joseph Larmor (1857–1942) was a physicist and mathematician who achieved innovations in the understanding of electricity, dynamics, thermodynamics, and the electron theory of matter. His most influential work was *Aether and Matter*, a theoretical physics book published in 1900.

Since we are dealing with a two-level system, we can use the results found in Chap. 2. We take into consideration that in this case $W_1 = W_k - \gamma \hbar B_{0z}/2$, $W_2 = W_k + \gamma \hbar B_{0z}/2$ (see (3.7)), and $\omega_{21} = \gamma B_{0z} = \omega_H$. Equations for elements of the density matrix can be obtained from (2.13). If we take into account the analogy of the influence of Hamiltonians, then we have

$$
\begin{aligned}
\frac{d\rho_{11}}{dt} &= j\frac{\gamma}{2} \left[\rho_{12}(B_x - jB_y) - \rho_{21}(B_x + jB_y) \right] - \frac{\rho_{11} - \rho^e_{11}}{T_1}; \\
\frac{d\rho_{22}}{dt} &= -j\frac{\gamma}{2} \left[\rho_{12}(B_x - jB_y) - \rho_{21}(B_x + jB_y) \right] - \frac{\rho_{22} - \rho^e_{22}}{T_1}; \\
\frac{d\rho_{12}}{dt} &= j\frac{\gamma}{2}(\rho_{11} - \rho_{22})(B_x + jB_y) - \frac{\rho_{12}}{T_2} - j\omega_H \rho_{12}; \\
\frac{d\rho_{21}}{dt} &= -j\frac{\gamma}{2}(\rho_{11} - \rho_{22})(B_x - jB_y) - \frac{\rho_{12}}{T_2} - j\omega_H \rho_{12}.
\end{aligned}
\tag{3.11}
$$

Now it is necessary to connect the substance magnetization with the elements of the density matrix. It is known from lecture courses on general physics that magnetization $\mathbf{M}$ represents the sum of the magnetic dipole moments of the substance in the volume unit:

$$
\mathbf{M} = \frac{1}{V} \sum_{i=1}^{N} <\mathbf{m}>_i = n <\mathbf{m}>,
\tag{3.12}
$$

where V is a volume occupied by N particles, n is its concentration, and $<\mathbf{m}>_i$ is the average value of the observable quantity of the magnetic moment of the i-th particle. In contrast to the case of polarization determination (2.15), we should not now average the particles' orientation. The thing is that the permanent magnetic field unambiguously specifies the coordinate system. Let us express the average value of the observable quantity of magnetic moments through elements of the density matrix, taking (3.3) into account:

$$
<\mathbf{m}> = -\gamma \mathrm{Tr}\left(\widehat{\rho J} \right).
$$

It follows from here that taking (3.8) and (3.9) into account, we have

$$
\begin{aligned}
<m_x> &= -\frac{\gamma\hbar}{2}\mathrm{Tr}\left[\begin{pmatrix} \rho_{11} & \rho_{12} \\ \rho_{21} & \rho_{22} \end{pmatrix} \begin{pmatrix} 0 & 1 \\ 1 & 0 \end{pmatrix} \right] = -\frac{\gamma\hbar}{2}(\rho_{12} + \rho_{22}); \\
<m_y> &= j\frac{\gamma\hbar}{2}(\rho_{12} - \rho_{21}); \\
<m_z> &= \frac{\gamma\hbar}{2}(\rho_{11} - \rho_{22}).
\end{aligned}
\tag{3.13}
$$

We now notice that the substance magnetization along the z axis $M_z = n\frac{\gamma\hbar}{2} \times (\rho_{11} - \rho_{22}) = \frac{\gamma\hbar}{2}n_{12}$ is proportional to the population difference of the lower (1) and upper (2) levels. Hence, the magnetization along the direction of the permanent magnetic field is the measure of the energy accumulated by the magnetic two-level system. As for magnetizations M_x and M_y, they are connected with nondiagonal elements of the density matrix that describe quantum jumps between the lower and upper levels.

Taking (3.12) and (3.13) into consideration, we move on from (3.11) to the equation system for components of the vector $\mathbf{M}$:

$$\frac{dM_x}{dt} + \frac{M_x}{T_2} = \gamma B_y M_z - \omega_H M_y;$$
$$\frac{dM_y}{dt} + \frac{M_y}{T_2} = -\gamma B_x M_z + \omega_H M_x; \tag{3.14}$$
$$\frac{dM_z}{dt} + \frac{M_z - M_z^e}{T_2} = -\gamma (B_x M_y - B_y M_x).$$

Here, M_z^e is the longitudinal equilibrium magnetization of the substance, equal to $\frac{\gamma\hbar}{2}n_{12}^e$.

If we take into account that $\omega_H = \gamma B_{0z}$, we may notice that the right parts of (3.14) are components of the vector product $[\mathbf{B}_\Sigma \times \mathbf{M}]$, where $\mathbf{B}_\Sigma = \mathbf{B} + B_{0z}\mathbf{1}_z$ is the total magnetic field. Then the equations in (3.14) can be written in a more compact form:

$$\frac{dM_x}{dt} + \frac{M_x}{T_2} = \gamma [\mathbf{B}_\Sigma \times \mathbf{M}] \mathbf{1}_x;$$
$$\frac{dM_y}{dt} + \frac{M_y}{T_2} = \gamma [\mathbf{B}_\Sigma \times \mathbf{M}] \mathbf{1}_y; \tag{3.15}$$
$$\frac{dM_z}{dt} + \frac{M_z - M_z^e}{T_2} = \gamma [\mathbf{B}_\Sigma \times \mathbf{M}] \mathbf{1}_z.$$

These equations are called the Bloch[7] equations. Given their significance, they warrant detailed discussion.

The Bloch equations, as well as Eqs. (2.24) and (2.25) for electric dipole transitions, represent the material equations for a medium consisting of magnetic particles. They determine the connection between $\mathbf{M}$ and $\mathbf{H}$. Really, $\mathbf{B} = \mu_0(\mathbf{H} + \mathbf{M})$. Since $[\mathbf{M} \times \mathbf{M}] = 0$, the right part of (3.15) can be written in the form $\gamma\mu_0[\mathbf{H}_\Sigma \times \mathbf{M}]$.

Let us examine the limiting case when the times of spin–spin and spin–lattice relaxation are very long, $T_{1,2} \to \infty$. Then (3.15) takes the form

---

[7]Felix Bloch (1905–1983) was an American physicist and a member of the US National Academy of Sciences. He was born in Switzerland, but from 1934 onward he worked in the USA. He laid the foundation for the theory of crystals and low-temperature ferromagnetism. He developed theoretical fundamentals and performed the first experiments on nuclear magnetic resonance. He introduced the concept of spin waves. In 1952 he received the Nobel Prize in Physics.

$$\frac{d\mathbf{M}}{dt} = \gamma\mu_0\left[\mathbf{H}_\Sigma \times \mathbf{M}\right]. \tag{3.16}$$

It appears that this equation can be obtained from very simple considerations based on the concepts of classical physics. We examine a certain charged top (rotating electron), which has the impulse moment $\mathbf{L}$ and, corresponding to it, the magnetic moment $m = -\gamma L^2$. The main equation for the dynamics of rotational movement has the view

$$\frac{d\mathbf{L}}{dt} = \mathbf{M}_F, \tag{3.17}$$

where $\mathbf{M}_F$ is the moment of force $\mathbf{F}$. As we know, force moment $\mathbf{M}_F = [\mathbf{m} \times \mathbf{B}]$ acts on the loop with a current, with magnetic moment $\mathbf{m}$ and placed in magnetic field $\mathbf{B}$. Substituting this equation in (3.17) and multiplying by n, we obtain the equation

$$\frac{d\mathbf{M}}{dt} = \gamma[\mathbf{B} \times \mathbf{M}],$$

which is completely similar to Bloch equation (3.16) in which we do not take into account relaxation items.[8]

What character does the magnetization vector movement have under the action of a magnetic field? It follows from (3.16) that

$$2\mathbf{M}\frac{d\mathbf{M}}{dt} = \frac{dM^2}{dt} \equiv 0.$$

This means that in the absence of relaxation, the length of the magnetization vector does not change in time. Hence, the vector $\mathbf{M}$ end moves on the sphere. Moreover, the velocity vector of magnetization variation at $\mathbf{B} = B_{0z}\mathbf{1}_z \neq 0$ is directed along the normal to the plane passing through the z axis and vector $\mathbf{M}$. Therefore, in the absence of an alternate field the magnetization vector performs movement along the circle around the z axis, i.e., makes a precession (Fig. 3.3), and the precessional frequency is equal to $\gamma B_{0z} = \omega_H$. Quantum interpretation of this fact consists in splitting of the energy level into two sublevels with the distance between them being $\hbar\omega_H = \gamma\hbar B_{0z}$, i.e., $\omega_{21} = \omega_H = \gamma B_{0z}$.

Now we will examine the influence of relaxation processes on the behavior of the magnetization vector in the absence of an alternate field. If at $T_{1,2} \to \infty$, quantity M_z remains constant in time, under the action of processes of the spin–lattice relaxation, the longitudinal magnetization component, in accordance with (3.14), exponentially

---

[8]This equation is called the Landau–Lifshitz equation. It was devised in 1935 when Lev Landau was 27 years old and Evgeny Lifshitz was 20 years old. The relativistic Dirac equation, in which the spin is included, was devised in 1927–1928. In this period, Pauli matrices appeared. So, at that time, application of the classical model was fully accepted. Moreover, the final answer was correct!

Fig. 3.3 Precession of
magnetization vector **M**
around a static magnetic
field. From the Bloch
equations it follows that in
the movement process the
length of vector **M** should
not change. Therefore, the
vector end moves on the
sphere. Similar properties
are typical of the mechanical
moment of a rotating top,
whose precession we
observed in childhood. Such
motion is called a
precession. In the absence of
external forces we deal with
free precession

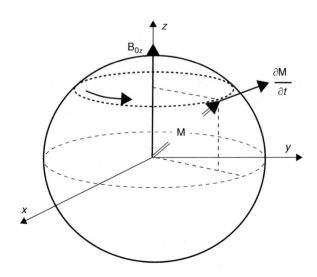

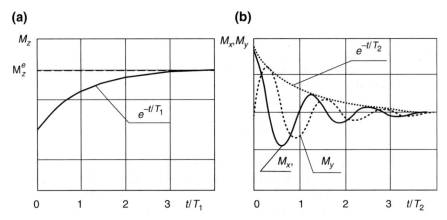

Fig. 3.4 Relaxation of longitudinal (**a**) and transverse (**b**) components of a magnetization vector.
Relaxation of the longitudinal component relates to energy exchange with a thermostat, while the
transverse component decreases as a result of disturbances of the precessional synchronism, which
is caused by all types of "collisions." The longitudinal relaxation is described by time T_1 and the
transverse relaxation by T_2

aspires to the equilibrium value of magnetization M_z^e (Fig. 3.4a). The time constant
T_1 corresponding to this process is called the longitudinal relaxation time. Transverse components of magnetization $M_{x,y}$ under the action of spin–spin relaxation
processes exponentially damp down to zero, performing oscillations with frequency
ω_H (Fig. 3.4b).

Damping time T_2 is often called the time of transverse relaxation. As we know
from the previous chapter, the time of transverse relaxation is defined by the intensity
of the energy exchange processes between magnetic particles and the lattice, and the
time of transverse relaxation is defined by how fast the synchronism of the magnetic

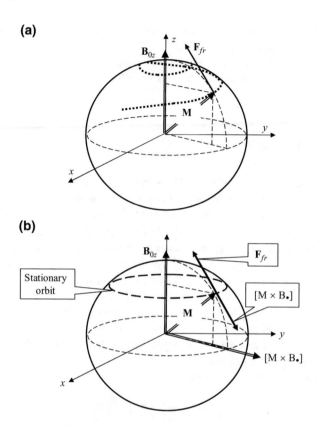

Fig. 3.5 Motion of magnetization vector **M**. (**a**) The relaxation processes, in a certain sense, are similar to friction force $\mathbf{F}_{fr}$, which is directed toward the vertex of the sphere. As a result, the precession damps. The precessional angle decreases and vector **M** passes by the spiral in the vertical position. This is exactly how magnetization of magnetic materials happens in a static magnetic field—for instance, during recording on a magnetic tape. This takes a time of about $T_{1,2}$ (about 1 μs). (**b**) If not only a static field but also an alternate field $\mathbf{B}_\sim$ is present, force $[\mathbf{M} \times \mathbf{B}_\sim]$ arises, which is able to take vector **M** out of the vertical position. This type of precession is called a forced precession. The precessional angle takes a stationary value under condition $[\mathbf{M} \times \mathbf{B}_\sim] = \mathbf{F}_{fr}$. To support equilibrium, the external field needs to perform work on the medium. This is manifested in energy absorption of the alternate field. Energy transfer is maximal for synchronous movement of vectors **M** and $\mathbf{B}_\sim$

dipole precession is disturbed. The spatial pattern of the magnetization vector movement is illustrated in Fig. 3.5.

Let us now discuss what the presence of an alternate magnetic field acting on the plane (x, y) leads to. It is clear enough that if this field has circular polarization and rotates with frequency ω_H in the same direction as vector **M**, then, moving synchronously with this vector, it will aspire to deflect it from the z axis, i.e., increase the precessional angle and oscillation amplitude M_x, M_y. If the field frequency $\omega \neq \omega_H$ or the rotational direction does not coincide with the precessional direction, we may expect a rather weak influence of the alternate field on the substance magnetization because of the absence of synchronism.

Before moving on to solution of the Bloch equations, we will discuss the issue of the equilibrium value of magnetization vector M_z^e.

3.5 Paramagnetism

From the last equation in system (3.15) it follows that in the stationary state ($d\mathbf{M}/dt = 0$), the longitudinal component of the magnetization vector is equal to some value M_z^e, even if field $\mathbf{B} = 0$. The Bloch equations themselves do not permit us to determine the numerical value of M_z^e. Hence, system (3.15) is incomplete and additional consideration is required for determination of the equilibrium magnetization value. It should be noted that we have already encountered the same situation during solution of equation system (2.24) and (2.25). The thing is that, as in Chap. 2, equations describing the thermostat dynamics are absent.

For $J = \frac{1}{2}$ $M_z^e = \frac{\gamma\hbar}{2}n_{12}^e$ it is necessary to find the equilibrium population difference of the two-level system whose energy diagram is presented in Fig. 3.2. We already know that for this it is necessary to use the Boltzmann distribution law. In this case it will appear as follows.

The population of the upper level

$$n_2^e = n_1^e \exp\left(-\gamma\hbar B_{0z}/k_B T\right).$$

If $n_1 + n_2 = n$, then

$$n_1^e = n/(1 + \exp\left(-\gamma\hbar B_{0z}/k_B T\right))$$

and

$$n_{12}^e = n\,\frac{1 - \exp\left(-\gamma\hbar B_{0z}/k_B T\right)}{1 + \exp\left(-\gamma\hbar B_{0z}/k_B T\right)} = n\,\tanh\left(\frac{\gamma\hbar B_{0z}}{2k_B T}\right).$$

Thus, the equilibrium value of the magnetization vector is

$$M_z^e = n\frac{\gamma\hbar}{2}\,\tanh\left(\frac{\gamma\hbar B_{0z}}{2k_B T}\right) = n\frac{\gamma\hbar}{2}\,\tanh\left(\frac{\gamma\hbar\mu_0 H_{0z}}{2k_B T}\right). \qquad (3.18)$$

The function of M_z^e versus H_{0z} is presented in Fig. 3.6. It can be seen that with the growth of the magnetic field intensity, M_z^e aspires to the limit M_{sat}:

Fig. 3.6 Characteristics of the equilibrium magnetization value of a paramagnet for several temperature values. With a temperature increase the populations of the magnetic sublevels become even and the magnetization reduces. For $H_{0z} \to \infty$ the magnetization aspires to the limit, which depends only on the number of particles

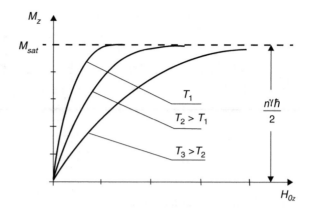

$$M_{\text{sat}} = n\frac{\gamma\hbar}{2} = n\frac{e\hbar}{2m_e} = n\mu_B.$$

Let us estimate the numerical value of M_{sat} for a diluted gas of weakly interacting particles (in this case, (3.18) is certainly true). If $n \approx 10^{22}\ m^{-3}$, then $M_{\text{sat}} \approx 0.1$ A/m. It is essentially less than the Earth's magnetic field intensity.

Now, using (3.18), we can determine the magnetic permeability. It is known that for magnetic materials, magnetic inductance **B** is

$$\mathbf{B} = \mu_0(\mathbf{H} + \mathbf{M}).$$

If the connection between **M** and **H** is linear, then

$$\mathbf{M} = \chi_M\mathbf{H}; \quad \mathbf{B} = \mu_0(1 + \chi_M)\mathbf{H} = \mu_a\mathbf{H},$$

where $\mu_a = \mu_0(1 + \chi_M) = \mu\mu_0$ is the absolute magnetic permeability. In the equilibrium state, according to (3.18), the connection of **M** with **H** is nonlinear. However, even if $T \approx 4$ K (the temperature of liquid helium), and $B_{0z} = 1$ T (one tenth of the maximal value obtained in practice), then

$$M_z^e \approx n\left(\frac{\gamma\hbar}{2}\right)^2 \mu_0 \frac{H_{0z}}{k_B T} = \chi_M H_{0z}. \tag{3.19}$$

The value of magnetic susceptibility:

$$\chi_M = n\left(\frac{\gamma\hbar}{2}\right)^2 \frac{\mu_0}{k_B T}$$

for $T = 4$ K, $n \approx 10^{22} m^{-3}$ is only $2 \cdot 10^{-8}$. Even if $n \approx 10^{28} m^{-3}$ (the concentration of particles in the solid body), then χ_M is merely 0.02. Hence, the relative magnetic

Table 3.1 Spectroscopic fission factor values

Iron ions			Lanthanide ions		
Ion	Calculated value	Experimentally observed value	Ion	Calculated value	Experimentally observed value
Ti^{3+}	1.55	1.8	Pr^{3+}	3.58	3.5
Fe^{3+}	5.92	5.9	Nd^{3+}	3.62	3.5
Fe^{2+}	6.70	5.4	Sm^{3+}	0.84	1.5
Co^{2+}	6.63	4.8	Gd^{3+}	7.94	8.0
Ni^{2+}	5.59	3.2	Dy^{3+}	10.63	10.6
Cu^{2+}	3.55	1.9	Ho^{3+}	10.60	10.4

permeability is a value hardly greater than 1. Magnetic materials with such properties are called paramagnets.

In quantum electronics the paramagnetic properties of interstitial ions in the lattice of a nonmagnetic material are of significant interest. In this case a number of levels equal to $2J + 1$ may essentially differ from 2, and to find the paramagnetic susceptibility we need to provide summation over all magnetic sublevels by using (1.64). This gives the following result:

$$\chi_M = n(\gamma\hbar)^2 \frac{J(J+1)}{3} \frac{\mu_0}{k_B T} = \left(\frac{eg\hbar}{2m}\right)^2 \frac{J(J+1)}{3} \frac{\mu_0}{k_B T} = (g_{eff}\mu_B)^2 \frac{\mu_0}{3k_B T}.$$

Here, $g_{eff} = g\sqrt{J(J+1)}$ is an effective factor of spectroscopic fission. Its values for some elements are presented in Table 3.1.

The magnetic susceptibility of rare-earth element ions is approximately one order greater than that of a single electron. However, in this case its value is small.

Although the magnetic properties of even rare-earth elements are very weakly expressed, paramagnets attract attention as materials for quantum amplifiers. In 1956 their use opened the way to the era of quantum electronics.

3.6   Ferromagnetism

When a magnetic field is absent, a paramagnet, as follows from (3.18), has zero magnetization. The ancient Greeks knew that iron could attract a stone, which they called a magnet, and nearly 4500 years ago, the Chinese began to use the natural magnetism of the Earth. Now every schoolkid knows that an iron rod, under certain conditions, can have a permanent magnetic moment. Materials with spontaneous magnetization are called ferromagnets, and two other varieties are also known: antiferromagnets and ferrimagnets.

Thus, in some cases, the magnetic moments of separate atoms, in spite of chaotic influence from a thermostat, will form parallel to each other. Everybody agrees that the forces of exchange interaction are responsible for this (see Chap. 1) but nobody can accurately say exactly how it happens in each specific case. To date, there is no strict quantitative theory of ferromagnetism. Therefore, there is nothing shameful in using the quite trustworthy classical model offered in 1907 by Pierre-Ernst Weiss.[9] He knew nothing of quantum mechanics; therefore, he offered the assumption that inside a ferromagnet there is some kind of internal magnetic field (the Weiss field):

$$H_{\text{int}} = \lambda M, \tag{3.20}$$

where λ is the Weiss constant. This field forms magnetic moments parallel to each other. Now it is clear that in real life there is no additional magnetic field and the forming force has an electric origin. But if we cannot calculate this force, then how is the situation better than (3.20) with an unknown value of λ?

We substitute (3.20) in (3.18) and obtain, as a result,

$$M_z^e = n\frac{\gamma\hbar}{2} \tanh\left(\frac{\gamma\hbar}{2}\frac{\mu_0}{k_B T}\lambda M_z^e\right). \tag{3.21}$$

This is the transcendent equation for determination of spontaneous magnetization, and it is convenient to present it in the following form:

$$x = \tanh\left(\frac{x}{t}\right), \tag{3.22}$$

where

$$x = \frac{2M_z^e}{n\gamma\hbar}; \quad t = \frac{4k_B T}{n(\gamma\hbar)^2\mu_0\lambda}.$$

The numerical solution of (3.22) does not present any difficulty, but we would like to use the graphic method. For this we draw the left and right parts of the equation so we can find its intersection point (Fig. 3.7). We see well that with an increase in temperature the solution point moves to zero and for $t > 1$ the solution is absent. Thus, at a temperature T value less that the Curie temperature,

$$T_{\text{Curie}} = n\left(\frac{\gamma\hbar}{2}\right)^2\frac{\mu_0\lambda}{k_B},$$

ferromagnetic materials have spontaneous magnetization, and at higher temperatures they becomes paramagnets (Fig. 3.8).

---

[9]Pierre-Ernest Weiss (1865–1940) was a French physicist. He investigated magnetic phenomena and discovered the magnetocaloric effect and the law of the temperature dependence of ferromagnetic susceptibility. He predicted the existence of magnetons, and he designed powerful electromagnets and a series of devices for magnetic and electric measurements.

Fig. 3.7 Solution of
ferromagnetic equation
(3.22). With growth of t the
solution point gradually
displaces and for $t = 1$ the
point will be in the
coordinate system origin. At
$t > 1$ the solution is absent

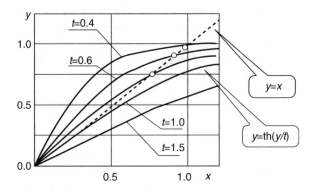

Fig. 3.8 Function of
spontaneous magnetization
of a ferromagnet versus
temperature. The Curie
point corresponds to the
value $t = 1$ in Fig. 3.7. At a
temperature higher than
T_{Curie} the ferromagnet
passes into a ferromagnetic
state

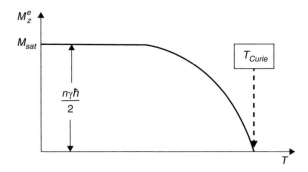

At zero temperature all magnetic moments are oriented parallel to each other and
magnetization achieves the maximal value. For particles with $J = 1/2$,

$$M_{\text{sat}} = n\frac{\gamma\hbar}{2} = n\frac{e\hbar}{2m_e} = n\mu_B,$$

but ferromagnetic substances are formed from atoms and molecules with a large
proper magnetic moment, i.e., with a large value of the quantum number J. Therefore, for ferromagnets we have

$$M_{\text{sat}} = n_B n \mu_B,$$

where n_B is the effective number of Bohr[10] magnetons. In Table 3.2 the values of
saturation magnetization, the Curie temperature, and the effective number of Bohr
magnetons are presented for some ferromagnetic and ferrimagnetic materials.

---

[10]This number should not be confused with the effective factor of spectroscopic fission. Here, we
are talking about a number of magnetic moments that are parallel to each other and fitted for
each atom.

Table 3.2 Characteristics of some magnetic materials

Substance	M_{sat} (кA/m)	T_{Curie} (K)	n_B	Substance	M_{sat} (кA/m)	T_{Curie} (K)	n_B
Fe	138.5	1043	2.22	GdMn$_2$	17.1	303	2.8
Co	115.1	1388	1.72	Gd$_3$Fe$_5$O$_{12}$[a]	46.1	564	16
Ni	40.6	627	0.606	Y$_3$Fe$_5$O$_{12}$[a]	15.9	560	5.0
Gd	159.9	292	7.10	CrO$_2$	41.0[b]	386	2.03
Dy	232.4	85	10.0	MnOFe$_2$O$_3$[a]	32.6[b]	573	5.0
Cu$_2$MnAl	43.8	710	4.0	FeOFe$_2$O$_3$[a]	38.2[b]	858	4.1
MnAs	69.2	318	3.4	CoOFe$_2$O$_3$[a]	31.8[b]	793	3.7
MnBi	54.1	630	3.52	NiOFe$_2$O$_3$[a]	21.5[b]	858	2.4
MnB	12.0	587	1.92	CuOFe$_2$O$_3$[a]	10.7[b]	728	1.3
EuO	152.8	69	6.8	MgOFe$_2$O$_3$[a]	8.8[b]	713	1.1

[a]Ferrite
[b]Value at room temperature

Using the known Curie temperature we can find the Weiss constant:

$$\lambda = \frac{4k_B T_{Curie}}{n(\gamma\hbar)^2 \mu_0}.$$

If we assume that $n \approx 10^{29} \text{m}^{-3}$ and $T_{Curie} = 10^3$ K, then $\lambda \cong 10^3$. Is that a lot or a little? To answer this question, we estimate the Weiss field value. If we suppose that all magnetic moments are oriented parallel to each other, spontaneous magnetization will have a value of about 10^6 A/m. This corresponds to the intensity of the internal magnetic field:

$$H_{int} = \lambda M_{sat} = 10^9 \text{ A/m}$$

or to magnetic induction:

$$B_{int} = \mu_0 H_{int} \approx 10^3 \text{ T}.$$

This is two orders greater than has sometimes been achieved in laboratory conditions. You see that quantum mechanical forces are much greater than any other magnetic forces.

Now we can move on to determination of the magnetic permeability of ferromagnets. In (3.21) we add the field $B = \mu_0 H$:

$$M_z^e = n\frac{\gamma\hbar}{2} \tanh\left(\frac{\gamma\hbar}{2}\frac{\mu_0}{k_B T}(H + \lambda M_z^e)\right).$$

Solution of this equation gives the functional coupling of **M** and **H**. A graphic illustration of this solution is presented in Fig. 3.9. In the paramagnetic state the

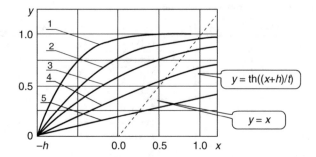

Fig. 3.9 Solution of the ferromagnetic state equation in a magnetic field for (1) $t = 0.5$, (2) $t = 1.0$, (3) $t = 1.5$, (4) $t = 2.5$, and (5) $t = 5.0$. With temperature growth the solution point moves to zero. In the paramagnetic state ($t > 1$) at a relatively high temperature (see the curve for $t = 5.0$) the tangent can be replaced by its argument to obtain an approximate solution from which the Curie–Weiss law follows, describing the temperature function of magnetic susceptibility

($T \gg T_{\text{Curie}}$) tangent can be replaced by its argument. Then the transcendent equation is transformed into the algebraic equation:

$$M_z^e = n \left(\frac{\gamma \hbar}{2}\right)^2 \frac{\mu_0}{k_B T} \left(H + \lambda M_z^e\right) = \frac{C}{T} \left(H + \lambda M_z^e\right),$$

from which it follows that

$$\chi_M = \frac{M}{H} = \frac{C}{T - C\lambda} = \frac{C}{T - T_{\text{Curie}}}.$$

This functional coupling is known as the Curie–Weiss law. It splendidly describes the function of susceptibility versus temperature in the area $T > T_{\text{Curie}}$. The temperature function of the inverse susceptibility of nickel, obtained by Weiss and published by him in 1926 (see Fig. 3.10), is a good confirmation.

At the Curie temperature the magnetic susceptibility becomes infinite. Hence, in the vicinity of this point our approximate calculations become wrong. This is connected with the fact that for $T = T_{\text{Curie}}$ a so-called phase step[11] happens. In our model the spontaneous magnetization in the ferromagnetic phase has no breaks, changing smoothly. It vanishes at the point of T_{Curie}. Such a behavior of material properties is typical for phase steps of the second kind.

In some materials the spontaneous magnetization rises by a step and the factual temperature of the phase jump does not coincide with the Curie temperature. Such

---

[11] Phase steps are steps between different macroscopic states of a multiparticle system with variation of external parameters (temperature, pressure, intensity of electric and magnetic fields, feedback coefficient, etc.). Phase steps are cooperative phenomena connected with the appearance or disappearance of mutual correlations in particle behavior. In lecture courses on the electron profiles of the process of oscillation excitation in the oscillator with variation of the feedback coefficient, the soft excitation mode is a step of the second kind but rigid excitation is a step of the first kind.

Fig. 3.10 Temperature
function of the inverse
magnetic susceptibility of
nickel (1 g) obtained
experimentally by Weiss
and published in 1926. The
Curie temperature is 358 °C.
The dotted curve
corresponds to the Curie–
Weiss law, which describes
the temperature function in
the region of high
temperatures well

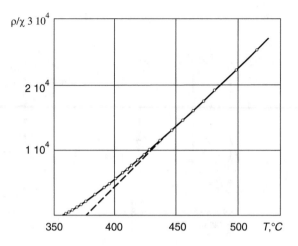

steps can be involved in phase steps of the first kind. With the temperature function
$M_z^e(T)$ a hysteresis is observed for steps of the first kind: the point of the phase step
with heating does not coincide with the point of the step at cooling. So, for instance,
for manganese–bismuth alloy (MnBi), $T_{Curie} \approx 630$ K $= 357°$ C. However, with
heating the step is at $T_{Curie} \approx 360°$ C, but with the cooling it is at $T_{Curie} \approx 345°$ C.
This difference is used for information recording and erasure with the help of laser
emission on magnetic thin-film disks.

Determination of magnetic susceptibility in the ferromagnetic phase represents a
more complicated problem. From our everyday experience we know well that under
normal conditions, real bodies made of ferromagnetic materials do not have a
permanent magnetic moment. Hence, the process of magnetization and the arising
of a spontaneous magnetic moment requires more careful examination.

3.7 Domains and the Hysteresis Curve

It is well known that immediately after smelting, iron has no permanent magnetic
properties. If at consolidation an external magnetic field is absent, after cooling
below the Curie temperature, spontaneous magnetization cannot be observed in any
way.[12] Why does this happen?

---

[12]In fact, the Earth's magnetic field is always present and weak residual magnetization is nevertheless observed. This phenomenon is used for dating of archeology samples. The thing is that the
orientation of the Earth's magnetic field changes over time. The regulation of this change is known.
At the time of baking of ceramic articles or simply cooling of the furnace in which they were baked,
magnetic moments were formed parallel to the Earth's magnetic field. Comparison of the magnetic
moment orientation with the geomagnetic field orientation allows us to determine, with sufficient
accuracy, the date of the last cooling of samples under investigation. This is a beautiful solution,
isn't it?

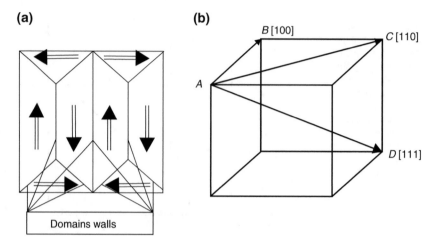

Fig. 3.11 (**a**) Domain structure of a ferromagnet. The domains are separated by walls, inside which the orientation of vector **M** changes by 180° (vertical walls) or by 90° (inclined walls). (**b**) Cubic structure of a ferromagnetic crystal. The field of anisotropy leads to the appearance of directions of easy AB ([100]) magnetization and hard AC ([110]) and AD ([111]) magnetization. The domains are oriented along the axes of easy magnetization

With surprising astuteness, Weiss postulated the existence of domains—finite regions with similarly oriented magnetic moments. Theoretical substantiation of the existence of domains was achieved in 1935 by Lev Landau[13] and Eugene Lifshitz.[14] They proved that formation of a domain structure is a consequence of joint action of exchange interaction forces and anisotropy forces.

Domains orient themselves so the external magnetic field is absent, as this is more profitable from the energy point of view. For instance, the orientation presented in Fig. 3.11a is possible. It is quite reasonable to ask the question: to what degree can a medium be broken into pieces of domains? Everything will stop when the energy spent on formation of the domain wall exceeds the gain from the decrease in the energy of the external field. Felix Bloch introduced the concept of the domain wall. The domain wall is the region within which vector **M** gradually changes its direction. An illustration of magnetization rotation between two domains is shown in Fig. 3.12.

---

[13]Lev Landau (1908–1968) was a Soviet physicist and academician (1946). In 1927 he was sent to Denmark to study with Niels Bohr and introduced the concept of the density matrix. In 1937 he created the general theory of the phase step of the second kind. He made a major contribution to the development of theoretical physics. In 1962 he was awarded the Nobel Prize in Physics. He was a member of the national academies of sciences in the USA, Denmark, the UK, France, and the Netherlands. The Institute of Theoretical Physics of the Russian Academy of Sciences was named in his honor.

[14]Eugene Lifshitz (1915–1985) was a Soviet physicist and academician (1979). His areas of scientific research were solid-body physics, cosmology, and the theory of gravity. He was a pupil of Lev Landau.

Fig. 3.12 Spin direction variation in the Bloch wall between domains with opposite directions of magnetization. The thickness of the intermediate layer in iron is about 300 constants of the lattice. The rotation of vector **M** in the domain wall is the result of anisotropy force action, which promotes rotation, and the forces of exchange interaction, which contradict it

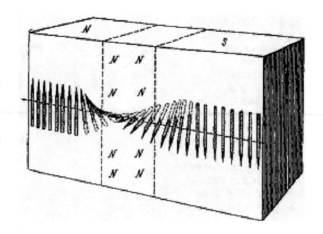

Walls at 90° and 180° are most typical. The wall thickness is of the order of 1 μm. Therefore, the domain size is not less than 1 μm.

Let us now examine the magnetization process of the ideal crystal.

Since crystal forms a structurally ordered space lattice, chemical bend forces are strictly oriented in the space. These forces, on the one hand, aspire to form magnetic moments parallel to each other and, on the other hand, orient them in a definite manner in the space. In connection with this we can speak about "forming forces" (exchange forces, the Weiss field) and anisotropy forces (the anisotropy field). Iron, for example, has a cubic structure and anisotropy fields are directed along the cube facets (Fig. 3.11b). These directions are called axes of easy magnetization (axes [100], [010], and [001]), and the direction AE is the axis of hard magnetization (axis [111]). If the crystal is not magnetized, domains are formed along directions of easy magnetization as a result of anisotropy force action. In the domain walls the magnetization vector rotates from one direction of easy magnetization to the other one. Forming forces aspire to make this rotation smoother, but anisotropy forces aspire to make it faster. With weak anisotropy forces the wall thickness reaches 10 μm, and with strong anisotropy forces it decreases to 0.3 μm.

If the external field is oriented along the axis of easy magnetization, with growth of this field the domain walls move so the domains oriented along the field increase. Several stages of this process are shown in Fig. 3.13. If the field does not coincide with the axis of easy magnetization and is oriented, for instance, along the diagonal of the AC facet, the walls move so the domains that are the closest in direction to the magnetization field grow (see Fig. 3.13). This will continue until only two types of domains remain. Then the essential energies against the anisotropy forces will require rotation of the magnetization vectors of the domains. Figure 3.14 shows the functions of magnetic induction versus the intensity of the magnetic field for three directions of an iron crystal. It can be seen that the iron is magnetized easily enough until the domain walls move, and rotation of the magnetic moments is essentially more difficult. During the magnetization process, work is performed for

Fig. 3.13 (**a**) Consistent stages of ferromagnetic magnetization. (**b**) Ferromagnetic magnetization in an arbitrary direction. Magnetization in relatively weak fields happens at the expense of domain wall movement. If the magnetic field does not coincide with the easy magnetization axis, in strong fields, rotation of the magnetization vector takes place in the domains

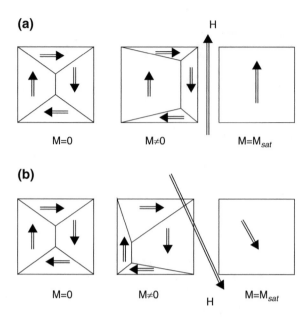

Fig. 3.14 Magnetization curves of a crystalline ferromagnet (iron). The initial part corresponds to relatively magnetic permeability of the order of 10^6. The process of rotation of **M** requires much more effort. The slightly sloping segments of curves AC and AD correspond to this process. Direction AD [111] is the hard magnetization direction. In the ideal crystal these processes are reversible and hysteresis is absent

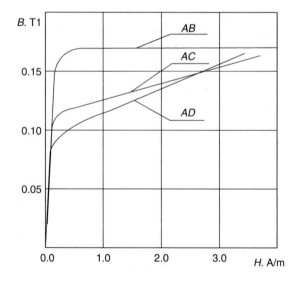

wall movements and rotation of magnetization. This work is fully converted into magnetic energy in the magnetized body equal to $\mathbf{M} \cdot \mathbf{H}$. If field **H** decreases to zero, everything will be repeated in the inverse order. Thus, magnetization of the ideal crystal lattice is a reversible process. Nevertheless, it is well known that real ferromagnetic materials are magnetized absolutely differently in other manners (Fig. 3.15). What is the matter here?

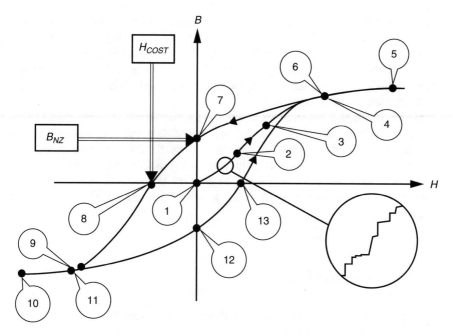

Fig. 3.15 Magnetization curve of a real ferromagnet. The process of domain wall movement happens in jumps with conversion of part of the energy into heat. At the initial point (*1*) the internal magnetic energy $W_M = BM$ is equal to zero. The hysteresis is connected with irreversibility of the heat processes. The energy of heat losses is equal to the accumulated energy of the magnetic field at point *7*; $W_M \sim B_{res}M_{res}$

Real crystals always have internal defects, which, in particular, lead to variations in the easy magnetization direction. Technical ferroalloys are polycrystalline materials in which all possible orientations of the separate crystallite axes are equiprobable. Therefore, firstly, magnetization always occurs in a direction that does not coincide with axes of easy magnetization, and the magnetization curve always has a segment in which domain magnetization vector rotation happens (segments 4, 5, 6 and 9, 10, 11 in Fig. 3.15). The process of **M** rotation is reversible. Secondly, movement of domain walls happens somewhat differently from that in an ideal crystal. The presence of defects and separate crystallites opposes free wall movement. Wall movement may be delayed in the defect regions until the forces acting on them from the field do not exceed the value necessary for further movement. As a result, the walls move in steps (jumps). This can be seen with careful examination of the magnetization curve. Wall steps (jumps) are accompanied by acoustic wave generation.[15] It is important that a part of the energy converts into another form and, at the end, it simply converts into heat, which is dissipated into the surrounding

---

[15]More meticulous readers can try to listen to the wall steps. For this it is necessary to wind a coil of copper wire around a magnetized rod and connect it to the input of a sensitive amplifier. But it is even more interesting to *hear* the domain steps with the help of a microphone.

space. There is no reverse process. The process cannot be reversible; the walls will not occupy their initial positions after the magnetic field is switched off. This is the reason of hysteresis. Let us identify two significant characteristics of engineering ferromaterials: the residue induction (points 7 and 12 in Fig. 3.15) and the coercive force (points 8 and 13 in Fig. 3.15). The square of the hysteresis curve determines the losses on magnetization reversal.

Materials that are used in transformers and electric motors should have a hysteresis loop that is as narrow as possible. Such materials are called magnetically-soft materials. On the other hand, for permanent magnets, residual magnetization is important; therefore, we must provide the maximum possible square of the hysteresis curve. Such materials are called magnetically-hard materials. Experts in material science can perform unbelievable tricks to obtain ferroalloys with the required properties.

3.8   Magnetic Resonance

Now we can perform the solution of the Bloch equation. Let

$$B_{\Sigma} = B_x(t)\mathbf{1}_x + B_y(t)\mathbf{1}_y + B_{0z}\mathbf{1}_z.$$

Then, the Bloch equations will take the form shown in (3.14). We can find out the solution of (3.14) with linear approximation. For this it is sufficient to assume that $M_z = M_z^e$. In this case we commit an error proportional to the square of transverse magnetization as $M_z = \sqrt{(M_z^e)^2 - M_x^2 - M_y^2}$. The remaining equations take the form

$$
\begin{aligned}
\frac{dM_x}{dt} + \frac{M_x}{T_2} &= \gamma B_y M_z - \omega_H M_y; \\
\frac{dM_x}{dt} + \frac{M_x}{T_2} + \omega_H M_y &= \gamma B_y M_z^e; \\
\frac{dM_y}{dt} + \frac{M_y}{T_2} - \omega_H M_x &= -\gamma B_x M_z^e.
\end{aligned}
\tag{3.23}
$$

We shall assume that

$$
\begin{aligned}
B_{x,y} &= \mathrm{Re}\left(\dot{B}_{x,y}\exp\left(j\omega t\right)\right); \\
M_{x,y} &= \mathrm{Re}\left(\dot{M}_{x,y}\exp\left(j\omega t\right)\right).
\end{aligned}
$$

Then, from (3.23) we obtain the system of equations for complex amplitudes:

$$j\omega\dot{M}_x + \frac{1}{T_2}\dot{M}_x + \omega_H\dot{M}_y = \gamma M_z^e \dot{B}_y;$$

$$-\omega_H\dot{M}_x + j\omega\dot{M}_y + \frac{1}{T_2}\dot{M}_y = -\gamma M_z^e \dot{B}_x,$$

which has the following solution:

$$\dot{M}_x \frac{\gamma M_z^e \left[\dot{B}_y \left(j\omega + T_2^{-1}\right) + \omega_H \dot{B}_x\right]}{\left(j\omega + T_2^{-1}\right)^2 + \omega_H^2};$$

$$\dot{M}_y \frac{\gamma M_z^e \left[-\dot{B}_x \left(j\omega + T_2^{-1}\right) + \omega_H \dot{B}_y\right]}{\left(j\omega + T_2^{-1}\right)^2 + \omega_H^2}.$$

Now the connection between the alternate magnetic field and the magnetization vector can be written through the susceptibility tensor as $\mathbf{M} = \dot{\chi}_M \mathbf{B}$,[16] where

$$\dot{\chi}_M = \begin{vmatrix} \dot{\chi} & \dot{\chi}_a & 0 \\ -\dot{\chi}_a & \dot{\chi} & 0 \\ 0 & 0 & 0 \end{vmatrix}; \quad \dot{\chi} = \frac{\gamma M_z^e \omega_H}{\omega_H^2 - (\omega - j/T_2)^2}; \quad \dot{\chi}_a = \frac{\gamma M_z^e \left(j\omega + T_2^{-1}\right)}{\omega_H^2 - (\omega - j/T_2)^2}.$$

It is easy to see that the interaction has a resonance character. This is especially distinctly demonstrated if as $\mathbf{M}$ and $\mathbf{B}$ we use the linear combinations $\dot{M}_\pm = \dot{M}_x \pm j\dot{M}_y$ and $\dot{B}_\pm = \dot{B}_x \pm j\dot{B}_y$, which correspond to fields with a circular polarization. For these fields,

$$\dot{M}_+ = \dot{M}_x + j\dot{M}_y = \frac{\gamma M_z^e}{\omega_H - \omega + j/T_2} \dot{B}_+ = \dot{\chi}_+ \dot{B}_+;$$

$$\dot{M}_- = \dot{M}_x - j\dot{M}_y = \frac{\gamma M_z^e}{\omega_H + \omega - j/T_2} \dot{B}_- = \dot{\chi}_- \dot{B}_-.$$

The rotational direction of field $\dot{B}_+$ coincides with the precessional direction of vector $\mathbf{M}$. Therefore, such a field intensively interacts with the medium, especially if $\omega = \omega_H$. Field B_- rotates in the opposite direction and cannot effectively influence the movement of vector $\mathbf{M}$. The imaginary parts of the susceptibilities equal to

$$\chi'_\pm = \mp \frac{\gamma M_z^e T_2}{1 + (\omega_H \mp \omega)^2 T_2^2}$$

describe the energy absorption. Functions $|\chi'_+|$ and $|\chi'_-|$ versus frequency are shown in Fig. 3.16. Media that have the above-mentioned properties are called gyrotropic,[17]

---

[16]Pay attention to the fact that here, χ_M connects the magnetic moment with the magnetic induction. If we understand $\mathbf{B}$ as a local field, then $\mathbf{B} = \mu_0 \mathbf{H}$ (see (3.16)). For weak magnetics we can neglect the difference between the external field and the local field. Ferromagnetic calculation of the local field represents a problem that is not so simple.

[17]The term "gyrotropic" comes from the Greek word *guros*, meaning "circle" or "ring."

Fig. 3.16 Function of the imaginary part of magnetic susceptibility, which is responsible for losses in the medium, versus the frequency for fields with different types of circular polarization. Energy absorption is observed for the field whose rotational direction coincides with the precessional direction

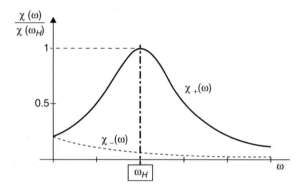

and the phenomenon of the frequency-sensitive energy absorption in magnetic media is called magnetic resonance.

We can distinguish between paramagnetic, ferromagnetic, and nuclear magnetic resonances depending on what the magnetic properties of the medium are caused by. Magnetic spin resonance plays a large and important role in solid-body physics.

Publications often use the following abbreviations for the main types of magnetic resonance: NMR for nuclear magnetic resonance, EPR for electron paramagnetic resonance (electron spin resonance), and FMR for ferromagnetic resonance.

3.9 Electron Paramagnetic Resonance

Electron paramagnetic resonance is observed on frequencies higher than 1 GHz ($\gamma = 28$ GHz/T1). This phenomenon was discovered in 1944 by Eugene Zavoiskiy.[18]

Since M_z^e is small in such media, we should cool paramagnets down to the temperature of liquid nitrogen (77 K) or even that of liquid helium (4 K). The paramagnetic particles defining the paramagnetism may be electrons, atoms, molecules, complex compositions, or crystal defects if they have a nonzero magnetic moment.

For EPR observation, specific spectrometers (radio spectrometers) have been developed, which operate in the centimeter and millimeter wavelength ranges. A sample with a volume of several cubic millimeters is placed in the resonator region, where the component of magnetic field intensity is maximal. The resonator is located

---

[18]Eugene Zavoiskiy (1907–1976) was a Soviet physicist and the founder of the Kazan school of thought. His discovery of the phenomenon of electron paramagnetic resonance established a new section of applied physics. In 1945 he defended his Sr.Sci. thesis, devoted to electron paramagnetic resonance. Under his supervision an investigation cycle was performed on application of electronic–optical transducers for investigation of high-speed (10^{-9}–10^{-14} s) light and other processes. Together with his colleagues, he created and further developed the method of thermonuclear plasm generation.

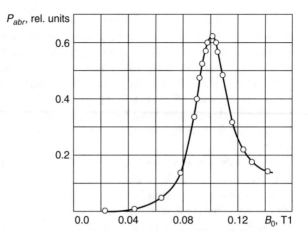

Fig. 3.17 Electron paramagnetic resonance in an $MnSO_4$ crystal at 298 K on a frequency of 2.75 GHz, observed by Eugene Zavoiskiy

between the poles of the electromagnet, which is the source of a permanent magnetic field. The resonance condition is usually achieved by means of variation of the magnetic field intensity at a fixed value of the generator frequency. The value of the magnetic field at resonance depends in the general case upon the orientation of the **H** vector with respect to the sample. Figure 3.17 shows the results of Zavoiskiy's experiments.

All free electrons in the main state are, of course, paramagnetic ones and in gases consisting of such atoms, paramagnetic resonance should probably be observed. However, the density of gases usually is so low that magnetic resonance cannot even be detected. All electrons in the most neutral molecules are paired (independently upon the aggregate state of a substance), so these molecules are nonmagnetic. However, in several molecule gases—namely, O_2, NO, NO_2, Cl_2O, and F_2O—the molecules have an unpaired electron and therefore they are paramagnetic. The paramagnetic resonance in these gases has been studied, but it is not of interest from a practical point of view.

Strong absorption under conditions of paramagnetic resonance can be observed using free radicals. Free radicals are a group of coupled atoms of organic composition containing an electron with an unpaired spin. In most free radicals, the unpaired electron behaves as a free electron with a 1/2 spin. In an external magnetic field, two energy sublevels are formed and the transitions between them are observed as paramagnetic resonance. The lines of paramagnetic resonance are usually narrow—of the order of 1 mT (about 30 MHz). The small line width of free radicals is connected to the rather strong exchange interaction between adjacent spins, which prevents precessional misphasing of separate spins.

One of the most frequent used radicals is hydrazyl—an abbreviated form of the names diphenylpicrylhydrazyl or diphenyl-trinitrophenyl-hydrazyl ($(C_6H_5)_2$–N–N–C_6H_2–$(NO_3)_2$). This substance, which is easy to obtain in a polycrystalline form and easy to grow in the form of small crystals, has an unusually narrow line (about 10 MHz). It gives a strong signal of magnetic resonance, which is easy to observe at room temperature even using the simplest instruments. A small crystal of hydrazyl is often introduced into the equipment for observation of magnetic resonance with the aims of magnetic field calibration and equipment adjustment.

During irradiation of crystals, plastics, and many other substances by different types of radiation (for example, x-rays, γ-rays, neutrons, electrons, and ultraviolet radiation), various defects may form in them—namely, defects of the crystalline lattice, disrupted chemical couplings, etc. In many cases in places where defects are located, unpaired electrons remain or appear, allowing paramagnetic resonance to be observed.

Some insignificant internal defects in crystals (for example, a missing atom or an excess atom located in the interstice) are called color centers, as variations in optical properties connected with them may cause coloration of a colorless crystal. So-called *F*-centers (or Farbe centers)[19] are most frequently encountered. An F-center is a type of crystallographic defect in which an anionic vacancy in a crystal is filled by one or more electrons. Electrons in such a vacancy tend to absorb light in the visible spectrum, such that a material that is usually transparent becomes colored. F-centers are often paramagnetic and can be studied by EPR techniques.

Among the substances that can be studied with the help of the electron spin resonance method are quartz and various plastics, which can be irradiated by x-rays; glass and ice, which can be irradiated by γ-rays; and diamonds, which can be irradiated by neutrons and electrons.

Of most practical interest to us are the paramagnetic properties of atoms located in the crystalline lattice of a solid body. In the first approximation, the levels of such atoms are similar to the levels of free atoms. Nevertheless, to decrease the mutual influence of atoms, which leads to widening of the magnetic resonance line, paramagnetic atoms are embedded into the substance in small quantities with nonmagnetic ions that have the same valency. The crystal of this type used in the creation of the first EPR maser in 1956 was a ruby, which has a crystalline lattice of corundum (Al_2O_3) with an impurity of a small quantity (nearly 0.1%) of chromium.

An isolated Cr^{3+} ion has an electron configuration of $3d^3$, which corresponds to the main level $S = 3/2$, $L = 3$, $J = 3/2$. The free ion has $2J + 1 = 4$ Zeeman levels; relatively, the ion in a ruby lattice has the same levels. However, in the ruby the chromium ion has a more complicated system of Zeeman levels. Energy level splitting is observed even for a zero (absent) magnetic field and its value is strong anisotropic. Figure 3.18a shows the energy levels of this ion in a magnetic field oriented along the symmetry axis of the ruby crystal. In a zero magnetic field the levels are divided into two doublets corresponding to $m = \pm 3/2$ and $m = \pm 1/2$; the energy difference between them corresponds to frequency transitions of 11.4 GHz. Figure 3.18b, c shows how the energy levels of the ruby vary when a permanent magnetic field forms angle θ with the crystal axis equalling, relatively, 0°, 30°, and 54° 44″. For θ ≠ 0 the eigenfunctions get mixed up, the energy levels do not intersect, and all forbidden transitions becomes more or less allowed. At the angle

---

[19]In German, *Farbe* means "color" and *Farbzentrum* means "color center."

Fig. 3.18 Energy diagram of magnetic sublevels in a ruby monocrystal (Al_2O_3: Cr^{3+}). The sublevels are related to the magnetic moments of the chromium ions embedded in the corundum lattice. The sublevel position in the monocrystal depends on the orientation of the permanent magnetic field with respect to the crystallographic axes. (**a**) The level splitting in a zero field is equal to ≈ 11 GHz. (**b**) At a nonzero angle the energy states are "mixed". (**c**) At an angle of 54° 44″ the energy diagram is symmetric with respect to the zero line

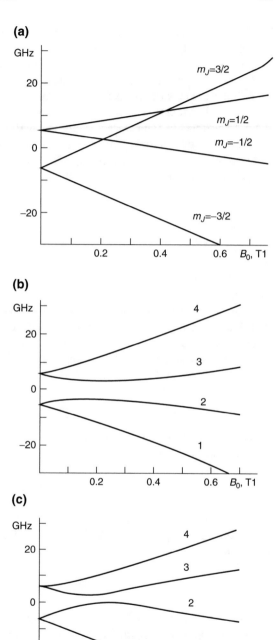

of $54° 44''$ $\left(\cos\left(\theta\right) = 1/\sqrt{3}\right)$, for reasons that we will not discuss there, a specific property occurs in which the corresponding energy levels turn out to be symmetric with respect to the imagined average line. This feature is very suitable for implementation of a quantum paramagnetic amplifier.

EPR has found wide applications in various areas of physics, chemistry, geology, biology, and medicine. It is intensively used for studying solid-body surfaces and phase transitions. In semiconductor physics, with the help of EPR, point impurity centers, free charge carriers, carrier–impurity pairs and complexes, radiation effects, dislocations, structural defects, interlayer formations (the Si–SiO_2 type of boundary), recombination processes, photoconductivity, and other phenomena are investigated.

3.10   Nuclear Magnetic Resonance

In 1948, Edward Purcell, Felix Bloch, and colleagues observed NMR for the first time.

NMR can be observed on frequencies of not more than 1 GHz. It is connected with the fact that geomagnetic constant γ for nuclei is of a lesser order than for electrons. According to (3.4), a nucleus has magnetic moment $\mathbf{m}$ and linear momentum, which can be expressed in $\hbar$ units ($\mathbf{JN} = \hbar\mathbf{I}$). If the nucleus is in a static magnetic field $\mathbf{B_0} = B_0\mathbf{1_z}$, then the interaction energy will be

$$W = -m_z B_0 = -\gamma\hbar I_z B_0.$$

We know well that in the stationary state the projection of the impulse moment $I_z = m_I \check{z}$, where $m_I = I,\ I - 1,\ \ldots,\ -I$; hence, the allowed energy values are

$$W = -m_I \gamma\hbar B_0.$$

In the magnetic field, the nucleus, for which $I = \frac{1}{2}$, can be in one of two energy states corresponding to $m_I = \pm\frac{1}{2}$. The energy difference of these two states defines the NMR frequency:

$$\omega_H = \gamma B_0.$$

For the proton, $\gamma = 2.675 \cdot 10^8 \frac{rad/s}{Tl}$ or $\gamma = 42.58 \frac{MHz}{Tl}$.

Nucleus spins with an odd mass number (total number of protons and neutrons) have half-integer values of m_I. In nuclei with an even mass number, either there is no spin at all if the nucleus charge is even, or m_I takes an integer value. For this reason, NMR is observed only in some atomic isotopes. For observation of NMR, the abundance of the isotope in the natural material is of practical importance. So, for instance, ^{59}Co isotopes constitute 100% of natural cobalt, and ^{57}Fe isotopes constitute somewhat more than 2% of natural iron.

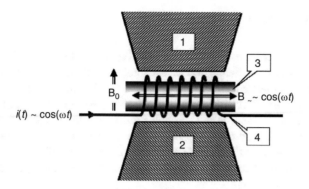

Fig. 3.19 Scheme of an experiment observing nuclear magnetic resonance (NMR). By changing the frequency of the sinusoidal current or the magnetic field intensity over time, we may detect a loss increase in the coil at the point of NMR. Usually, to increase sensitivity, the oscillator current variation is registered in an oscillating circuit connected to the sample by a coil. An NMR signal in water can have a width of the order of hertz. The line width in liquids will be narrower because of particle heat movement. *1* and *2* denote the magnetic poles, *3* denotes the sample, and *4* denotes the coil creating the alternate field

For nuclei of different elements in a field of 1 T the resonance frequency is in the range of 1–10 MHz. This difference allows identification of atomic isotopes.

NMR is observed by variation of losses in the inductance coil that surrounds the sample (Fig. 3.19). However, the NMR signal in most cases is weak. Therefore, we need to use highly sensitive equipment and specific methods of signal processing. It is much more convenient to observe NMR in ferromagnets. Spontaneous magnetization allows it to be done without an external magnetic field. For example, in cobalt the induction of the equivalent internal field is 2.1 T. In this field, NMR is observed on a frequency of 213.1 MHz.

NMR methods have found wide applications in organic chemistry. With the help of NMR, the structure and chemical composition of compounds and also the dynamics of some chemical reactions can be studied. Spectra of fine NMR lines are characterized by a so-called chemical shift, which is caused by the surrounding atoms. A paramagnetic hydrogen nucleus, which is included in different organic molecules, has a maximal magnetic moment value in comparison with other nuclei and is a suitable object for NMR observation. The values of the chemical shift allow information on the character of the hydrogen bond to be obtained. Besides hydrogen atoms, ^{19}F, ^{14}N, ^{15}N, ^{31}P, ^{13}C, and ^{29}Si can be used.

The most widely known application of NMR is the study of the properties of protein molecules. NMR is used for diagnosing pathological variations in living tissue and is also used together with computerized tomography to obtain volumetric images of organs in living organisms. As we can see, there is just one step from quantum mechanics to the altruistic work of human health care!

3.11 Ferromagnetic Resonance

In principle, spin resonance in ferromagnets[20] is similar to NMR. The term "ferromagnetic resonance" is often extended include magnetic resonance in ferromagnets as well. This is connected with the fact that in a weak static field ($\mathbf{B}_0$) a ferromagnet behaves as a ferromagnet with some magnetization ($\mathbf{M}_0$) and gyromagnetic constant γ_{eff}. Precessional frequencies in such fields fall within the range of centimeter and millimeter waves, and additional resonance can be discovered in the infrared range. Engineers usually do not want to draw attention to these details.

The FMR phenomenon has a series of unusual features, among which we would like to note the following:

1. The transverse susceptibility components are very large, since the ferromagnetic magnetization in a static magnetic field is much greater than the magnetization of an electron paramagnet and of the paramagnetic systems of nuclei in the same magnetic field.
2. The major part in the phenomenon is played by the sample shape. The thing is that since the magnetization is strong, the demagnetizing fields are strong as well. Usually, a sample in the form of an ellipsoid is considered, whose main axes coincide with the axes of the coordinate system. Internal static magnetic field $\mathbf{B}^i$ is connected with external static magnetic field $\mathbf{B}^e$ as follows:

$$B_x^i = B_x^e - N_x \mu_0 M_x; \quad B_y^i = B_y^e - N_y \mu_0 M_y; \quad B_z^i = B_z^e - N_z \mu_0 M_z.$$

Here N_x, N_y, and N_z are demagnetizing factors, taking into consideration the shape. Variation of the internal field intensity, first of all, is manifested in the FMR frequency value:

$$\omega_H^2 = \gamma^2 \left[B_0 + (N_y - N_z)\mu_0 M \right] \left[B_0 + (N_x - N_z)\mu_0 M \right].$$

For a sphere, $N_x = N_y = N_z$; therefore, $\omega_H = \gamma B_0$. For other forms the FMR frequency essentially differs from this value. So, for instance, for a plate with a magnetic field perpendicular to its surface, $N_x = N_y = 0$, $N_z = 1$:

$$\omega_H = \gamma [B_0 - \mu_0 M].$$

3. The strong exchange bond between ferromagnetic electrons manifests a tendency to suppress the dipole contribution to the line width; therefore, the FMR lines under favorable conditions may be very sharp ($\Delta f < 1$ MHz). The function of the

---

[20]The phenomenon of ferromagnetic resonance was observed for the first time in 1946 by James Griffiths. However, the phenomenon of ferromagnetic resonance had been predicted in 1913 by Vladimir Arkadiev (1884–1953), who was a Russian physicist and a corresponding member of the Academy of Sciences of the USSR.

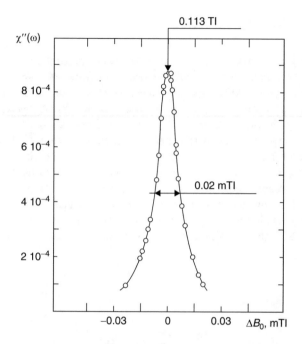

Fig. 3.20 Magnetic resonance in a yttrium ferrite garnet, showing the function of the imaginary part of the susceptibility versus the magnetic field intensity. The frequency of the electromagnetic field is 3.33 GHz. The line width is only 0.6 MHz, which corresponds to a spectral line Q factor of about 5600. The sample is a polished sphere 1 mm in diameter, manufactured from a yttrium ferrite garnet monocrystal. The precession is accompanied by excitation of coherent magnetic–elastic waves (spin waves). On the rough surface these waves are dispersed and, at the end, their energy is transformed into heat (noncoherent elastic waves (phonons)). This process increases the relaxation speed

imaginary part of the magnetic susceptibility of a yttrium ferrite garnet ($Y_3Fe_5O_{12}$) sample is shown in Fig. 3.20. The sample is spherical with a diameter of about 1 mm and a polished surface. The resonance frequency is 3.33 GHz.

4. The spin system of the ferromagnet is rather stable, and it is impossible to control it as the system of nuclei spins; the magnetization of M_z cannot be reduced to zero or change its direction to the opposite. However, the FMR excitation is disintegrated into separate spin wave modes until the magnetization vector is somewhat noticeably deflected from its initial direction. Therefore, saturation effects are observed at relatively low power levels of the external electromagnetic field.

FMR is easily observed at room temperature on the same frequencies as EPR. The main FMR application is connected with development of nonreciprocal devices, or gates (i.e., devices that pass the electromagnetic wave in one direction but do not pass it in the opposite direction). The gate is an essential part of devices in the ultrahigh-frequency range. It excludes the influence of waves reflected from a load upon the generator.

3.12   The Faraday Effect

Let us examine, in detail, the task of plain wave propagation in a gyrotropic medium—for example, in a dielectric located in a permanent magnetic field. If the direction of magnetic field $\mathbf{B}$ is specified, it is necessary to specify the direction of the wave vector. Let us concentrate on the solution of the longitudinal wave propagation task, which, in spite of its relative simplicity, has an important practical significance. This case was investigated experimentally by Michael Faraday.[21] He observed the phenomenon of light polarization plane rotation in optical glasses placed in a permanent magnetic field. This phenomenon became known as the Faraday effect.

Let a plane electromagnetic wave with linear polarization propagate as shown in Fig. 3.21a. It is known that any plane wave can be presented as a sum of two waves of circular polarization with opposite rotational directions. If, for example, at point $z = 0$ (Fig. 3.21b) the complex amplitude of the electric field vector takes the form

$$\dot{\mathbf{E}} = E_{0x}\mathbf{1}_x,$$

the complex amplitudes of these waves are

$$\dot{\mathbf{E}}_{\pm} = \frac{E_{0x}}{2}\left(\mathbf{1}_x \pm j\mathbf{1}_y\right).$$

In a gyrotropic medium, waves with right and left rotation propagate with different velocities (V_+ and V_-, relatively). Therefore, at any point,

$$\dot{\mathbf{E}}_{\pm} = \frac{E_{0x}}{2}\exp\left(-j\frac{\omega}{v_{\pm}}z\right) = \frac{E_{0x}}{2}e^{j\psi_{\pm}(z)},$$

and coordinate components of the appropriate wave with the linear polarization vary in the space as

$$\dot{E}_x(z) = \frac{E_{0x}}{2}\left[e^{j\psi_{\pm}(z)} + e^{j\psi_-(z)}\right];$$

$$\dot{E}_y(z) = j\frac{E_{0x}}{2}\left[e^{j\psi_+(z)} - e^{j\psi_-(z)}\right].$$

---

[21]Michael Faraday (1791–1867) was a Britain physicist. At the age of 14 years he became an apprentice to the owner of a bookshop and bindery, where he practiced self-education. He was a talented experimenter with good scientific intuition. Faraday's ideas about electric and magnetic fields were a major influence on the development of all physics. In 1832 Faraday stated the idea that propagation of electromagnetic interactions is a wave process taking place at a finite speed. Laws, phenomena, units of physical quantities, etc. were subsequently named after him.

(a)

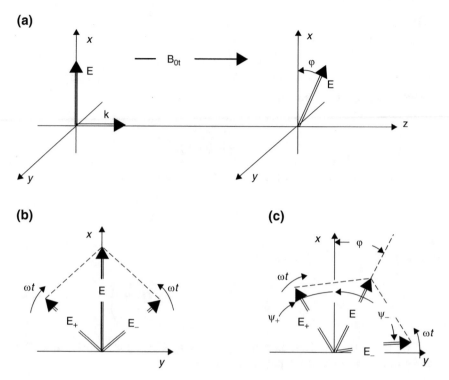

(b)   **(c)**

Fig. 3.21 (**a**) The Faraday effect. (**b**) Vector diagram of the origin. Two rotating vectors form the total electric field vector with linear polarization directed along the x axis. At propagation along a permanent magnetic field, partial waves with circular polarization obtain different phase shifts Ψ_+ and Ψ_- and the total field has a linear polarization directed along the bisector of the angle $\Psi_+ + \Psi_-$. This means that the electric field polarization changes by angle φ (**c**)

The total vector forms the angle φ with the x axis (Fig. 3.21c), such as

$$\frac{\dot{E}_y}{\dot{E}_x} = \text{tg}(\varphi) = \text{tg}\left(\frac{\Psi_- - \Psi_+}{2}\right).$$

The polarization plane rotation on way L will have a value equal to

$$\varphi = \frac{\omega L}{2}\left(\frac{1}{V_-} - \frac{1}{V_+}\right).$$

Let us use the equation for the phase velocity of the plane wave, which for waves with circular polarization in magnetic materials takes the form

$$V_\pm = \frac{c}{\sqrt{\varepsilon \mu_\pm}}.$$

Then,

$$\varphi = \pi \frac{L}{\lambda_0} \sqrt{\varepsilon} \left(\sqrt{\mu_-} - \sqrt{\mu_+} \right). \tag{3.24}$$

From Sect. 3.8 it follows that

$$\mu_\pm = 1 + \frac{\gamma \mu_0 M_z^e}{\omega_H \mp (\omega - j/T_2)}.$$

Usually, the Faraday effect is observed for $\frac{1}{T_2} < \omega >> \omega_H$. Therefore,

$$\mu_\pm \cong 1 \mp \frac{\gamma \mu_0 M^e}{\omega}.$$

Substituting $\mu_\pm$ in (3.24), we have

$$\varphi = \pi \frac{L}{\lambda_0} \sqrt{\varepsilon} \left(\sqrt{1 + \frac{\gamma \mu_0 M_z^e}{\omega}} - \sqrt{1 - \frac{\gamma \mu_0 M^e}{\omega}} \right). \tag{3.25}$$

If we take into consideration that $\mu_\pm \approx 1$, we may obtain the following formula for the rotational angle of the polarization plane:

$$\varphi = \pi \frac{L}{\lambda_0} \sqrt{\varepsilon} \frac{\gamma \mu_0 M_z^e}{\omega} = L \sqrt{\varepsilon} \frac{\gamma \mu_0 M_z^e}{2c}. \tag{3.26}$$

Here, c is the light speed. From (3.26) it follows that the rotational angle of the polarization plane in this approximation does not depend on the frequency.

Now we estimate the value of φ for the ferrite with $\varepsilon = 10$, $\mu_0 M_z^e = 0.1\ T$, $\gamma = 28$ GHz/T1. Substituting numerical values in (3.26), for $L = 1$ cm we obtain $\varphi = 57^0$.

As we can see, the rotational angle of the polarization plane is quite a noticeable quantity and we may, evidently, find some useful practical application for this phenomenon. The Faraday effect has been used for a long time, and with great success, on ultra-high frequencies in nonreciprocal devices such as isolators, which exclude the harmful influence of wave reflections in transmission lines. A sketch of a wave guide isolator is shown in Fig. 3.22a).

The device contains two segments of a wave guide with a rectangular section rotated around the longitudinal axis at an angle of 45° with respect to each other, and a wave guide segment with a circular section, connected by smooth junctions to the rectangular wave guide segments. In the circular wave guide there is a ferrite rod, which is bias magnetized by an external magnetic system. To improve the isolator parameters, the absorbing plates are mounted in junctions.

Fig. 3.22 (**a**) Isolation of the Faraday effect: *1* and *6* denote the rectangular wave guides, *2* and *5* denote the wave guide junctions, *3* denotes a circular wave guide, *4* denotes the ferrite, and *7* and *8* denote the absorbers. (**b, c**) Configuration of the field in the direct (**b**) and inverse (**c**) propagation directions. The fact that changing the propagation direction does not change the rotation of the polarization plane is important

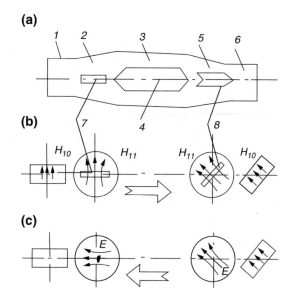

If the wave propagates from the left to the right, it passes through the full device practically without difficulty. The length and diameter of the rod, and the value of the bias magnetization field, are chosen so the wave polarization at the end of the segment of the circular wave guide rotates by 45°. The wave converted by this manner passes without difficulty through the smooth junction to the rectangular wave guide having rotated also by 45°. For a wave propagating from the right to the left, everything principally happens differently. Since the sign of the rotational angle of the polarization plane does not depend on the direction of propagation, the wave will have polarization perpendicular to the vertical axis of the device. A field with such polarization cannot propagate in the left segment of the rectangular wave guide; moreover, it is intensively absorbed by the plate. As a result, in the left part of the path the reflected waves are practically absent.

In the optical range in an isolator with the Faraday effect, the crossed wave guides are replaced by a pair of crossed polaroids. The role of the intermediate sections is played by lenses. An optical isolator can be used for joining a quantum generator to a fiber-optic communication line.

We know of other examples of application of the Faraday effect—for instance, in circulators, phase shifters, optical modulators, controllable optical stencils, optical magnetic devices for information storage, etc. Unfortunately, it is not possible for us to go into more detail here regarding the variety of interesting ideas directed at practical application of the various physical phenomena. We will encounter many of them during practical activities. We can hope that when it is necessary, everybody will, without great difficulty, assimilate the details of specific practical implementations of some of these ideas relying on the obtained fundamental knowledge.

3.13   Problems for Chapter 3

3.1. From (3.8) and equation $\left[\hat{J}_i \hat{J}_j\right] = j\hat{J}_k$, obtain equations for two other spin Pauli matrices (3.9).

Instruction: Look at Sect. 1.3.

3.2. A paramagnetic crystal contains 10^{23} m^{-3} ions with a spin value of 1/2. Determine the magnetic moment of saturation. Calculate the linear magnetic susceptibility at temperatures of $T = 300$ K and $T = 4$ K.

3.3. The iron density is 7.87×10^3 kg/m^3 and the molecule weight is 55.85. Calculate the saturation magnetization and compare it with the value presented in Table 3.2.

3.4. Using (3.14), obtain the equation system for slowly changing complex amplitudes of the magnetization vector.

Assistance: Assume that

$$\mathbf{B}(t) = \text{Re}\left\{\mathbf{B}(t)\exp\left(j\omega_H t\right)\right\}; \mathbf{M}(t) = \text{Re}\left\{\mathbf{M}(t)\exp\left(j\omega_H t\right)\right\}.$$

3.5. Prove that under the conditions of magnetic resonance $\omega = \omega_H$ from the Landau–Lifshitz equation for an alternate field with circular polarization $\widetilde{\mathbf{B}} = \widetilde{B}\cos\left(\omega_H t\right)\mathbf{1}_x + \widetilde{B}\sin\left(\omega_H t\right)\mathbf{1}_y$, the equality $M_z(t) = M_0\cos\left(\gamma\widetilde{B}t\right)$ follows.

3.6. The population difference of particle levels with a spin value of 1/2 is 10^{23} m^{-3}. The EPR frequency is 10 GHz and the width of the magnetic resonance line is 30 MHz. Find damping factor α. The relative dielectric permeability $\varepsilon = 9$.

3.7. For observation of NMR a test tube filled with water is placed inside an induction coil ($L = 2$ µH) of an oscillating circuit (Fig. 3.19). Loss resistance $r_L = 5$ Ω. The NMR frequency is 30 MHz (the field of bias magnetization is about 0.7 T). The proton concentration is $n = 10^{27}$ m^{-3}, the time of transverse relaxation is 0.1 s, and the temperature is 300 K. Determine the parameters of the equivalent circuit of the coil on the NMR frequency. Determine the variation of the circuit's Q factor with NMR.

Assistance: If we neglect the leakage field, then $\widetilde{L} = (1 + \dot{\chi}_M)L$.

3.8. Investigate the function of complex conductivity of a serial oscillating circuit versus frequency under the conditions of the previous task. Varying the numerical values of the initial data, determine the influence of physical parameters on the character of the frequency function.

Instruction: Use a computer to perform the calculations and draw the diagrams.

3.9. The device described in task 3.7 is used as a magnetometer to measure the magnetic field of the Earth. For this, by changing the current direction in the electromagnet, we can determine two values of the NMR frequency. The vertical component of the induction of the Earth's magnetic field is near 70 µT. Find the frequency shift of NMR of protons.

3.10. Suppose that in the device described in task 3.7 the nuclei spin concentration changes along the coil axis according to the law $n(x)$ and the field of bias magnetization $B_{0z}(x) = B_0 + kx$. Write the expression for determination of the resistance inserted into the oscillating circuit. If $T_2 \to \infty$, what will the function of the losses versus the frequency be?

Instruction: Perform the numerical simulation on a computer.

3.11. Using the Bloch equation system, obtain the formula for longitudinal magnetization, taking into account the saturation effect in a form similar to (2.41).

Instruction: Assume that $\omega = \omega_H \gg 1/T_2$.

3.12. Using the result of the previous task, obtain the saturation induction and the power flow Π_{sat} under EPR conditions. For metal ions in a crystalline lattice of dielectric for liquid nitrogen temparature, the longitudinal relaxation time $T_1 \approx 10^{-3}$ s and the transverse relaxation time $T_2 \approx 10^{-8}$ s. The relative dielectric permeability of the lattice is 9.

3.13. Thin ferromagnetic films are divided into domains whose walls are perpendicular to the plane. Under definite conditions, domains are formed in the form of cylinders with a radius of the order of the film thickness (cylindrical domains). A plane of light polarization passing through such a domain rotates. Determine the rotational angle in a film with a thickness of 1 µm for $\mu_0 M_{sat} = 0.07$ TL (MnBi alloy).

3.14. A linear polarized wave falls from a vacuum to the boundary of a magnetic medium. A permanent magnetic field is perpendicular to the boundary. Find the field of the reflected wave if $\omega \gg \omega_H > 1/T_2$.

Chapter 4
Field Interaction with "Free Charges"

4.1 Introduction: The Interzone and Fundamental Absorption

While studying Chap. 1 you probably paid attention to the fact that as far as complication of a microparticle's internal structure is concerned, its energy spectrum becomes more and more dense. In a hydrogen atom the energy of stationary states depends only upon the main quantum number. For a lithium atom, stationary states with different orbital number values become distinguishable. If several electrons are located in the external orbit, the energy value begins to depend upon the complete mechanical moment of these electrons. When transferring to molecules, we must add vibrational and rotational forms of motion. As a result, vibrational–rotational energy sublevels appear for each electron level. The dimensions of operator matrices quickly increase and the problem of interaction of a particle and emission becomes very bulky.

Nevertheless, sooner or later, the fundamental law about quantity transition to quality begins to act. In a solid body the level number is so large (see Sect. 1.7) that, at least inside the energy zones, the stationary state's spectrum can be considered a continuous one. This essentially changes the situation. Firstly, summation over levels can be replaced by integration; secondly, continuous variation of energy is typical for classical physics and there may be a possibility of a classical explanation for these phenomena.

A classical gas of nonstructural particles is an example of a continuous spectrum system, which has been well studied for a long time. Real atoms and molecules can be considered elementary particles if their interaction energy with the field or with each other is not high enough to ensure a transition from one discrete energy level to another. Nonstructural particles, which are not subject to external influence, have kinetic energy only with a continuous spectrum of eigenvalues. A gas of such particles will interact with an electromagnetic field if the particles have an electric charge. An electrically neutral gas of positively and negatively charged particles is

© Springer Nature Switzerland AG 2020
V. V. Shtykov, S. M. Smolskiy, *Introduction to Quantum Electronics and Nonlinear Optics*, https://doi.org/10.1007/978-3-030-37614-7_4

called plasma. This term was introduced in 1929 by Irving Langmuir.[1] Electronics engineers deal with plasma in connection with solution of the problems of radio wave propagation in the ionosphere.

Since electrons and holes, which move inside the conduction zone and the valence zone, have a practically continuous energy spectrum, the plasma model can rather well describe the electromagnetic properties of a solid body. We should not forget to use, in all formulas, the effective mass of electrons m_e* and holes m_h* instead of the real mass of electrons m_e.

Taking into consideration the universality of the plasma model, we start with examination of electromagnetic wave interactions with plasma. The mathematical approach describing this phenomenon and the physics process itself will be absolutely different from the case of the discrete energy spectrum, which we dealt with in the two previous chapters.

Further, for simplicity, we shall not consider the issues of which way is the definite concentration of charged particles kept and of how ionization and recombination processes happen in plasma. We shall simply assume that some external energy source maintains the charged particle balance, and we shall try to find out the response of the system of these particles to an external electromagnetic field. Such a problem statement is far from a self-consistent one, since we may certainly expect that an electric field will promote ionization and thereby change the balance of charged particles in the plasma. However, the accounting for this factor would essentially complicate our lecture course, and so we shall neglect it, assuming that the field is rather weak.

As we shall see, intraband motion of carriers plays a noticeable role in semiconductors. However, in connection with the wide application of these materials in quantum electronics, we cannot disregard the fundamental absorption that arises when the energy of fallen photons exceeds the width of the forbidden band. It is clear that investigation of optical properties in this frequency region serves as an invaluable instrument for studying the zone structure. Moreover, almost all photoelectric phenomena representing the fundamental physical interest and significant from the applied positions are caused by such interband excitations of charge carriers. Lastly, semiconductor quantum generators operate exclusively on the basis of interband transitions.

The absorption coefficient connected to these processes depends upon the probability (normalized to a time unit) that the electron under the action of radiation with frequency ω will perform a jump from the valence band to the conduction band. The value of this jump probability represents a purely quantum problem, as the structure

---

[1]Irving Langmuir (1881–1937) was an American chemist and physicist. His most noted publication was his famous 1919 paper titled *The Arrangement of Electrons in Atoms and Molecules*, in which he outlined his "concentric theory of atomic structure," building chemical bonding theory. His investigations of electric discharges in gases, studying thermal–electric emissions, were used for designing gas-filled electric filament lamps and electronic lamps in electronics. He was awarded the 1932 Nobel Prize in Chemistry for his work in surface chemistry. He was the first industrial chemist to become a Nobel laureate.

of electron energy levels in the crystal is completely left out of the "kingdom" of classical physics. Below (and without performing detailed calculations), we shall touch on the most important and interesting features of this process.

4.2   The Kinetic Equation

If no external impact is applied to the plasma, under the influence of relaxation processes it will, sooner or later, transfer to the state of thermodynamic equilibrium. With that, the plasma will have spatial homogeneity and will be characterized by the equilibrium distribution function of particles over impulses $f_e(\mathbf{p})$. As was shown in Sect. 1.12, in the case of nondegenerate plasma, $f_e(\mathbf{p})$ represents the Maxwell distribution (1.73).

What will happen to the plasma if a electromagnetic field affects it? Evidently, the thermodynamic equilibrium will be disturbed and the distribution function will change in some manner. If the field appears to be heterogeneous, we may expect that the plasma will be heterogeneous in its properties. It is clear that we already cannot describe it using the distribution function (1.69) depending upon the impulse only. The nonequilibrium distribution function should depend upon both the impulse and the coordinate. It should be defined so that the value $f(\mathbf{r}, \mathbf{p})d\mathbf{r}d\mathbf{p}$ represents the average number of particles in the element of the phase space with a center at the point $(\mathbf{r}, \mathbf{p})$ and a volume equal to $d\mathbf{r}d\mathbf{p}$. This function is closely connected with the "classical" probability density $P(\mathbf{r}, \mathbf{p})$ of the particle presence in the phase space cell, which was used in Sect. 1.11 to introduce the density matrix concept for systems with a continuous energy spectrum (see (1.59)) as

$$f(\mathbf{r}, \mathbf{p}) = NP(\mathbf{r}, \mathbf{p}). \tag{4.1}$$

Here, N is the particle number in the plasma. We would like to note that $\int f(\mathbf{r}, \mathbf{p}) d\mathbf{p} = n(\mathbf{r})$, where n is the particle concentration in the plasma. Of course, the density matrix $\rho(\mathbf{p}, \mathbf{p}')$ describes in-depth the nonequilibrium plasma, and we could obtain all results of interest to us by starting from Eq. (1.57). Nevertheless, $\rho(\mathbf{p}, \mathbf{p}')$ is a less visual characteristic of the plasma for physical system than the distribution function.

Let us determine the connection between the density matrix and the distribution function. First of all, we would like to note once more that wave functions of a free particle localized in a cell $(\mathbf{r}_0, \mathbf{p}_0)$ are connected by the Fourier transformation (1.58) (see Sect. 1.11) in coordinate $\psi_{\mathbf{r}_0\mathbf{p}_0}(\mathbf{r})$ and impulse $c_{\mathbf{r}_0\mathbf{p}_0}(\mathbf{r})$ representations. Hence, according to the known shift theorem, the wave function in the impulse representation can be written as

$$c_{\mathbf{r}_0\mathbf{p}_0}(\mathbf{r}) = C(\mathbf{p} - \mathbf{p}_0)e^{-j\frac{\mathbf{p}\mathbf{r}_0}{\hbar}}, \tag{4.2}$$

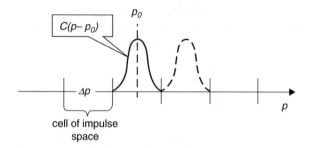

Fig. 4.1 Packet of the wave function localized in cell Δp of the impulse space. In the coordinate space the particle will be localized in cell Δr. The connection between impulse and coordinate representations is described by the Fourier transformation. In the signal theory a similar connection exists between a pulse and its spectrum. A pulse with duration τ has a bandwidth of about $1/\tau$

where $C(\mathbf{p})$ represents a finite function concentrated near zero, whose extension does not exceed the cell size of the impulse space $\Delta \mathbf{p}$ (Fig. 4.1) and whose form is determined by the localization character in the coordinate space. Then the equation for the density matrix element (1.59) takes the form

$$
\rho(\mathbf{p}, \mathbf{p}') =
$$
$$
= \iint P(\mathbf{r}_0, \mathbf{p}_0) C^*(\mathbf{p}' - \mathbf{p}_0) \, C(\mathbf{p} - \mathbf{p}_0) e^{-j\frac{(\mathbf{p}-\mathbf{p}')\mathbf{r}_0}{\hbar}} d\mathbf{r}_0 d\mathbf{p}_0. \tag{4.3}
$$

Since the required distribution function (4.1) is connected with $P(\mathbf{r}_0, \mathbf{p}_0)$, we need to express this quantity through the density matrix. For this we assume that $\mathbf{p}' = \mathbf{p} - \mathbf{k}$. Quantity $\mathbf{k}$ should be less than $\Delta \mathbf{p}$, otherwise the wave functions will not overlap. Then (4.3) can be transformed into

$$
\rho(\mathbf{p}, \mathbf{p} - \mathbf{k}) =
$$
$$
= \iint P(\mathbf{r}_0, \mathbf{p}_0) C^*(\mathbf{p} - \mathbf{p}_0 - \mathbf{k}) C(\mathbf{p} - \mathbf{p}_0) e^{-j\frac{\mathbf{p}\mathbf{r}_0}{\hbar}} d\mathbf{r}_0 d\mathbf{p}_0 \cong
$$
$$
\cong \int C^*(\xi - \mathbf{k}) C(\xi) d\xi \int P(\mathbf{r}_0, \mathbf{p}) e^{-j\frac{\mathbf{k}\mathbf{r}_0}{\hbar}} d\mathbf{r}_0. \tag{4.4}
$$

In (4.4) we divided the integration over $\mathbf{p}_0$ and $\mathbf{r}_0$ because the integrand is nonzero within the cell of the impulse space only, where $|\xi| = |\mathbf{p} - \mathbf{p}_0| < \Delta p$, and the probability density is the slow function of $\mathbf{p}_0$. Therefore, $P(\mathbf{r}_0, \mathbf{p}_0) \approx P(\mathbf{r}_0, \mathbf{p})$. It changes smoothly and in the coordinate space; in any case, the typical scale of the plasma heterogeneity Δr_0 is much greater than the cell size in coordinate space Δr. Hence, the second integral in (4.4) has a noticeable value if the value of $|\mathbf{k}|$ is much less than cell size Δp. Then, $C(\xi - \mathbf{k}) \approx C(\xi)$ and the first integral in (4.4) can be considered as 1, since the wave function must satisfy the normalization condition. As a result,

$$\rho(\mathbf{p}, \mathbf{p} - \mathbf{k}) = \int P(\mathbf{r}_0, \mathbf{p}) e^{-j\frac{\mathbf{k}\mathbf{r}_0}{\hbar}} d\mathbf{r}_0.$$

Having taken the inverse Fourier transformation and replacing $\mathbf{r}_0$ with $\mathbf{r}$, we obtain the required connection between the nonequilibrium distribution function and the density matrix:

$$f(\mathbf{r}, \mathbf{p}) = \frac{N}{(2\pi)^3} \int \rho(\mathbf{p}, \mathbf{p} - \mathbf{k}) e^{j\frac{\mathbf{k}\mathbf{r}}{\hbar}} d\mathbf{k}. \qquad (4.5)$$

A distribution function introduced in such a manner is called the Wigner function. As we can see, when thermodynamic equilibrium is absent, the essential role in determination of the distribution function is played by nondiagonal elements of the density matrix.

Function (4.5) depends upon both the coordinate and the impulse, which, according to quantum theory, cannot be measured simultaneously with arbitrary accuracy. That is why the distribution function is a classical object rather than a quantum one.

If $f(\mathbf{r}, \mathbf{p})$ is known, we may obtain the distribution function over the impulses only:

$$f(\mathbf{p}) = \int f(\mathbf{r}, \mathbf{p}) d\mathbf{r} = N \int \rho(\mathbf{p}, \mathbf{p} - \mathbf{k}) \delta(\mathbf{k}) d\mathbf{k} = N\rho(\mathbf{p}, \mathbf{p}) = N\rho(\mathbf{p}).$$

The obtained result coincides completely with definition (1.69), which was introduced earlier.

4.2.1 Statistics of Nonequilibrium States

Thus, we have decided to describe plasma subjected to disturbing field action with the nonequilibrium distribution function (4.5). It is reasonable to obtain the equation that determines the behavior of this function in time. The density matrix equation (1.57) may be the starting point for us and, together with (4.4), yields the following equation:

$$\frac{df}{dt} = \frac{N}{(2\pi)^3} \int \left[\hat{H}, \hat{\rho} \right]_{\mathbf{p}, \mathbf{p} - \mathbf{k}} e^{j\frac{\mathbf{k}\mathbf{r}}{\hbar}} d\mathbf{k}.$$

The Hamiltonian is as before:

$$\widehat{H} = \widehat{H}_0 + \widehat{H}_{\text{rel}} + \widehat{H}_{\text{int}}.$$

However, the matrix elements of the Hamiltonian $\widehat{H}_{\text{rel}}$ in the impulse representation are, in fact, continuous functions of two variables. In this case, in accordance with the rules of matrix multiplication, the commutator included in (1.57) will have the following matrix elements:

$$\left[\widehat{H}, \widehat{\rho}\right]_{\mathbf{pp'}} = \int \left[H_{\mathbf{pp''}}\rho(\mathbf{p''}, \mathbf{p'}) - \rho(\mathbf{p}, \mathbf{p''})H_{\mathbf{p''p'}}\right]d\mathbf{p''}, \tag{4.6}$$

i.e., summation on the index is replaced by integration.

If we apply transformation (4.5) to Eq. (1.57), which connects the density matrix with the distribution function, in the left part of the equation we have the derivative df/dt. And what will the right part be? First of all, let us consider the item containing the undisturbed Hamiltonian H_0. If we examine free electrons as plasma particles, the matrix element in the impulse representation owing to (1.48) is equal[2]:

$$H_{0\mathbf{pp'}} = \int e^{-\frac{\mathbf{p'r}}{\hbar}}\left(-\frac{\hbar}{2m_e}\nabla_r^2\right)e^{\frac{\mathbf{p'r}}{\hbar}}d\mathbf{r} = \frac{p^2}{2m_e}\delta(\mathbf{p} - \mathbf{p'}),$$

i.e., the undisturbed Hamiltonian has a diagonal form. Let us substitute this expression in (4.6) to find the matrix element of the commutator:

$$\left[\widehat{H}_0, \widehat{\rho}\right]_{\mathbf{p,p-k}} = \frac{1}{2m_e}\int\left[p^2\rho(\mathbf{p''}, \mathbf{p} - \mathbf{k})\delta(\mathbf{p} - \mathbf{p''}) - p''^2\rho(\mathbf{p}, \mathbf{p''})\delta(\mathbf{p''} - \mathbf{p} + \mathbf{k})\right]d\mathbf{p''} =$$
$$= \frac{\mathbf{pk}}{m_e}\rho(\mathbf{p}, \mathbf{p} - \mathbf{k}).$$

Having applied the Fourier transformation (4.5) to this equation, we obtain, according to the theorem on differentiation,

$$\frac{\mathbf{p}}{m_e}\frac{N}{(2\pi)^3}\int \mathbf{k}\rho(\mathbf{p}, \mathbf{p} - \mathbf{k})e^{\frac{\mathbf{kr}}{\hbar}}d\mathbf{k} = -\frac{\mathbf{p}}{m_e}j\hbar\frac{\partial f}{\partial \mathbf{r}}. \tag{4.7}$$

Here, $\partial f/\partial r$ represents the gradient of the distribution function in coordinate space $\nabla_r f$.

We have successfully examined the first item in the right part of Eq. (1.57). The matter is more complicated with the item containing the disturbance Hamiltonian

---

[2]Gas discharge plasma consists of negatively charged particles—electrons and positive ions. The electron subsystem is described by its distribution function, and the ion subsystem by its own distribution function. In particular, at thermodynamic equilibrium in "cold" gas discharge plasma the electron temperature is higher than the ion temperature owing to the greater inertial properties of ions.

$\widehat{H}_{int}$. We assume now, for simplicity, that the electron gas is under the action of a field of potential forces $\mathbf{F}(\mathbf{r}) = -e\mathbf{E} = e\nabla_r\varphi$, where φ is the potential of this field. Then $\widehat{H}_{int} = e\varphi(\mathbf{r})$, and the matrix element of the disturbance Hamiltonian is equal in accordance with (1.48):

$$\widehat{H}_{int\,\mathbf{p}\,\mathbf{p}'} = -e\int\varphi(\mathbf{r})\,e^{-j\frac{(\mathbf{p}-\mathbf{p}')\mathbf{r}}{\hbar}}d\mathbf{r} = -e\Phi\left(\frac{\mathbf{p}-\mathbf{p}'}{\hbar}\right), \tag{4.8}$$

where $\Phi(\mathbf{k})$ is the three-dimensional Fourier image of potential $\varphi(\mathbf{r})$. Now we must find the matrix element of the commutator, which, owing to (4.6), is equal:

$$\left[\widehat{H}_{int},\widehat{\rho}\right]_{\mathbf{p},\mathbf{p}-\mathbf{k}} = -e\int\left[\Phi\left(\frac{\mathbf{p}-\mathbf{p}''}{\hbar}\right)\rho(\mathbf{p}'',\mathbf{p}-\mathbf{k}) - \rho\left(\mathbf{p},\mathbf{p}''\right)\Phi\left(\frac{\mathbf{p}-\mathbf{p}''+\mathbf{k}}{\hbar}\right)\right]d\mathbf{p}''.$$

This integral can be presented in the form of two integral difference and to make the variable changing in them so that Φ is the function of integration variable $\mathbf{p}'$ only. Taking into consideration that integration is provided in infinite limits, we can again unite both integrals into one:

$$\left[\widehat{H}_{int},\widehat{\rho}\right]_{\mathbf{p},\mathbf{p}-\mathbf{k}} = e\int\mathbf{p}'\Phi\left(\frac{\mathbf{p}'}{\hbar}\right)\frac{\rho(\mathbf{p},\mathbf{p}'-\mathbf{k}) - \rho(\mathbf{p}-\mathbf{p}',\mathbf{p}-\mathbf{k})}{p'}d\mathbf{p}'. \tag{4.9}$$

To move further, it is necessary to determine the peculiarities of the behavior of the density matrix element (4.3) as a function of two variables. If one of these variables gets an increment and the second remains unchanged, then ρ sharply changes depending on this increment within the limits of one cell of the impulse space. This is a consequence, firstly, of the finiteness of the $C(\mathbf{p})$ function and, secondly, of the slow function $P(\mathbf{r}_0,\mathbf{p}_0)$ of $\mathbf{r}_0$. But if both variables get the same increment, ρ changes depending on this increment rather slowly (as slow as probability P as a function of $\mathbf{p}_0$). The Fourier image of the potential $\Phi(\mathbf{p}/\hbar)$ is a much more fast-changing function than $\rho(\mathbf{p},\mathbf{p}+const)$, as $\varphi(\mathbf{r})$ changes slowly. This fact allows us to replace the fraction in the integral (4.9) with the derivative:

$$\left[\widehat{H}_{int},\widehat{\rho}\right]_{\mathbf{p},\mathbf{p}-\mathbf{k}} = -ej\hbar\int j\frac{\mathbf{p}'}{\hbar}\Phi\left(\frac{\mathbf{p}'}{\hbar}\right)\frac{\partial\rho(\mathbf{p},\mathbf{p}-(\mathbf{k}-\mathbf{p}'))}{\partial\mathbf{p}}d\mathbf{p}'. \tag{4.10}$$

The obtained integral represents the convolution of two Fourier images. The first two items are the Fourier transformation of the potential derivative over the coordinate, and the last multiplier, according to (4.5), is connected to the derivative of the distribution function of the impulse. From this, according to the convolution theorem, we obtain

$$\frac{N}{(2\pi)^3} \int \left[\hat{H}_{\text{int}}, \hat{\rho}\right]_{\mathbf{p},\mathbf{p}-\mathbf{k}} e^{\frac{\mathbf{k}\mathbf{r}}{\hbar}} d\mathbf{k} = -ejh\frac{\partial\varphi}{\partial\mathbf{r}}\frac{\partial f}{\partial\mathbf{p}} = -jh\mathbf{F}\frac{\partial f}{\partial\mathbf{p}}.$$

Here, we must understand the derivative on the impulse as the gradient of the distribution function in impulse space $\nabla_{\mathbf{p}}f$.

To finally obtain the equation for the distribution function, it remains for us to examine that part of the Hamiltonian that describes relaxation processes. As you will remember, we always had three problems with determination of $\hat{H}_{\text{rel}}$ and we were limited by the fact that we had written the relaxation items in the equation starting from common sense. In this case we shall act in a similar way. Moreover, we shall say nothing yet about the form of the relaxation item; we shall designate it only by a letter, S, and we shall call it the collision integral (collisions set the plasma in the equilibrium state). Then, taking into consideration (4.7) and (4.10), the equation for the density matrix (1.57) after application of the transformation (4.5) takes the form

$$\frac{\partial f}{\partial t} + \frac{\mathbf{p}}{m_e} \cdot \nabla_{\mathbf{r}}f + \mathbf{F} \cdot \nabla_{\mathbf{p}}f = S. \tag{4.11}$$

This equation is called a kinetic equation or the Boltzmann equation, and it describes the nonequilibrium distribution function of a gas containing particles with mass m_e, which are subjected to the action of force $\mathbf{F}$. It turns out that (4.11) is true for any but not only potential forces. Below, we shall make sure of it.

4.2.2 The Classical Approach to Derivation of the Kinetic Equation

We obtained the Boltzmann equation for the nonequilibrium distribution function (4.11). It does not contain the Planck constant, and the distribution function itself has a classical sense. Therefore, there is nothing surprising in the fact that (4.11) can be derived from purely classical considerations without using the density matrix concept.[3]

Let us examine some small volume of the phase space presented naturally for the two-dimensional case in Fig. 4.2. The number of particles in this volume is $f\Delta r\Delta p$, where f is an average value of the distribution function in this volume. We are interested in how the particle number will vary in time. To begin we assume that particle collisions with each other and with a thermostat are absent. Then, variation of this number can be caused by the fact that particles with some motion velocity $\partial r/$

---

[3]The derivation of the kinetic equation from the equation for the density matrix is given to demonstrate the possibility of a unified approach to description of a variety of manifold physical phenomena. The tendency to develop a universal approach of such a kind often acts as a driving force in the development of physics.

Fig. 4.2 Impulse coordinate phase representation used for classical derivation of the kinetic equation. The *arrows* show the directions of particle motion. If the number of entering particles is not equal to the number of outgoing particles, the number of particles in the cell changes. This permits us to form a difference equation, which is then transformed into a differential one

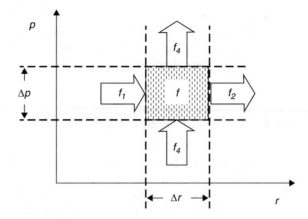

∂t in the coordinate space come into this volume and go out of it. The number of entering particles, which is proportional to the value of distribution function f_1, is not equal to the number of outgoing particles, which is proportional to f_2. Just like this, particles may move in the impulse space with some velocity $\frac{\partial p}{\partial t}$. If the number of entering particles f_3 is unequal to the number of outgoing particles f_4, then it will lead to variation of the particle number in the volume. Let the variation of the particle number Δf occur during time Δt. Then,

$$\frac{\Delta f}{\Delta t} = -\frac{f_2 - f_1}{\Delta r}\frac{\Delta r}{\Delta t} - \frac{f_4 - f_3}{\Delta p}\frac{\Delta p}{\Delta t}.$$

Rushing Δt and the volume sizes of the phase space to zero, we obtain the speed of the particle concentration variation in the phase space:

$$\frac{\partial f}{\partial t} = -\nabla_r f \frac{\partial \mathbf{r}}{\partial t} - \nabla_p f \frac{\partial \mathbf{p}}{\partial t}. \tag{4.12}$$

If we remember that the particle number variation in the element of the phase space can occur also as a result of collisions, we must add to the right part of (4.12) the collision integral $S = \left(\frac{\partial f}{\partial t}\right)_{\text{coll}}$. At last, transferring from a two-dimensional phase space to a six-dimensional one, and taking into account that $\frac{\partial \mathbf{r}}{\partial t} = \frac{\mathbf{p}}{m_e}$, $\frac{\partial \mathbf{p}}{\partial t} = \mathbf{F}$, we obtain from (4.12) nothing other than the kinetic equation (4.11).

Since we are interested in the interaction of the particle gas with the electromagnetic field, we consider charged particles. Then for an electron gas we should understand the Lorentz force as force $\mathbf{F}$, and (4.11) takes the form

$$\frac{\partial f}{\partial t} + \frac{\mathbf{p}}{m_e} \cdot \nabla_r f - e\left(\mathbf{E} + \frac{1}{m_e}[\mathbf{p} \times \mathbf{B}]\right) \cdot \nabla_p f = S. \tag{4.13}$$

Variation of the number of electrons in the cell of the phase space according to (4.13) is caused by (1) variation of their coordinates, i.e., the diffusion; (2) variation of their impulses, i.e., acceleration under action of the Lorentz force; and (3) collisions. Now we shall investigate the last factor.

4.2.3 The Collision Integral

There are many different ways to write collision integrals—for bosons, for fermions, and for classical particles and quasiparticles. It was shown in Chap. 1 that in practically all important cases a particle gas can be considered a classical one, and we shall do this. We shall also assume that the plasma has a small density. Therefore, it is enough to take into account pair collisions only (writing S in the Boltzmann form). Thus, we will not take into consideration electron–electron collisions, and we will limit our discussion to collisions of electrons with heavy particles—ions and neutral molecules in the gas discharge plasma—or with impurity ions and phonons (vibrations of the lattice) in the semiconductor.

Let us examine the following model: let the electron have mass m_e and impulse $\mathbf{p}$, and let the heavy scattering particles have mass M and impulse $\mathbf{p}_1$ (Fig. 4.3). We assume that after collision the electron has impulse $\mathbf{p}'$ and the particle has impulse $\mathbf{p}_1'$. The probability of such an scattering act is characterized by the scattering differential section $\sigma(\theta, u)d\Omega$ of equal area, passing through which the electron, after the collision, will be deflected at angle θ between the vectors of mutual velocities $\mathbf{v} = \frac{\mathbf{p}}{m_e} - \frac{\mathbf{p}_1}{M}$ and $\mathbf{v}' = \frac{\mathbf{p}'}{m_e} - \frac{\mathbf{p}_1'}{M}$ at spatial angle $d\Omega$. This value is found either with the help of quantum theory or from experimental data.

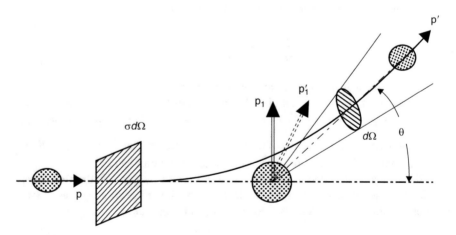

Fig. 4.3 Electron scattering on the heavy particle. As a result of the collision, the impulse changes. The collision process in the particle ensemble has a probability character. The velocity of the incoming electron is connected with heat motion. Therefore, the collision frequency is a function of the temperature

To write $S = \left(\frac{\partial f}{\partial t}\right)_{coll}$, it is necessary to determine the number of particles performing collisions in the time unit with variation of vector $\mathbf{p}$ to $\mathbf{p}'$. It is clear that this number will be larger for a larger concentration of electrons with impulse $\mathbf{p}$, which is proportional to $f(\mathbf{p})$ for a larger concentration of scattering particles, which is proportional to the distribution function of these particles $F(\mathbf{p}_1)$ for larger their mutual velocity and the scattering section. We are interested in the complete number of collisions independent of scattering angle θ and the impulse of the scattering particles; therefore, we need to fulfill integration over $\mathbf{p}_1$ and the spatial angle $d\Omega = \sin(\theta)d\varphi d\theta$. As a result, we obtain

$$S = \iint \left[f(\mathbf{p})F(\mathbf{p}_1) - f(\mathbf{p}')f(\mathbf{p}_1')\right]|\mathbf{v}|\sigma(\theta, u)d\mathbf{p}_1 d\Omega. \tag{4.14}$$

The second item in the square brackets under the integral has been included to take into consideration not only the decrease in electrons in the cell of the phase space but also its filling due to the inverse transition $\eth' \to \eth$ for collisions.

As we see, (4.13), together with (4.14), represents a rather complicated integral–differential equation. To solve it we need to make some simplifying assumptions.

We shall assume that the collisions are elastic and as $M \gg m_e$, the scattering particles can be considered fixed in the first approximation. Then $\mathbf{p}' = \mathbf{p}$ and $f(\mathbf{p}')$ will be the function of impulse value p and the scattering angle, i.e., $f(\mathbf{p}') = f(p, \theta)$. The mutual velocity $|\mathbf{v}|$ will be transferred to the electron velocity $v = \frac{p}{m_e}v$, and the scattering section will depend only on scattering angle θ and the electron impulse value p. In this case, integration over $\mathbf{p}_1$ can be fulfilled independently, which gives a multiplier equal to the concentration of scattering particles n_m. As a result, (4.14) will be transformed into the form

$$S = -n_m v \int [f(\mathbf{p}) - f(p, \theta)]\sigma(\theta, p)d\Omega =$$
$$= -2\pi n_m v \left\{ f(\mathbf{p}) \int_0^{2\pi} \sigma(\theta, p) \sin(\theta)d\theta - \int_0^{2\pi} f(p, \theta)\sigma(\theta, p) \sin(\theta)d\theta \right\}. \tag{4.15}$$

Further, any function of a single variable can be presented as a sum of even and odd functions (Fig. 4.4). The matter with the function of three variables $f(\mathbf{p})$ is more complicated. Nevertheless, we shall assume in the future that the distribution function can also be presented as the sum of the isotropic part $f_0(p)$ depending upon the impulse value and the directional part $f_1(\mathbf{p})$:

$$f(\mathbf{p}) \approx f_0(p) + f_1(\mathbf{p}) = f_0(p) + f_1(p, \theta). \tag{4.16}$$

By the way, it corresponds to the first two items of expansion in the series of spherical harmonics.

Fig. 4.4 Representation of distribution functions as the sum of even (isotropic, undirected) and odd (directed) parts. The directed part of the function is the vector. Under conditions of heat equilibrium, the distribution function is isotropic in a three-dimensional impulse space. Asymmetry is a result of external influence

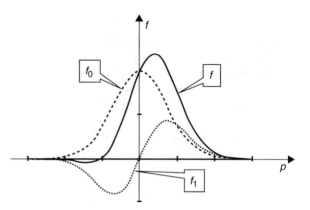

Let us introduce the concept of the scattering transport section:

$$\sigma_T(p) = 2\pi \int\limits_0^{2\pi} \sigma(\theta, p) \sin(\theta) d\theta, \tag{4.17}$$

which characterizes the integral probability of electron scattering by a particle. Substituting (4.16) and (4.17) in (4.15), and taking into account that $f_1(\mathbf{p})$ is the odd function of the cosine of the scattering angle and σ is the even function, we obtain

$$S = -n_m v \sigma_T(p) f_1(\mathbf{p}). \tag{4.18}$$

As we can see, (4.18) does not contain function $f_0(p)$, which, as we shall see later, characterizes the energy of the electron gas. This relates to the fact that with absolutely fixed scatterers the electron energy due to elastic collisions does not transfer to them; hence, energy relaxation is absent. In fact, the energy of scattering particles at collisions, nevertheless, is changed, and its relative variation has the order $\chi = \partial_{\tilde{a}} / \tilde{I} \ll 1$. To take the energy exchange into consideration, we must add into (4.18) the item $\chi \sigma_T(p)\left(f_0(p) - f_0^e(p)\right)$, where $f_0^e(p)$ is the equilibrium distribution function. As a result, the collision integral will be transformed into the form

$$S = -n_m v \sigma_T(p)\left\{ f_1(\mathbf{p}) + \chi\left[f_0(p) - f_0^e(p)\right]\right\}. \tag{4.19}$$

Equation (4.19) corresponds to the so-called relaxation time approximation. The matter is that the value

$$v(p) = n_m v \sigma_T(p) \tag{4.20}$$

is the frequency of elastic collisions of particles with impulse p; the inverse quantity $\tau(p) = \nu^{-1}(p)$ has a sense of the relaxation time of the directional part of the distribution function (the analogue of time T_2), and the quantity $\chi^{-1}\nu^{-1}(p)$ is the relaxation time of the isotropic part of the distribution function (the analogue of time T_1). In this approximation, the Boltzmann equation takes the form

$$\frac{\partial f}{\partial t} + \frac{\mathbf{p}}{m_e}\nabla_r f - e\left(\mathbf{E} + \frac{1}{m_e}[\mathbf{p} \times \mathbf{B}]\right)\nabla_\mathbf{p} f +$$
$$+\nu(p)f_1(\mathbf{p}) + \chi\nu(p)\left(f_0 - f_0^e\right) = 0 \tag{4.21}$$

According to (4.21), in the absence of an external impact in the homogeneous plasma, as a result of relaxation processes, f_1 during time $\tau(p)$ aspires exponentially to zero, and f_0 achieves f_0^e for much more time $\frac{\tau(p)}{\chi}$.

4.3 Current Density and Average Energy of Charge Carriers in Plasma

We are interested in the plasma reaction to a high-frequency electromagnetic field. When we considered the field effect on the coupled charges, we found substance polarization as such a reaction. According to the accepted traditions, when studying the field effect on the free charges, we may find not polarization but current density $\mathbf{J}$. The difference in approaches is not so major since the current density $\mathbf{J} = \frac{\partial \mathbf{P}}{\partial t}$. For electrons the current density is $\mathbf{J} = -en\langle\mathbf{v}\rangle$, where $\langle\mathbf{v}\rangle$ is the average particle velocity in the plasma, which can be determined using impulse averaging over the distribution function. Therefore,

$$\mathbf{J} = -\frac{e}{m_e}\int \mathbf{p}f(\mathbf{p})d\mathbf{p}. \tag{4.22}$$

If we substitute (4.16) in (4.22), the integral from the isotropic part of the distribution function gives zero because of the oddness of the integrand, and (4.22) can be transformed into the following form:

$$\mathbf{J} = -\frac{e}{m_e}\int \mathbf{p}f_1(\mathbf{p})d\mathbf{p}. \tag{4.23}$$

Thus, the average impulse in the plasma—and, hence, the density of the electron current—is determined by the directional part of the distribution function.

Studying Chaps. 2 and 3, we made sure that the substance reaction to the field depends upon the energy accumulated in this substance. We may expect that the plasma will not be excluded from the rule and besides the equation for $\mathbf{J}$ we should derive the equation for the average electron energy in the plasma. How great is this quantity? Energy $Q = <p^2/2m_e> n$ corresponds to the volume unit. We shall find the average energy value using the distribution function. As a result, we obtain

$$Q = -\frac{e}{m_e} \int p^2 f(\mathbf{p}) d\mathbf{p} = \frac{1}{2m_e} \int p^2 f_0(p) dp. \qquad (4.24)$$

Thus, the electron energy in the plasma is determined by the isotropic part of the distribution function. By the way, if we substitute the Maxwell distribution function (1.73) instead of f_0 in (4.24), we obtain the equilibrium value of the energy accumulated in the volume unit:

$$Q^e = \frac{3}{2} k_B T n, \qquad (4.25)$$

which is well known to us from general physics. The concept of temperature T, generally speaking, is defined for the thermodynamic equilibrium only. But according to (4.25), we may generalize it, considering that temperature is a measure of the average energy of the electron gas in the nonequilibrium condition as well:

$$T(\mathbf{r}) = \frac{2}{3} \frac{Q(\mathbf{r})}{n(\mathbf{r}) k_B}. \qquad (4.26)$$

If, in a homogeneous (in terms of concentration) plasma, the average electron energy depends upon the coordinate, we usually say that nonuniform heating of electrons takes place.

Now, as in Chaps. 2 and 3, it is expedient to transfer from equations for $f(\mathbf{p})$ to equations for the current density and the average electron energy in the volume unit of the plasma.

4.3.1   Equation for Current Density

Let us multiply the kinetic equation (4.21) according to (4.22) by minus $\mathbf{p}e/m_e$ and integrate each equation item over all of the impulse space. The first item included in the left part of (4.21) gives us the time derivative of the current density. The second item leads to the integral

$$\int \mathbf{p}\left(\mathbf{p}\frac{\partial f}{\partial \mathbf{r}}\right)d\mathbf{p} = \frac{\partial}{\partial x}\int\limits_{-\infty}^{\infty}\int \int p_x\mathbf{p}f(\mathbf{p})dp_xdp_ydp_z+$$

$$+\frac{\partial}{\partial y}\int\limits_{-\infty}^{\infty}\int \int p_y\mathbf{p}f(\mathbf{p})dp_xdp_ydp_z + \frac{\partial}{\partial z}\int\limits_{-\infty}^{\infty}\int \int p_z\mathbf{p}f(\mathbf{p})dp_xdp_ydp_z.$$

Let us analyze the integrands, each of which can be written in the form $p_{x,y,z}\mathbf{p}f_1(p) + p_{x,y,z}\mathbf{p}f_0(p)$. The first item, being the odd function of $p_{x,y,z}$, gives zero at integration, and in the second item a nonzero result will be given by the item $p_{x,y,z}^2f_0(p)1_{x,y,z}$ only. Taking into account the isotropic character of $f_0(p)$ and hence the fact that the average values of $\left\langle p_{x,y,z}^2\right\rangle$ are equal to each other, the above-mentioned integral can be written as

$$\frac{\partial}{\partial x}\left\langle p_x^2\right\rangle 1_x + \frac{\partial}{\partial x}\left\langle p_y^2\right\rangle 1_y + \frac{\partial}{\partial x}\left\langle p_y^2\right\rangle 1_z = \frac{1}{3}\mathbf{grad}\left(\left\langle p^2\right\rangle\right).$$

As a result, the second item in Eq. (4.21), taking account of (4.24), is

$$-\frac{e}{m_e^2}\int \mathbf{p}(\mathbf{p}\nabla_r f)d\mathbf{p} = -\frac{2e}{3m_e}\mathbf{grad}Q. \qquad (4.27)$$

The third item in (4.21) containing the electric field at integration by parts reduces to the equation

$$\frac{e^2}{m_e}\mathbf{E}\int \mathbf{p}\nabla_p f(\mathbf{p})d\mathbf{p} = \frac{e^2}{m_e}\mathbf{E}\left(\oint_S \mathbf{p}f(\mathbf{p})ds - \int f(\mathbf{p})d\mathbf{p}\right),$$

in which the surface integral is calculated in the impulse space along the surface of an infinitely large radius. Taking into consideration the fact that the distribution function in infinity tends exponentially to zero, we obtain the final equation for the third item:

$$-\frac{e^2}{m_e}\mathbf{E}\int f(\mathbf{p})\,d\mathbf{p} = -\frac{ne^2}{m_e}\mathbf{E}. \qquad (4.28)$$

In a similar way, we can show[4] that the fourth item containing the magnetic field at integration by parts gives

---

[4]We hope the reader can perform this small mathematical exercise independently.

$$\frac{e^2}{m_e} \int \mathbf{p}([\mathbf{p} \times \mathbf{B}] \nabla_{\mathbf{p}} f) d\mathbf{p} = -\frac{e^2}{m_e} \int f[\mathbf{p} \times \mathbf{B}] d\mathbf{p} = -\frac{e}{m_e}[\mathbf{B} \times \mathbf{J}]. \qquad (4.29)$$

The integral generated by the fifth item in (4.21) can be calculated exactly only in the case that we know the function of the collision frequency versus the impulse. For many models of collisions this function is really well known. We shall not go into details here; we will simply use the theorem about the average value, in accordance with which, as well as taking account of (4.23),

$$-\frac{e}{m_e} \int v(p) \mathbf{p} f_1(\mathbf{p}) d\mathbf{p} = -\frac{e}{m_e} \nu(p_{\text{aver}}) \int \mathbf{p} f_1(\mathbf{p}) d\mathbf{p} = \tilde{\nu}(T) \mathbf{J}. \qquad (4.30)$$

Here, $\tilde{\nu}$ is some average frequency of elastic collisions being the function of the average particle energy, i.e., their temperature (see (4.26)). It can be understood what the character of this function is, knowing how the transport scattering section depends on the impulse. If we are speaking about electron collisions with neutral heavy particles (atoms and molecules), then σ_T does not depend on the impulse and according to (4.20), $\nu(p) \sim p$; hence, $\tilde{\nu}$ is proportional to $\sqrt{T}$. At collision with ions in the gas discharge plasma, $\sigma_T \sim p^{-4}$; therefore, $\nu(p) \sim p^{-3}$ and $\tilde{\nu} \sim T^{-3/2}$. The similar function of the average collision frequency versus temperature takes place at scattering of electrons and holes on impurity ions in semiconductors. With regard to their scattering on the lattice vibrations, in the case of acoustic phonons $\tilde{\nu} \sim T_1 T^{1/2}$, for optical phonons[5] $\tilde{\nu} \sim T_1 T^{-1/2}$, where T_1 is the temperature of the lattice. It is clear that the specific values of the average collision frequency depend upon the concentration and type of neutral particles and impurity ions, and change in the wide limits.

Let us return to our equation. The last sixth item included in its left part gives zero at integration as the integrand function after multiplication by $\mathbf{p}$ appears to be odd function. As a result, taking into account (4.27), (4.28), (4.29), and (4.30), we obtain the following equation for the current density:

$$\frac{\partial \mathbf{J}}{\partial t} - \frac{2e}{3m_e} \text{grad} Q + \tilde{\nu}(T) \mathbf{J} = \frac{e^2 n}{m_e} \mathbf{E} + \frac{e}{m_e}[\mathbf{B} \times \mathbf{J}]. \qquad (4.31)$$

Equation (4.31) describes a variety of manifold effects taking place in the gas discharge and solid-body plasma, and we consider some of these effects slightly later. Now we note that according to Eq. (4.31) the current variation is caused by

• Displacement of charged particles due to heterogeneity of their temperature or concentration (diffusion and thermal–electric current)

---

[5]Phonons are the quantum of elastic oscillations $\hbar\Omega$. Oscillation movements of lattice atoms are high-frequency acoustic waves—acoustic phonons. In lattices formed by ions of different kinds the elastic waves are accompanied by excitation of polarization waves—optical phonons. This issue will be discussed again more in Sect. 4.6.

- The presence of an electric or magnetic field
- Relaxation processes of collisions of charged particles with molecules and ions, which result in loss of the directional component of the particle velocity.

At field absence in the homogeneous plasma, the current is exponentially damped in time, and the speed of its damping is defined by the average frequency of elastic collisions of electrons with heavy particles.

4.3.2 Equation for Electron Energy in Plasma

According to (4.24), we multiply the kinetic equation (4.21) by $p^2/2m_e$ and integrate it over all of the impulse space. As a result, the first item in the left part of the equation gives us the time derivative of the average energy. Before transferring to the second item, we introduce into consideration one more physical quantity—the flow density (or the heat flow) $\mathbf{I}_T$, i.e., the average energy value passing through the normal unit section in a time unit. It follows from the definition of this quantity that

$$\mathbf{I}_T = \int \frac{\mathbf{p}}{m_e} \frac{p^2}{2m_e} f(\mathbf{p}) d\mathbf{p}. \tag{4.32}$$

The heat flow is directed from the more heated plasma region into a less heated region. The energy flow density is related to the gradient of the electron (hole) temperature through the coefficient of the electron (hole) heat conductivity.[6] The heat flow arises also in the absence of a temperature gradient under the action of the electric field or the gradient of the charge carrier concentration. In principle, from the kinetic equation we may obtain the equation for $\mathbf{I}_T$ and describe the mentioned phenomena, but that is not included in our tasks in this lecture course. We shall be limited by the fact that using (4.32) we shall write the second item in the equation for the electron average energy in the form

$$\int \frac{\mathbf{p}}{m_e} \frac{p^2}{2m_e} \nabla_r f(\mathbf{p}, \mathbf{r}) d\mathbf{p} = \operatorname{div} \mathbf{I}_T. \tag{4.33}$$

The third item in (4.21), during integration by parts and taking account of (4.22), gives

---

[6]Besides electron and hole heat conductivity in a solid body, there is also lattice heat conductivity, in which heat flow is transferred from a more heated place to a less heated one as a result of lattice oscillations (phonons).

$$-\frac{e}{2m_e}\mathbf{E}\int p^2\nabla_\mathbf{p}f(\mathbf{p},\mathbf{r})d\mathbf{p} = \frac{e}{m_e}\mathbf{E}\int \mathbf{p}f(\mathbf{p},\mathbf{r})d\mathbf{p} = -\mathbf{EJ}. \qquad (4.34)$$

The obtained result is not a surprise; the time variation of the average electron energy in the plasma is defined by the power that is spent by the electric field creating the electric current. The magnetic field, as we know, does not ever cause an energy variation in a charged particle. Therefore, not calculating the integral, we may assert that the fourth item of (4.21) does not make any contribution to the equation of the average electron energy. However, we immediately make sure in this trying to calculate the appropriate integral by parts. The fifth item at integration will also give zero, as after multiplication by p^2 the integrand will be an odd function. At last, the last sixth item can be transformed into the form using the mean value theorem:

$$\frac{1}{2m_e}\chi\int v(p)p^2\left(f_0 - f_0^e\right)d\mathbf{p}$$
$$= \chi v(p_{aver})\frac{1}{2m_e}\int p^2\left(f_0 - f_0^e\right)d\mathbf{p} = \chi\tilde{v}(T)(Q - Q^e). \qquad (4.35)$$

Of course, p_{aver} in (4.35) most likely does not coincide with p_{aver} in (4.30). Therefore, strictly speaking, it is impossible to use the same designation in (4.35) for the average frequency $\tilde{v}$ as in (4.30). But we allow this small inaccuracy, taking into consideration that the initial data necessary for theoretical calculations are often known to us with accuracy to the order.

According to the obtained Eqs. (4.33), (4.34), and (4.35), from the kinetic equation, the following equation for the average electron energy in the plasma follows directly:

$$\frac{\partial Q}{\partial t} + \chi\tilde{v}(Q - Q^e) = \mathbf{EJ} - \text{div}\,\mathbf{I}_T. \qquad (4.36)$$

Equation (4.36) describes the energy balance in the plasma. The power spent by the electric field is consumed in the average energy variation that is accumulated in the electron movement and dissipated in the thermostat as a result of relaxation processes and is moved by heat flows into other parts of the plasma if it is heterogeneous. We would like to note that in the absence of an electric field, the average electron energy in the homogeneous plasma tends to its equilibrium value Q^e as a result of energy exchange between electrons and heavy particles, but this process occurs much more slowly than the process of current damping as $\chi \ll 1$.

Note that Eq. (4.36) looks like the equation for the concentration difference of the two-level system (2.25). This is not surprising, because both $n_{12}\check{z}\omega_{21}$ and Q are the volumetric density of the system energy. Absence in (2.25) $\text{div}(n_{12})$ is connected with uniform distribution of particles in the space.

Thus, we have two equations ((4.31) and (4.36)) at our disposal, which are quite sufficient to describe some very important effects of electromagnetic field interaction with plasma.

4.4 Linear Interaction of an Electric Field with Plasma

Plasma properties are described by the system of nonlinear material equations (4.31) and (4.36). We have already encountered the same situation in Chaps. 2 and 3. For a wide class of tasks it is enough to use linear approximation. In this approximation for homogeneous isotropic plasma in the absence of a permanent magnetic field, only one simple equation remains:

$$\frac{\partial \mathbf{J}}{\partial t} + \tilde{\nu}(T)\mathbf{J} = \frac{e^2 n}{m_e}\mathbf{E}. \tag{4.37}$$

If the plasma is in a weak alternate electric field $\mathbf{E}(t) = \mathrm{Re}\,(\mathbf{E}e^{j\omega t})$, the complex amplitude of the alternate component of the current density will, according to (4.37), be

$$\dot{\mathbf{j}} = \frac{e^2 n/m_e}{\tilde{\nu}(T) + j\omega} \tag{4.38}$$

The current density can be represented as the sum of the conduction current and the polarization current:

$$\dot{\mathbf{j}} = \sigma\dot{\mathbf{E}} + j\omega\dot{\mathbf{P}} = \sigma\dot{\mathbf{E}} + j\omega\varepsilon_0(\varepsilon - 1)\dot{\mathbf{E}}.$$

This allows transfer from (4.38) to the linear electric parameters of the plasma. The relative dielectric permeability is

$$\varepsilon = 1 - \frac{e^2 n}{m_e\left(\omega^2 + \tilde{\nu}^2\right)\varepsilon_0} = 1 - \frac{\omega_0^2\tau^2}{1 + \omega^2\tau^2} \tag{4.39}$$

and the specific conductance is

$$\sigma = \frac{e^2 n\tilde{\nu}}{m_e\left(\omega^2 + \tilde{\nu}^2\right)} = \frac{e^2 n\tau}{m_e(1 + \omega^2\tau^2)} = \frac{\sigma_0}{(1 + \omega^2\tau^2)}. \tag{4.40}$$

Here, σ_0 is the specific conductance on DC, and $\omega_0 = \sqrt{\frac{e^2 n}{m_e\varepsilon_0}}$ is the so-called plasma oscillation frequency. This concept was introduced for the first time by Irving Langmuir during study of plasma oscillations in gas discharge. The functions of the

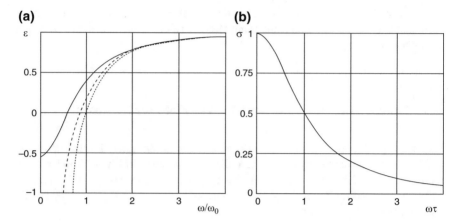

Fig. 4.5 Functions of relative dielectric permeability versus frequency (**a**) and normalized specific plasma conductivity versus frequency (**b**). The *solid curve* corresponds to $\omega_0\tau = 1.25$, the *dashed curve* to $\omega_0\tau = 2$, and the *dotted curve* to $\omega_0\tau = 10$. At frequencies of less than ω_0, the dielectric permeability is negative. At frequencies much greater than the collision frequency, the plasma behaves as a dielectric with low losses (see Fig. 4.5b)

dielectric permeability and the specific plasma conductivity versus the frequency are shown in Fig. 4.5.

In a long-wave (low-frequency) limit, $\omega\tau \ll 1$, and Eqs. (4.39) and (4.40) can be simplified. They can frequently be written in the form

$$\varepsilon \cong 1 - \omega_0^2\tau^2 \cong 1, \quad \sigma \cong \vec{\sigma}_0 = e\frac{e\tau}{m_e}n = e\mu n, \tag{4.41}$$

where μ is the mobility of charge carriers in the plasma, through which the average velocity of the charge drift in the field $\mathbf{E}(\mathbf{v} = \mu\mathbf{E})$ can be expressed. The collision frequency of the charge carriers in metals and semiconductors is of the order of 10^{12}–10^{13} s^{-1}. Therefore, (4.41) describes the properties of these materials from DC to the far-infrared ranges.

If $\omega\tau \gg 1$, we can speak about the short-wave (high-frequency) limit. Then

$$\varepsilon \cong 1 - \frac{e^2n}{m_e\omega^2\varepsilon_0} = 1 - \frac{\omega_0^2}{\omega^2}, \quad \sigma \cong \frac{e^2n}{m_e\omega^2\tau} = \frac{\omega_0^2\varepsilon_0}{\omega^2\tau}. \tag{4.42}$$

Since $\omega\tau \gg 1$ corresponds to $\omega\tau \gg \nu$, plasma under such conditions is often called collisionless plasma. For collisionless plasma,

$$\varepsilon = 1 - \frac{\omega_0^2}{\omega^2}, \quad \sigma \cong 0. \tag{4.43}$$

In the expressions in (4.42), plasma frequency ω_0 acts as some threshold radiation frequency ω. If $\omega < \omega_0$, the plasma dielectric permeability is less than zero

(Fig. 4.5a). In this medium the electromagnetic wave cannot propagate, since the phase constant $\beta = \omega\sqrt{\mu_0\bar{\varepsilon}}$ is a purely imaginary quantity. Therefore, on frequencies less than the plasma frequency, the electromagnetic wave will effectively reflect from the boundary with the vacuum.

The specific plasma conductivity for $\omega\tau \gg 1$ decreases proportionally to the square frequency (Fig. 4.5b). This is connected with the fact that during the time between collisions, the electron has time to perform several oscillations under the action of the alternate field. The wave propagation will be accompanied by damping. The damping exponent of the power flow is

$$\alpha_p = \sigma\sqrt{\frac{\mu_0}{\varepsilon_0}} = \sigma Z_0 = \frac{\omega_0^2\varepsilon_0}{\omega^2\tau} = \frac{\omega_0^2\lambda^2}{c^3\tau}. \tag{4.44}$$

Figure 4.6 shows the function of α_Π versus the square wavelength. Experimental results agree well with the theoretical law $\alpha_p \sim \lambda^2\alpha$.

All considerations can be successfully extended to metals and semiconductors. Considering electrons and holes as a gas of charge carriers with definite density, temperature, charge, mass, and the appropriate distribution function, we can write equations of type (4.41) for each kind of particle. We omit this procedure here and shall take into consideration, in the future, that when it is necessary, we shall introduce some evident changes in our formulas. In particular, as was mentioned in Sect. 4.1, when the inertial masses of charge carriers differ from the electron mass (for instance, in semiconductors) in formulas (4.38), (4.39), (4.40), (4.41), (4.42), and (4.43), instead of the electron mass m_e, we should use the effective mass of electrons m_e^* or holes m_h^*. Some value of plasma frequency will correspond to each kind of charged particle. For metals and semiconductors in (4.42) we must also take

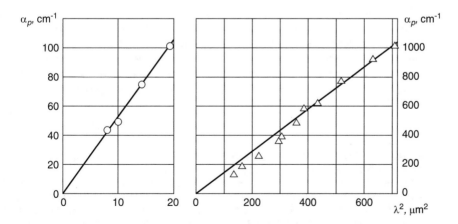

Fig. 4.6 Function of the damping factor of the power flow in indium antimonide versus the square wavelength. In the short-wave limit the specific conductivity is inversely proportional to the frequency; therefore, $\alpha_p \sim \lambda^2$. The *experimental points* are well laid on the *straight line*

into account the contribution to the dielectric permeability of the crystal lattice ($\varepsilon_0 \to \varepsilon_p \varepsilon_0$). In semiconductors $\varepsilon_p \approx 16$, and in metals $\varepsilon_p \approx 1$.

In semiconductors at room temperature, the collision frequency of electrons and holes is of the order of 10^{12}. Radiation in the far- and near-infrared ranges is concerned in the short-wave case. In metals the collision frequency is one order greater. The short-wave limit for metals falls at the long-wave limit of the optical range.

Let us estimate the value of the specific conductivity of metal. Substituting in (4.41) $n \sim 10^{28}$ m^{-3}, $m^* \sim m = 9.1 \times 10^{-31}$ kg, $\tau \sim 10^{13}$, we obtain $\sigma_0 \approx 3 \times 10^7$ Sim/m. This agrees well with the values for metal conductivity that are usually used in engineering electrodynamics (for copper, $\sigma_0 \approx 5.7 \times 10^7$ Sim/m).

Plasma properties cause the reflection of radio waves from the ionosphere. They are also responsible for electromagnetic wave reflection of both optical and radiofrequency ranges from a metal surface. However, both the ionosphere and metals are transparent to electromagnetic waves with oscillation frequency $\omega > \omega_0$ or wavelength $\lambda < \lambda_0 = \frac{2\pi c}{\omega_0}$. In the ionosphere, charged particles arise under the action of solar radiation. The values of the plasma frequency depend upon the part of 24 hours and the season, and oscillate within the 10^7 MHz limit ($\lambda_0 \approx 100$–10 m). The collision frequency in the ionosphere is about 10^6. In the short-wave range, (4.42) satisfactorily describes ionosphere plasma.

Since the electron density is many orders less in the ionosphere than in metals, the radiation wavelength limit at which metals are transparent is many orders less than the wavelength limit in the ionosphere. Using Eq. (4.42) and specifying a reasonable electron density in metal ($n \sim 10^{28}$ m^{-3}), we can easily find that λ_0 for metal is a tenth of a micrometer and can be confirmed experimentally.

Radiation absorption by "free" carriers in semiconductors is widely used for studying relaxation processes and for measurement of the effective mass of carriers. For example, in the region $\omega\tau \ll 1$, the relaxation time can be found for the known carrier concentration by comparing the experimental values of the dielectric permeability of two samples. With known relaxation time and concentration values we can determine the effective mass. When William Shockley[7] was starting to create the first semiconductor devices, he used this method to measure the semiconductor parameters. Afterward, the results of his measurements were confirmed in experiments on cyclotron resonance, which will be discussed in the next section.

---

[7]William Bradford Shockley (1910–1989) was an American physicist who developed the theory of the electron–hole junction. In 1948 he discovered the transistor effect. From 1955 onward he worked as the director of the Shockley Semiconductor Lab, which subsequently became the Shockley Transistor Corporation, of which he was the president and director. In 1956 he received the Nobel Prize in Physics. He held more than 90 patents. He received the Morris Leibmann Memorial Prize from the Institute of Radio Engineers, the Oliver E. Buckley Solid-State Physics Prize from the American Physical Society, and the Cornstock Prize in Physics from the US National Academy of Sciences. He was a member of the American Physical Society and the Institute of Electrical and Electronic Engineers. In his youth he was a keen mountain climber.

In injection laser diodes, the dielectric properties of the charge carriers lead to formation near the *p–n* junction of the planar optical wave guide, which is manifested in threshold current reduction and in narrowing of the radiation beam of the semiconductor quantum generator.

4.5 Cyclotron Resonance

Let us examine again a homogeneous isotropic plasma. We assume that permanent magnetic field **B** affects such a plasma. For distinctness, let the magnetic field be directed along axis *z*, i.e., $\mathbf{B} = B\mathbf{1}_z$. A charge placed in the magnetic field under the action of the Lorentz force will move along a circle in the plane perpendicular to the magnetic field (Fig. 4.7).

If, besides the magnetic field, there is an electric field,

$$\mathbf{E} = E_x\mathbf{1}_x + E_y\mathbf{1}_y, \qquad (4.45)$$

then, taking into account the above-mentioned, we can rewrite (4.31) as

$$\frac{\partial \mathbf{J}}{\partial t} - \frac{eB}{m_e}[\mathbf{1}_z \times \mathbf{J}] + \tilde{\nu}(T)\mathbf{J} = \frac{e^2 n}{m_e}\mathbf{E}. \qquad (4.46)$$

If E is the alternate electric field changing with frequency ω, the differential equation (4.46) can be written in the vector form

Fig. 4.7 An electron rotates around the vector of magnetic induction with angular velocity $\omega_H = \frac{eB}{m}$. This kind of movement is called cyclotron motion. The cyclotron orbit radius is determined by the value of the orbit velocity $r = \frac{v}{\omega} = \frac{vm}{eB}$

– cyclotron radius. The magnetic field does not change the modulus of the velocity vector, i.e., it does not change the electron energy

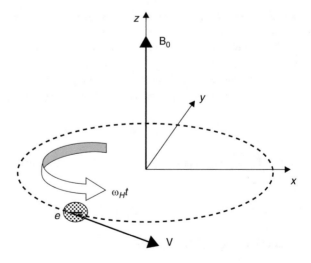

$$\dot{\mathbf{J}} = \frac{\omega_H}{\tilde{\nu} + j\omega}\left[\mathbf{1_z} \times \dot{\mathbf{J}}\right] + \frac{e^2 n}{m_e(\tilde{\nu} + j\omega)}\dot{\mathbf{E}}, \tag{4.47}$$

where $\omega_H = \frac{eB}{m_e}$. The vector equation of type

$$\mathbf{X} = [\mathbf{C} \times \mathbf{X}] + \mathbf{A} \tag{4.48}$$

has the solution

$$\mathbf{X} = \frac{\mathbf{A} + [\mathbf{C} \times \mathbf{A}] + \mathbf{C}(\mathbf{CA})}{1 + |\mathbf{C}|^2}. \tag{4.49}$$

We can make sure of this by direct substantiation. Comparing (4.47) and (4.48), we can, right away, write the solution to Eq. (4.47). Taking (4.49) and (4.45) into account, we obtain, after transformations,

$$\dot{\mathbf{J}} = \frac{e^2 n}{m_e\left[(\tilde{\nu} + j\omega)^2 + \omega_H{}^2\right]}$$
$$\times \left\{\left[(\tilde{\nu} + j\omega)\dot{E}_x - \omega_H\dot{E}_y\right]\mathbf{1}_x + \left[\omega_H\dot{E}_x + (\tilde{\nu} + j\omega)\dot{E}_y\right]\mathbf{1}_y\right\}. \tag{4.50}$$

Using (4.50) we can determine parameters ε and σ of the plasma placed in a permanent magnetic field. Without doing the further calculations (they can be completed by the reader in the same manner as above), we would like to note only that now both ε and σ are tensor quantities. This means that the current density and polarization do not coincide with the electric field direction that gives birth to them. Moreover, in the presence of a permanent magnetic field, other specific properties are observed in the plasma. In the first place, we can attribute to them the phenomenon of resonance absorption.

For revelation of typical properties of this phenomenon, we assume that an electric field with left or right circular polarization affects the plasma. The convenience of the transfer to fields with circular polarization will be quite clear if we look at Fig. 4.7. The complex amplitude of the electric field (4.45) can be presented as

$$\dot{\mathbf{E}}_{\pm} = \left(\mathbf{1_x} \pm j\mathbf{1}_y\right)E_m, \tag{4.51}$$

where the signs (+) and (−) correspond to two directions of vector E rotation. Let $\dot{E} = \dot{E}_-$, then

$$\dot{E}_x = E_m, \dot{E}_y = -jE_m. \tag{4.52}$$

Substituting (4.52) in (4.50), we obtain

$$\dot{\mathbf{J}}_- = \frac{e^2 n}{m_e} E_m \frac{(\tilde{\nu} + j\omega)(\mathbf{1}_x - j\mathbf{1}_y) + \omega_H(\mathbf{1}_y - j\mathbf{1}_x)}{(\tilde{\nu} + j\omega)^2 + \omega_H{}^2} =$$

$$= \frac{e^2 n}{m_e} \frac{\dot{\mathbf{E}}_-}{\tilde{\nu} + j(\omega - \omega_H)}. \tag{4.53}$$

Analyzing Eq. (4.53), it is easy to see that for $\omega = \omega_H$ the value of $\dot{\mathbf{J}}$ increases in a resonant manner. This phenomenon is called the cyclotron resonance. The physical reason for the cyclotron resonance is connected with the specific movement of electric charges in electric E and magnetic **B** fields. Since the Lorentz force (which is perpendicular to both the electric and magnetic fields) will affect the electric charge, in the polarization plane of vector E it will circumscribe a trajectory close to a circular arc. The rotational angular velocity of the electric charge is determined by its inertial mass and the magnetic field value. This is equal to the so-called cyclotron frequency:

$$\omega_H = \frac{eB}{m_e}. \tag{4.54}$$

If the direction of electric field vector E coincides with the direction of the radius vector of the rotated charge carrier, the additional movement in the direction of the electric field will be superimposed on the circular movement. In this case, the charge carrier will move on the trajectory of an expanding helix, effectively collecting the energy from electric field E. It is easy to see that these conditions are satisfied for electric field $\dot{\mathbf{E}}_-$.

If these considerations are true, for electric field $\dot{\mathbf{E}}_+$ with the opposite rotational direction to vector E, the cyclotron resonance must be absent. Really, let $\dot{\mathbf{E}} = \dot{\mathbf{E}}_+$. Then, in accordance with (4.51),

$$\dot{E}_x = E_m, \dot{E}_y = jE_m. \tag{4.55}$$

Substituting (4.55) in (4.50), we obtain

$$\dot{\mathbf{J}}_+ = \frac{e^2 n}{m_e} E_m \frac{(\tilde{\nu} + j\omega)(\mathbf{1}_x j\mathbf{1}_y) + \omega_H(\mathbf{1}_y - j\mathbf{1}_x)}{(\tilde{\nu} + j\omega)^2 + \omega_H{}^2} =$$

$$= \frac{e^2 n}{m_e} \frac{\dot{\mathbf{E}}_+}{\tilde{\nu} + j(\omega + \omega_H)}, \tag{4.56}$$

which confirms that at such polarization of electric field E the cyclotron resonance really is absent. Figure 4.8 shows the function of specific conductivity versus the frequency of the electric field for the opposite directions of its rotation. The power absorbed in the plasma will change similarly.

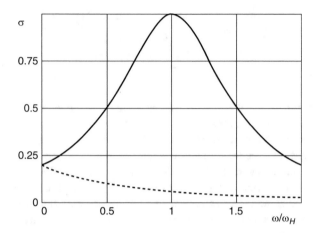

Fig. 4.8 Function of the specific conductivity of plasma normalized to its value at $B = 0$ and $\omega = 0$ for fields with different rotational directions. Cyclotron resonance is observed in the case of a field whose rotational direction coincides with the electron rotational direction in a permanent magnetic field. At the resonance point the conductivity is numerically equal to conductivity on DC for $B = 0$ (see Fig. 4.5b). The magnetic field decreases the conductivity value on DC more than four times. This phenomenon is used in magnetoresistors

As in the case of the plasma frequency, the cyclotron frequency of the charge carrier is determined by the inertial mass. Therefore, in the general case, the equation for the cyclotron frequency (4.54) should be changed to

$$\omega_H = \frac{eB}{m^*}, \tag{4.57}$$

where m^* is the effective mass of the electron m_e^* or the hole m_h^*.

Taking the above-mentioned into consideration, Eq. (4.57) can be used as calculated for experimental determination of the effective mass of the charge carrier. In the simplest case, the substance under investigation for which we need to measure m^* is placed in the wave guide system. If the oscillation frequency of electromagnetic radiation ω coincides with cyclotron frequency ω_H, we shall observe the maximal absorption of the field energy. Changing ω for a fixed **B** or changing **B** for a fixed ω and measuring the power that is passed, we can calculate m^* by using Eq. (4.57).

For the first time, cyclotron resonance was observed in germanium (Ge) with the specific resistance 38 Ohm · cm on the frequency 9.05 GHz at a temperature of 4 K. For an n-type substance the resonance field at 0.037 Tl corresponds to the effective mass of the electron $m^* = 0.11m_e$. For a p-type substance there are two types of holes—light and heavy—with effective masses of $m^* = 0.04m_e$ and $m^* = 0.03m_e$, respectively. The function of electromagnetic field absorption on a frequency of 24 GHz versus the intensity of a permanent magnetic field is presented in Fig. 4.9.

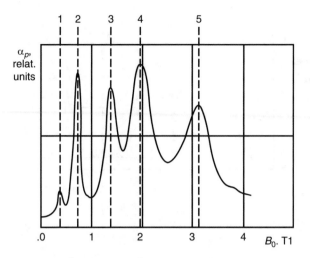

Fig. 4.9 Function of electromagnetic field absorption on a frequency of 24 GHz in germanium versus the magnetic induction value at $T = 4$ K. Germanium has a complex band structure (see Fig. 4.11). Carriers gather in several valleys (the vicinities of the local minimums of energy). The curvature of the valleys differs; therefore, carriers in different valleys have different effective masses. _1_ denotes holes $m_h^* \approx 0.46 m_e$, _2_ denotes electrons $m_e^* \approx 0.93 m_e$, _3_ denotes electrons $m_e^* \approx 1.63 m_e$, _3_ denotes electrons $m_e^* \approx 2.33 m_e$, _4_ denotes holes $m_h^* \approx 0.46 m_e$ and _5_ denotes holes $m_h^* \approx 3.61 m_e$

From the quantum point of view, the phenomenon of cyclotron resonance is connected with zone splitting into so-called "Landau levels." In this case the energy of electrons is

$$W_e(k_z, l_c) = W_c + \hbar \omega_H \left(l_c + \frac{1}{2} \right) + \frac{\hbar^2 k_z^2}{2 m_e^*}$$

and the energy of holes is

$$W_e(k_z, l_v) = W_v - \hbar \omega_H \left(l_v + \frac{1}{2} \right) - \frac{\hbar^2 k_z^2}{2 m_h^*}.$$

In Fig. 4.10 we can see zone splitting into one-dimensional subzones, each of which can be identified by orbital quantum number l. The function of permitted state density $N(W)$ is shown in Fig. 4.10b. Transitions between Landau levels lead to energy absorption of the alternate electric field. If we do not take into consideration the relativistic correction, the level system is identical to the levels of the quantum harmonic oscillator. Relativistic corrections disturb the levels' equidistance[8] and are manifested in nonlinear effects of a saturation effect type. Nevertheless, the

---

[8]In the classical interpretation we should substitute in (4.54) mass equal to $m / \sqrt{1 - v^2/c^2}$.

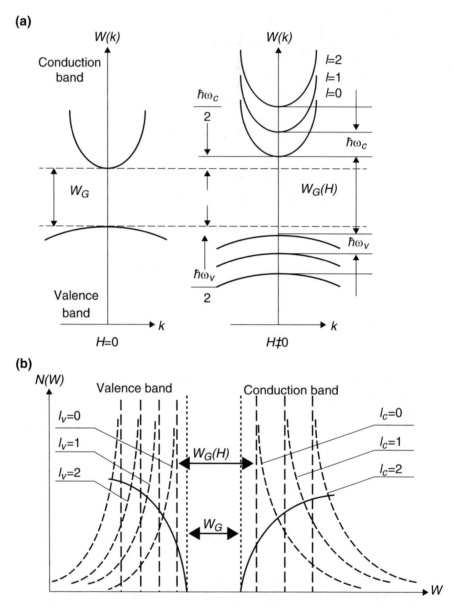

Fig. 4.10 Splitting of semiconductor bands into subbands in the presence of a permanent magnetic field. (**a**) For $k = 0$ the distance between adjacent subbands is $\hbar\omega_H$. The magnetic field changes the density of the permitted states. (**b**) Electrons concentrate near the bottom of each subband. This gives the basis for identification of movement in the plane perpendicular to the magnetic field with discrete levels (Landau levels)

corrections are small and for satisfactory description of the phenomenon of cyclotron resonance we must take into consideration about 1000 energy levels. For this reason, the quantum description becomes extremely ineffective, whereas the classical one (for the same reason) gives quite acceptable results in terms of accuracy.

4.6   Fundamental Absorption

If a radiation quantum is absorbed by a semiconductor and electrons in the valence band (taking additional energy that exceeds or is equal to the width of the forbidden band) perform transitions to the conduction band, such absorption is called intrinsic or fundamental absorption. When studying intrinsic absorption by a semiconductor we should take into account the structure of its energy bands.

The semiconductors that are known at present can be divided into two main classes depending upon the type and configuration of their energy bands. The first type has minimum energy in the conduction band and maximum energy in the valence band, and this minimum and maximum are at the same value of wave vector **k** (Fig. 4.11a). Indium antimonide is an example of a semiconductor with such a band structure. The second class of substances has extrema of the conduction band and the valence band at different **k** values (Fig. 4.11b). Most semiconductors, including germanium and silicon, are in the second class of substances.

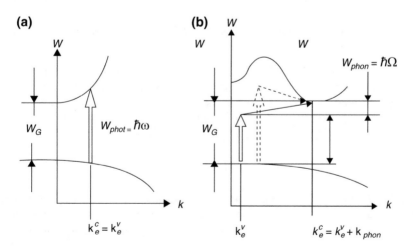

Fig. 4.11 Fundamental absorption due to transitions between bands. With direct transitions the impulse of the headway motion of the charge carrier does not change. Therefore, the wave vector of electron $\mathbf{k}_e^c$ passed to the conduction band is equal to the vector of electron $\mathbf{k}_e^v$ in the valence band. (**a**) Direct transitions are observed in semiconductors with a band configuration close to ideal (for example, indium antimonide). In multivalley semiconductors (for example, germanium and gallium arsenide), indirect transitions are observed (**b**) with participation of phonons (lattice oscillations). With such transitions the electron changes its forward movement velocity (impulse $\mathbf{k}_e^c = \mathbf{k}_e^v + \mathbf{k}_{phon}$)

Transitions of electrons through the forbidden band will happen, first of all, between energy states corresponding to the maximum of the valence band and to the minimum of the conduction band, i.e., at values of quasi-impulses ($\mathbf{p} = \hbar\mathbf{k}$) or wave vector $\mathbf{k}$ close to zero, as presented in Fig. 4.11a.

Really, in the first order of the theory of perturbations depending on time, probability W_{if} of the transition from initial state i to a group of final states f is

$$W_{if} = \frac{2\pi}{\hbar} \left| \widehat{H}_{if} \right|^2 N(W_f). \tag{4.58}$$

Here, $\widehat{H}$ is the matrix element of perturbation connecting the system states i and f; $N(W_f)$ is the final state density.

The energy of electron interaction in the semiconductor with the electric field of the plane linear-polarized electromagnetic wave,

$$\mathbf{E}(\mathbf{r}, t) = \mathrm{Re}\,(E \exp[j(\omega t - \mathbf{k}\mathbf{r})]\mathbf{1}_x),$$

is

$$W = \frac{eE_x}{2}\,\mathrm{Re}\,(\exp[j(\omega t - \mathbf{k}\mathbf{r})]).$$

Acting as in Chap. 2, we can write the matrix element of the perturbation operator

$$\widehat{H}_{vc}(\mathbf{k}_i, \mathbf{k}_f) = \frac{eE}{2} \int \dot{u}_v^*(\mathbf{k}_i, \mathbf{r}) x \dot{u}_c(\mathbf{k}_f, \mathbf{r}) \exp[j(\mathbf{k}_i - \mathbf{k}_f - \mathbf{k})\mathbf{r}] d\mathbf{r}, \tag{4.59}$$

where $\dot{u}_v(\mathbf{k}_i, \mathbf{r})$ and $\dot{u}_c(\mathbf{k}_f, \mathbf{r})$ are the proper wave functions of the valence band and the conduction band, respectively. The exponent in the integrand is a fast-changing function. Therefore, the value of $\widehat{H}$ is vanishingly small in all cases besides $\mathbf{k}_i - \mathbf{k}_f - \mathbf{k} = 0$. In the optical range, $|\mathbf{k}| \approx 10^7\,\mathrm{m}^{-1}$ while $|\mathbf{k}_i - \mathbf{k}_f| \approx |\mathbf{k}|_i \approx 10^{10}\,\mathrm{m}^{-1}$. Hence, we can write the condition of possible transitions in the form

$$\mathbf{k}_i = \mathbf{k}_f, \tag{4.60}$$

i.e., transitions occur without a change of direction and the intraband forward movement velocity. Transitions of such a type are called direct or vertical transitions.

For vertical transitions the matrix elements of the perturbation operator take the form

$$\widehat{H}_{vc}(\mathbf{k}) = \frac{eE}{2} \int \dot{u}_v^*(\mathbf{k}, \mathbf{r}) x \dot{u}_c(\mathbf{k}, \mathbf{r})] d\mathbf{r} = \frac{E}{2} d_{vc},$$

where $d_{vc} = e\langle u_v(\mathbf{k})|x|u_c(\mathbf{k})\rangle$ is the transition matrix element.

Now we must find the state density $N(W_f)$ appearing in (4.58). We remember that near the bottom of the conduction band the electron energy is $W_e = W_c + \frac{\hbar^2 k^2}{2m_e}$, and near the top of the valence band, $W_h = W_v + \frac{\hbar^2 k^2}{2m_h}$. Hence,

$$\hbar\omega = W_e - W_h = W_G + \frac{\hbar^2 k^2}{2m_r},$$

where $m_r^* = \frac{m_e^* m_h^*}{m_e^* + m_h^*}$ is the reduced effective mass and W_G is the width of the forbidden band. The number of permitted states in the volume unit according to [32] (see also (1.70)) is

$$N_e(W_e) = \frac{(2m_e^*)^{3/2}}{2\pi^2\hbar^3}(W_c - W_e)^{1/2}, \quad N_h(W_h) = \frac{(2m_h^*)^{3/2}}{2\pi^2\hbar^3}(W_v - W_h)^{1/2}.$$

For the model of the reduced mass we finally obtain

$$N(W_f) = \frac{(2m_r^*)^{3/2}}{2\pi^2\hbar^3}(\hbar\omega - W_G)^{1/2}. \tag{4.61}$$

To get the function of the absorption coefficient versus the energy near the edge of fundamental absorption, it is necessary, using (4.58), to determine the difference in the transition number per second from the valence band to the conduction band, and vice versa. For simplicity we assume that the semiconductor temperature is equal to zero. In this case the conduction band will be completely empty and transitions downward from the conduction band will be absent.

Using Eqs. (4.61) and (4.59) we can conclude that with permitted direct transitions for which $\widehat{H}_{if}$ does not depend upon k, the transition number from the valence band is

$$N_{vb} = \frac{(2m_r^*)^{3/2}}{\pi\hbar^4}\left|\widehat{H}_{vc}(\hbar\omega - W_G)\right|^2 (\hbar\omega - W_G)^{1/2}. \tag{4.62}$$

The power absorbed in the volume unit is

$$P_{abs} = N_{vb}\hbar\omega = \frac{\sigma E^2}{2}.$$

This equation allows determination of the specific conductivity describing the fundamental absorption:

$$\sigma(\omega) = \frac{(2m_r^*)^{3/2} d_{vc}^2}{2\pi\hbar^3} \omega(\hbar\omega - W_G)^{1/2}.$$

The value of the square of the matrix element for interband transitions is $d_{vc}^2 \approx \frac{e^2\hbar}{m\omega}$. Using this expression, we obtain

$$\sigma(\omega) \approx \frac{(2m_r^*)^{1/2} e^2}{\pi\hbar^3} (\hbar\omega - W_G)^{1/2}.$$

If the specific conductivity is known, we may find damping factor α_{II}, which is for a medium with small losses:

$$\alpha_p = \sigma Z_c \approx \frac{(2m_r^*)^{1/2} e^2}{\pi\hbar^2} (\hbar\omega - W_G)^{1/2} \sqrt{\frac{\mu_0}{\varepsilon\varepsilon_0}}.$$

If we assume that $\varepsilon = 16$ and $m_e^* = m_h^* = m_e$, for the energy in eV we have

$$\alpha_p \approx 2 \cdot 10^6 (\hbar\omega - W_G)^{1/2} \, \mathrm{m}^{-1}.$$

The experimental function of damping factor α_{II} versus the photon energy for indium antimonide is presented in Fig. 4.12a. This function is well described by the formula $\alpha_p \approx 1.5 \cdot 10^6 (\hbar\omega - W_G)^{1/2} \, \mathrm{m}^{-3}$. For semiconductors with a complicated band structure near the boundary $z\omega = W_G$, absorption is observed as a result of indirect transitions. As an example, Fig. 4.12b shows the function $\alpha_p(\hbar\omega)$ for GaAs. The dotted line corresponds to direct transitions.

If we use the Kramers–Kronig relations, we can, if necessary, determine the contribution from interband transitions to the dielectric permeability of the medium.

All of the above-mentioned relations are fulfilled within the limits of balance equations. You probably noticed that nowhere above did we speak about relaxation processes. This becomes possible for the following reason. In essence, fundamental absorption is similar to absorption under conditions of nonuniform widening. In this context, N_c and N_v act as the degeneration degree of energy levels, and $F(W)$ as the distribution function on transition frequencies. Thus, our results are obtained absolutely similarly to 2.7—namely, by changing the Lorentz spectral line to the δ function. A strict and consequent solution with the help of the density matrix can be found in the literature.

Fig. 4.12 Function of the damping factor of power flow α_p of pure indium antimonide (InSb), caused by direct interband transitions versus photon energy $\hbar\omega$. The frequency response of absorption is well described by the function $\alpha_p = 1.5 \cdot 10^6$ $\sqrt{\hbar\omega - W_G}\,\mathrm{m}^{-1}$. In other semiconductors the bands are more complicated and indirect transitions make a noticeable contribution to fundamental absorption. The contribution of indirect transitions to absorption is well seen in the function of α_p versus the frequency for GaAs (**b**)

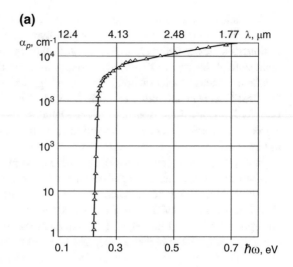

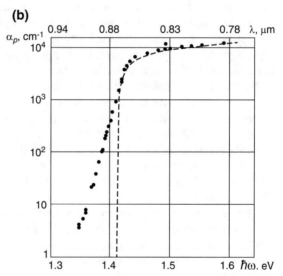

4.7   Other Mechanisms of Absorption

Interaction of radiation with free carriers and fundamental absorption happen on a background of a series of phenomena that also make a contribution to absorption. Let us examine some of them.

In investigation of the optical properties of solid bodies with high spectral resolution we can see sharp absorption lines located in the frequency slightly below the edge of direct transitions. These lines are connected with so-called excitonic absorption.

An exciton is a system from an electron and a hole coupled by the Coulomb force. The exciton is in some sense similar to the hydrogen atom. The energy of exciton stationary states is inversely proportional to the square of quantum number k. The main state of this pair is located at a distance of about 0.005 eV from the edge of the conduction band. Radiation breaks the coupling (ionizes) and leads to appearance of the "free" electron and the "free" hole in bands. The energy diagram is presented in Fig. 4.13a. Transitions from the valence band to the energy levels of the exciton correspond to absorption. The absorption spectrum of copper protoxide is presented in Fig. 4.13b.

In doped semiconductors we can observe absorption connected with transitions between impurity levels and semiconductor bands. Such absorption is called impurity absorption. The donor and acceptor levels are located within the forbidden band at a distance of about 0.1 eV from its boundaries. Therefore, bands of impurity absorption lie beyond the edge of fundamental absorption. Besides the main state, impurity centers have higher energy levels. Electron excitation leads to the appearance of absorption, which has a spectrum with several lines. The absorption spectrum of silicon doped with boron is shown in Fig. 4.14.

The crystal lattice can absorb radiation of definite frequencies as well. The absorption spectrum of the lattice is characterized by a series of peaks, which are superimposed on the absorption of free charge carriers. Absorption is connected with excitation of elastic lattice oscillations (phonons). Each absorption peak is connected with excitation of a definite type of phonons.

Semiconductor compositions containing atoms of different kind (e.g., GaAs) can be presented as a set of electric dipoles. The electromagnetic field excites the lattice oscillations at which atoms of different kinds perform out-of-phase oscillations (optical phonons). In homopolar crystals (for instance, Si) there are no dipoles and only acoustic phonons exist. In this case, interaction with electromagnetic radiation is connected with induced dipoles. That is why, for example, absorption is greater in GaAs than in Si.

The spectrum of lattice absorption of GaAs is shown in Fig. 4.15. The crystal lattice makes the main contribution to the "background" dielectric permeability. The relative dielectric permeability of a semiconductor is about 16. This value needs to be taken into consideration during use of formulas that contain absolute dielectric permeability ε_0.

A detailed description of optical phenomena in semiconductors can be found in the literature.

4.8   The Nonequilibrium State of Semiconductors

An external impact brings a semiconductor out of the equilibrium position. A semiconductor behaves slightly differently from a simple two-level system. The reason is that intraband relaxation processes pass much faster than intraband

(a)

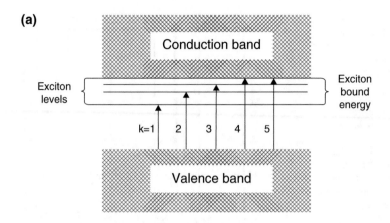

(b)

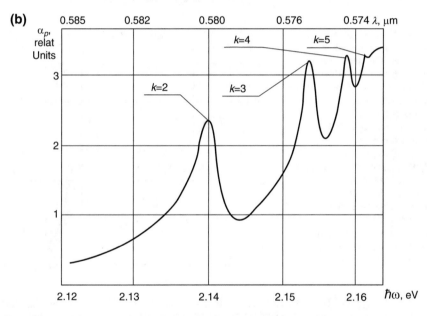

Fig. 4.13 Scheme of energy levels of an exciton with a fixed mass center (**a**) and the function of damping factor α_p versus photon energy for Cu_2O at 77 K (**b**). Optical transitions from the top of the valence band are shown by *arrows*. The longest corresponds to the exciton ionization potential; hence, this energy is equal to the width of the forbidden energy band. We see a series of peaks. The width of the forbidden band for Cu_2O is ≈ 2.17 eV. In physics, a **bound state** describes a system in which a particle is subject to a potential such that the particle has a tendency to remain localized in one or more regions of space. The potential may be either an external potential or the result of the presence of another particle

Fig. 4.14 Absorption by impurities. The function of the damping factor of power flow α_p in a silicon sample doped with boron (the donor) is presented. The peaks on the graphs correspond to donor transitions from the main state to the excited state. The absorption peaks gradually coincide with the donor ionization band—transitions to the conduction band, whose probability decreases fast

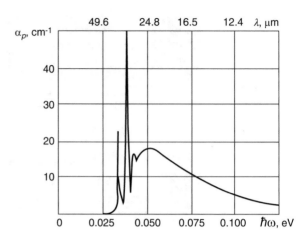

Fig. 4.15 Absorption by a crystal lattice of gallium arsenide. The function of the damping factor of power flow α_p versus photon energy is shown. It is connected with ion oscillations. As a result, elastic transverse and longitudinal waves (phonons) are excited in the crystal. Each peak corresponds to a definite type of phonon. Lattice absorption is superimposed on the absorption spectrum of free charge carriers (the rising branch on the left)

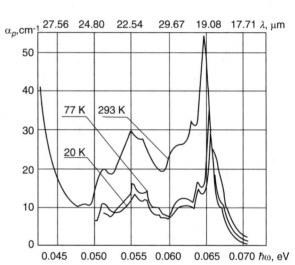

ones. The typical time for the former is about 10^{-12}–10^{-13} s, whereas the time for the latter is more than 10^{-9} s.[9] This difference allows us to speak about the existence of the quasiequilibrium distribution in each band. Such a distribution can be described by two Fermi functions:

$$F_c(W) = \left(1 + \exp\left(\frac{W - W_{Fc}}{k_{\text{B}}T}\right)\right)^{-1} ; \quad F_v(W) = \left(1 + \exp\left(\frac{W - W_{Fv}}{k_{\text{B}}T}\right)\right)^{-1} ,$$

where W_{Fc}, W_{Fv} are the Fermi quasilevels for the conduction and valence bands, respectively. The conception of quasilevels is widely used in the theory of

---

[9]This time acts as T_1 (see Chaps. 1 and 2).

semiconductor devices. The location of quasilevels at a given carrier concentration is
determined from the conditions of normalization:

$$n_e = \int_{Wc}^{\infty} N_c(W)F_c(W)dW; \quad n_h = \int_{\infty}^{Wv} N_v(W)F_v(W)dW,$$

which are essentially simplified at zero temperature T:

$$n_e = \int_{Wc}^{W_{Fc}} N_c(W))dW; \quad n_h = \int_{W_{Fv}}^{W_v} N_v(W))dW.$$

   Under these conditions, all energy levels up to the quasilevel are occupied and the
upper levels are free.
   Now we imagine that in some way we can create a semiconductor in which
quasilevel W_{Fc} lies in the conduction band and W_{Fv} lies in the valence band. Such a
situation is shown in Fig. 4.16. All of the calculations performed above can be
repeated. It is clear that the result will differ only by the sign. The sign change of the
specific conductivity will lead to a situation in which, instead of being damped, the
radiation in a such medium will be amplified. This idea was published by Nikolay
Basov as long ago as 1959 (i.e., before the appearance of the laser) and in 1961 he

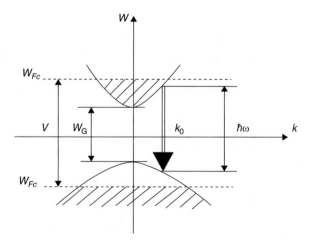

Fig. 4.16 Nonequilibrium electron distribution in semiconductor bands. The Fermi quasilevel for
electrons W_{Fc} is located inside the conduction band. At zero temperature, all permitted energy levels
below this level will be occupied. In the valence band, all permitted levels above the Fermi
quasilevel W_{Fv} will be free (occupied by holes). In such a nonequilibrium system the induced
absorption gives place to induced radiation and the medium-specific conductivity becomes negative

provided a method of amplification obtained in structures containing degenerate semiconductors.[10]

The situation in real degenerate semiconductors is more complicated. To place the Fermi level, for instance, in the conduction band, the donor impurity concentration should be very high. As a result, essential variations in energy bands arise in comparison with an intrinsic semiconductor, which, in particular, lead to the fact that transitions between any energy levels become permitted. Therefore, integration over energy will be necessary at determination of the transition number. Moreover, it is necessary to introduce some corrections into our considerations for nonzero temperatures.

To find out the number of transitions upward, we must take into account the number of particles really located in the valence band at the level with energy W_{Fv}. For this, we need to multiply (4.58) by the Fermi distribution function $F(W)$ (1.68). After that, because of the impossibility of more than one particle being in the same state, we must take into account the number of free places in the conduction band. Then, instead of (4.62), we obtain

$$N_{up} = \int \frac{(2m_r)^{3/2}}{\pi\hbar^4} \left|\widehat{H}_{vc}(W,\omega)\right|^2 F(W - \hbar\omega)N_v(W - \hbar\omega)(1 - F(W))N_c(W)dW$$

and similarly

$$N_{dw} = \int \frac{(2m_r)^{3/2}}{\pi\hbar^4} \left|\widehat{H}_{vc}(W,\omega)\right|^2 F(W)N_{\tilde{n}}(W)(1 - F(W - \hbar\omega))N_v(W - \hbar\omega)dW.$$

Now the absorbed power is determined by the difference in the transition numbers upward and downward. Amplification will happen under condition $N_{up} > N_{dw}$, which is satisfied if $W_{Fc} - W_{Fv} > \hbar\omega$. These rather simple considerations act as an initial point in creation of semiconductor quantum generators.

4.9   Problems for Chapter 4

4.1. Electrons in metal are well described by the model of a degenerate electron gas. Derive a formula for determination of the Fermi energy of the metal at specified values of the electron concentration and the effective mass.

Assistance: The density of permitted states of the electron gas $N(W) = \frac{(2m^*)^{3/2}}{2\pi^2\hbar^3} W^{1/2}$; the Fermi function at $T = 0$ is equal to 1 for $0 < W < W_F$.

---

[10]See (1.12) regarding a degenerate gas of fermions.

4.2. Using the result of task 4.1, find the electron concentration in copper on the basis of the known values of $W_F = 7\,eV$ and $m^* = 1.5 m_e$. Calculate the value of the plasma frequency and the corresponding wavelength.

4.3. At $T = 300$ K the specific conductivity of copper is 5.7×10^7 Sim/m. Using the value of the electron concentration obtained above, determine the effective frequency of the collisions.

4.4. A plain mirror is formed by evaporation of a copper film on a transparent substrate. Using the parameter values obtained in previous tasks, determine the reflection coefficient from the mirror at wavelengths of 10 μm, 0.5 μm, and 0.1 μm.

4.5. The electron mobility in pure GaAs is 0.85 m²/(V · s). The effective electron mass is 0.07 m. Determine the effective frequency of the collisions. Besides "light" electrons in GaAs, there are "heavy" ones with a mass of 1.2 m. Find the drift velocity of the two types of carriers in a field of 1 V/m. What will the specific conductivity be if all of the electrons become "heavy"?
   Note: For this, you will need energy of about 0.36 eV. The phenomenon you will discover is known as the Gunn effect.

4.6. Figure 4.9 shows the function of absorption in germanium on a frequency of 24 GHz versus the intensity of magnetic induction. Determine the effective masses of all kinds of charge carriers. Find the effective frequency of collisions.

4.7. Derive a formula for the specific conductivity on DC $\rho(\omega = 0, B)$ of some hypothetical n-type semiconductor located in a permanent magnetic field. In the absence of the magnetic field, $\rho(\omega = 0, B = 0) = 0.01$ Ohm × m, and the effective frequency of collisions is 10^{12}. Calculate $\rho(\omega = 0, B = 0.3$ Tl).
   Assistance: Use the electron gas model and, for simplicity, assume that ρ is a scalar function.

4.8. Discuss the possibility of using pure crystals of germanium and silicon as windows for a CO_2 laser. The widths of the forbidden bands are 0.7 eV for Ge and 1.14 eV for Si. The charge carrier concentrations at $T = 300$ K are 5×10^{19} and 2×10^{16}, respectively. The optical phonon energy of germanium is 0.034 eV and that of silicon is 0.12 eV. The relative dielectric permeability of the lattice is $\varepsilon = 16$.

4.9. Investigate the function of the reflection coefficient of radiation with a wavelength of 10 μm versus the concentration of free electrons. The relative dielectric permeability of the lattice is $\varepsilon = 16$.
   Instruction: Use the electron gas model. For performing calculations and drawing graphs, use a computer.

4.10. Solve task 4.9 for radiation with a wavelength of 0.9 μm falling on a silicon surface. The width of the forbidden band of Si is 1.14 eV.
   Assistance: Take into account the contribution to the complex dielectric permeability of fundamental absorption, for which $\alpha_{PO} \approx 2 \cdot 10^6 (\hbar\omega - W_G)^{1/2}\,m^{-1}$.

4.11. Hydrostatic pressure changes the width of the forbidden band. For silicon, $\frac{dW_G}{dP} \approx 1.5 \times 10^{-11}$ eV/Pa. A semiconductor is subjected to compression at a pressure of 10,000 atmospheres (10^{10} Pa). How much will the edge of the fundamental absorption width be shifted?

4.12. A permanent electric field causes inclination of the band edges, which is manifested in a shift in the width edge of fundamental absorption. This phenomenon is called the Keldysh–Frantz effect. The shift is approximately equal to

$$\Delta W = 1,5 \left(\frac{\hbar^2 E^2}{em} \right)^{1/3} eV.$$

Calculate the variation in damping factor α at a wavelength of 0.9 µm for silicon in a field of 10^6 V/m.

Chapter 5
Quantum Amplifiers and Generators

5.1 Introduction: Population Inversion

Under conditions of heat equilibrium, the number of particles with higher energy is always smaller than the number with lower energy. Therefore, $n_{12} > 0$, $\sigma > 0$, $P_{abs} > 0$ and a wave propagating in a medium is damped. Nevertheless, the equilibrium condition can be disturbed by use of an additional energy source (a pumping source). Under definite conditions the number of particles on the upper level becomes greater than in the lower level. As a result, $n_{12} > 0$, $\sigma > 0$, $P_{abs} < 0$ and instead of absorption we may observe electromagnetic wave amplification. Such a state is called population inversion or a state with a negative temperature. A medium with inversion is often called an active medium.

Originally, inversion was obtained in molecular beams by using particle assortment with the help of the heterogeneous space of an electric or magnetic field. In 1956, Nicolaas Bloembergen[1] offered a so-called three-level method for obtaining inversion.[2] Inversion was provided by particle transitions between additional levels under the action of a pumping source. A four-level system is often used for the operating transition inversion instead of a three-level one. The process of inversion

---

[1]Nicolaas Bloembergen (1920–2017) was a Dutch–American physicist. He is known for his findings on nuclear magnetic resonance and electron paramagnetic resonance, quantum electronics, and nonlinear optics. He shared the 1981 Nobel Prize in Physics with Arthur Schawlow and Kai Siegbahn for work on spectroscopy. He was awarded prizes by the American Physical Society and the Institute of Radio Engineers, the US President's National Medal of Science, and the Frederic Ives Medal from the Optical Society of America. He was a member of the US National Academy of Sciences and the Dutch Royal Academy of Sciences.

[2]In 1955, Nikolay Basov and Alexander Prokhorov offered a three-level pumping method in conformity with a molecule gas. However, it was not practically realized, apparently because of the absence of suitable generators to use as the pumping source. Generalization to a solid body was also not realized. In 1962 a gas laser was created, using cesium vapor, in which a three-level scheme with optical pumping from a helium gas discharge lamp was used. Generation was obtained at wavelengths of about 7 and 3 μm.

© Springer Nature Switzerland AG 2020
V. V. Shtykov, S. M. Smolskiy, *Introduction to Quantum Electronics and Nonlinear Optics*, https://doi.org/10.1007/978-3-030-37614-7_5

Fig. 5.1 Inversion of population levels at the operating transition. The pumping source fully equalizes the populations of levels *1* and *3*. As a result, population inversion occurs at the operating transition (1 ↔ 2). To find an accurate answer it is necessary to solve a system of three differential equations

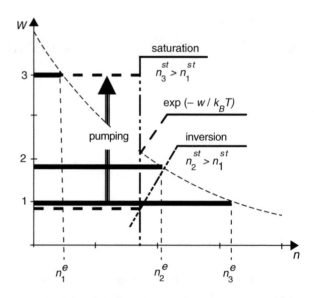

formation can be described by differential equations of population level variation in time. These equations are sometimes called balance equations. However, in essence, inversion at the operating transition is formed as a result of saturation (see Sect. 2.12) of the auxiliary transition, as shown in Fig. 5.1.

Bloembergen's suggestion was implemented in the centimeter range in parametric amplifiers. An additional coherent generator was used as the pumping source. There were initially no coherent light sources in the optical range; therefore, a xenon lamp served as the pumping source in the first optical quantum generator. After that, researchers began to use electron impacts, collisions of atoms and molecules, chemical reaction energy, and other mechanisms to obtain inversion.

Within the limits of our students' lecture course it is unnecessary to discuss the finer points of the process of inversion creation. To describe the process of population inversion development in time, we shall use the following simple model.

In the final analysis, independently of the process details in the multilevel quantum system, the evolution of the population level difference at the operating transition can be taken into consideration by adding some equivalent "source" in Eq. (2.25). This fictitious source has no direct influence on the field of the operating frequency and is not manifested at all during the solution of the Maxwell equations. Within the limits of such a model, the action of the pumping source with the specific energy W_{pump} can be described by

$$\frac{dn_{12}}{dt} + \frac{n_{12} - n_{12}^e}{T_1} = -\frac{2W_{\text{pump}}}{\hbar\omega_{21}}.$$

In the stationary mode the population difference takes the value

Fig. 5.2 The function of the level populations normalizes to the equilibrium value n_{12}^e versus normalized time. The initial condition corresponds to heat equilibrium. The pumping intensity is $n_{12}^{st} = -2n_{12}^e$. Inversion occurs after time approximately equal to $0.5 \cdot T_1$

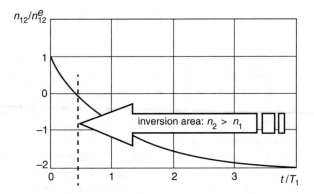

$$n_{12}^{st} = n_{12}^e - \frac{2T_1 W_{pump}}{\hbar \omega_{21}}.$$

If we assume that n_{12}^{st} has a specified value, the equation for the population difference can be written as

$$\frac{dn_{12}}{dt} + \frac{n_{12}}{T_1} = \frac{n_{12}^{st}}{T_1}. \tag{5.1}$$

This equation simulates the switching-on of the pumping source with constant intensity at the time moment $t = 0$. It must be solved for the initial condition $n_{12}(0) = n_{12}^e$. The transient process for the inversion condition is shown in Fig. 5.2, in which the function of the normalized population difference n_{12}/n_{12}^e versus normalized time t/T_1 is presented.

In the real case, the transients in the system as a whole will affect the process of inversion creation. This can be taken into consideration by modifying Eq. (5.1) as follows:

$$\frac{dn_{12}}{dt} + \frac{n_{12}}{T_1} = \frac{n_{pump}(t)}{T_1}.$$

Equation (5.1) is used everywhere below as the model of the pumping source. To emphasize the specific role of the population difference in quantum amplifiers and generators, we transfer from $n_{12} = n_1 - n_2$ to the quantity $n_{21} = n_2 - n_1$, which is positive in the condition of population inversion.

Taking into account the actions of the pumping source, the material equations (2.24) and (2.25) take the form

$$\frac{d^2 \mathbf{P}}{dt^2} + \frac{2}{T_2} \frac{d\mathbf{P}}{dt} + \omega_{12}^2 \mathbf{P} = -\frac{2\omega_{21} d^2}{3\hbar} n_{21} \mathbf{E}; \tag{5.2}$$

$$\frac{dn_{21}}{dt} + \frac{n_{21} - n_{21}^{st}}{T_1} = \frac{2}{\hbar \omega_{21}} \frac{d\mathbf{P}}{dt} \mathbf{E}. \tag{5.3}$$

Equation (5.2) needs no specific comments. Equation (5.3) describes the saturation effect that leads to equalization of the level population (see Sect. 2.12). Oscillations arising in the quantum generator contradict the pumping source. As a result, at some time after the start of its operation, a dynamic equilibrium will be established, which will define the stationary parameters of the generator. The process of the generator's evolution to the steady-state mode can be found during the solution of the Maxwell equations together with Eqs. (5.2) and (5.3) only. It is clear that this is a complicated mathematical problem. However, some characteristics of quantum devices can be determined from simpler models.

5.2   Linear Amplification of Electromagnetic Waves

Let us examine the process of amplification of electromagnetic waves by the inverted two-level system. We assume that along the whole length, the field intensity is so low that we can use the linear approximation. In this case it follows from (5.3) that $n_{12} = -n_{21}^{st}$. Substituting this quantity into the appropriate formulas in Sect. 2.6, we obtain the complex dielectric susceptibility of the active medium:

$$\dot{\chi}_a(\omega) = j \frac{\chi_a}{1 + j\xi}, \tag{5.4}$$

where $\chi_a = \frac{d^2 n_{21}^{st} T_2}{3\hbar}$. By analogy with (2.31) we introduce the conductivity of the active medium:

$$\sigma_a = \frac{\chi_a \omega}{1 + \xi^2}. \tag{5.5}$$

In the active medium the energy of the source is transferred to the electromagnetic field through the mechanism of quantum transitions and under definite conditions the natural wave damping transfers to its amplification.

The space region with length L, in which the plain wave propagates, can be considered the simplest model of the distributed quantum amplifier of the traveling wave. Let us determine the characteristics of such a device.

At propagation of the plain wave in the medium, the complex amplitude of the electric field changes in accordance with the law

$$\dot{E}(z) = \dot{E}(0) \exp(-j\dot{\gamma}z),$$

where $\dot{\gamma}$ is the propagation constant equal to

$$\dot{\gamma} = \omega\sqrt{\mu_0 \dot{\varepsilon}_a} = \beta - j\alpha.$$

Now we can write the complex dielectric permeability of the active medium as follows, using (5.4):

$$\dot{\varepsilon}_a = \varepsilon_0 + j\frac{\chi_a}{1 + j\xi}.$$

This results in the following form of the propagation constant:

$$\dot{\gamma} = \frac{\omega}{c}\sqrt{1 + j\frac{1}{\varepsilon_0}\frac{\chi_a}{1 + j\xi}},$$

where $c = \frac{1}{\sqrt{\varepsilon_0\mu_0}}$ is the light speed.

We introduce the complex transfer coefficient of the amplifier with the length:

$$\dot{K}(j\omega) = \frac{\dot{E}(L)}{\dot{E}(0)} = \exp(-j\dot{\gamma}L). \tag{5.6}$$

If $\varepsilon_0 \gg \chi_a$, we may use the approximated expression for the square root. In this case the propagation constant is

$$\dot{\gamma} = \frac{\omega}{c}\left(1 + j\frac{1}{2\varepsilon_a}\frac{\chi_a}{1 + j\xi}\right),$$

the phase constant is

$$\beta = \frac{\omega}{c}\left(1 + \frac{1}{2\varepsilon_0}\frac{\chi_a\xi}{1 + \xi^2}\right), \tag{5.7}$$

and the damping constant is

$$\alpha = -\frac{1}{2}\frac{\chi_a\omega}{1 + \xi^2}Z_{med}. \tag{5.8}$$

Here, $Z_{med} = \sqrt{\frac{\mu_a}{\varepsilon_a}}$ is the characteristic impedance of the medium.

Let us substitute (5.7) and (5.8) in (5.6) and determine the modulus of the transfer coefficient:

$$\left|\dot{K}(j\omega)\right| = \exp\left[\frac{1}{2} \cdot \frac{\chi_a \omega}{1+\xi(\omega)^2} Z_{med} L\right] \tag{5.9}$$

and its phase:

$$\Phi(\omega) = -\frac{\omega}{c}\left(1 + \frac{1}{2\varepsilon_0} \frac{\chi_a \xi(\omega)}{1+\xi(\omega)^2}\right) L. \tag{5.10}$$

During analysis of electromagnetic wave propagation it is often sufficient to track the active power flow $\mathbf{I} = \frac{1}{2}\operatorname{Re}\left(\dot{\mathbf{E}} \times \dot{\mathbf{H}}^*\right)$. At propagation of the plain wave, the power flow changes according to the law $I(z) = I(0)\exp[\alpha_p z]$, where $\alpha_p = 2\alpha$ is the amplification coefficient of the power. The transfer coefficient of the quantum amplifier is

$$K_p = \exp\left[\alpha_p L\right], \tag{5.11}$$

where $\alpha_p = \frac{\chi_a \omega}{1+\xi^2} Z_{med}$.

Figure 5.3a shows the amplitude–frequency characteristic (5.9) of the quantum amplifier on the ammonia molecules. The numerical values necessary for the calculations are taken from Chap. 2. The square of the transition dipole moment is accepted as being equal to $1.2 \cdot 10^{-59}\ C^2 \cdot m^2$, and relaxation time $T_2 = 6.2 \cdot 10^{-9}$ s. Calculations have been provided for $L = 1$ m for four values of population differences. The point $\xi = 0$ corresponds to the frequency $f_{21} = 23.866$ GHz or $\lambda_{21} = 1.26$ cm. The phase characteristic is presented in Fig. 5.3b as the function

$$\varphi(\xi) = -\frac{1}{2\varepsilon_0}\frac{\chi_a \xi}{1+\xi^2}\frac{2\pi L}{\lambda_{21}}.$$

The amplification coefficient of the transition frequency is units of percentage. This is connected with the fact that the gas is at low pressure and the molecule concentration at the operating transition is low. For the same reason, the phase deflection on the bandwidth boundary does not exceed 0.02 rads (about 1.2°). However, it is quite sufficient for creation of the first quantum generator.

The particle concentration is several orders higher in a solid body than in a gas. Therefore, amplification factor α_p can achieve a value of 1–$10\ m^{-1}$. However, there are examples of active media with large gains among gases. So, in a mixture of helium and neon at a pressure of about 1 kPa (about 10 mm of a mercury column) for pumping by an electric discharge at the wavelength 3.39 μm, $\alpha_p \approx 2 - 5\ m^{-1}$ ($K_p(1m) \approx 10 - 100$).

Fig. 5.3 Dependence of the modulus (**a**) and the phase (**b**) deflections from the resonance value of the transfer function of the amplifier on ammonia molecules versus generalized detuning for several values of population difference: *1* denotes $n_1^{st} = 5 \cdot 10^{18} \mathrm{m}^{-3}$, *2* denotes $n_1^{st} = 2.5 \cdot 10^{18} \mathrm{m}^{-3}$, *3* denotes $n_1^{st} = -2.5 \cdot 10^{18} \mathrm{m}^{-3}$, and *4* denotes $n_1^{st} = -5 \cdot 10^{18} \mathrm{m}^{-3}$. It can be seen how absorption transfers to amplification with the appearance of population inversion

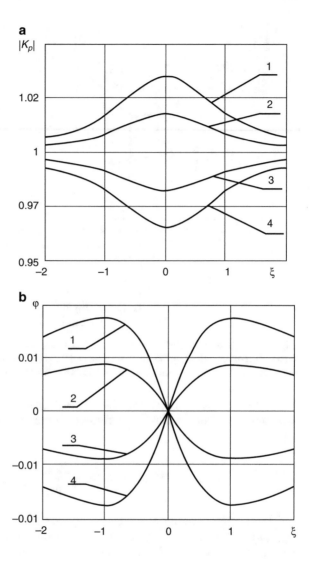

5.3   Regenerative Amplification

The gain may be increased if we place the active medium in a resonator. Really, the field in the resonator can be considered as a wave that propagates in the short-circuited segment of the transmission line reflecting from the "mirrors." Multiple passes of the wave through the active medium increase the gain. We shall find the most important characteristics of such an amplifier.

Let us examine an amplifier with a transmission resonator. A physical model of an amplifier with a transmission-type resonator is shown in Fig. 5.4a. We can associate the process of multiple reflections with the structural circuit shown in

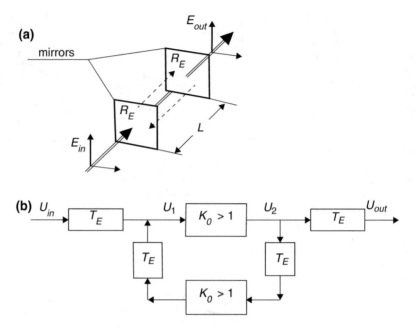

Fig. 5.4 Physical model of an amplifier with a transmission resonator (**a**) and its structural circuit (**b**). Mirrors reflecting the incident waves and the active medium act as elements of feedback. At $K_0 > 1$ the feedback is positive and the device's amplification factor increases

Fig. 5.4b. That circuit shows that the resonator quantum amplifier is a system with feedback. The complex transfer coefficient of the system surrounded by feedback is equal to

$$\dot{K}_{FB}(j\omega) = \frac{\dot{K}(j\omega)}{1 - \dot{K}(j\omega)\dot{k}_{FB}(j\omega)}.$$

The complex transfer function $\dot{K}(j\omega)$ has already been found above, and $\dot{k}_{FB}(j\omega) = R_E^2 \dot{K}(j\omega)$. We take into consideration that according to the energy conservation law, $R_E^2 = R$ is the reflection coefficient of the power and $T_E^2 = 1 - R$. From the structural circuit we can easily find that the complex transfer function of the resonator amplifier is

$$\dot{K}_p(j\omega) = \frac{(1 - R)\dot{K}(j\omega)}{1 - R\dot{K}^2(j\omega)}.$$

Substituting (5.6), we obtain

$$\dot{K}_p(j\omega) = \frac{(1-R)\exp(-j\dot{\gamma}L)}{1 - R\ \exp(-2j\dot{\gamma}L)}.$$

Taking (5.11) into account, we obtain the transfer coefficient of the power:

$$K_p = |K(j\omega)|^2 = \frac{(1-R)^2 K_p}{\left(1 - RK_p\right)^2 + 4RK_p^2 \sin^2(\beta L)}. \qquad (5.12)$$

The transfer function achieves its maximum when

$$\beta L = n\pi. \qquad (5.13)$$

This is the well-known condition of resonance. If it is satisfied, then

$$K_p = K_p\left[\frac{1-R}{1-RK_p}\right]^2. \qquad (5.14)$$

At $R = 0$ we go to the well-known result. In an active medium, $K_p > 1$; therefore, with growth of R the transfer function grows as the positive feedback partially compensates for the losses. This process is called regeneration. Thus, in a resonator amplifier we deal with the process of regenerative amplification.

At $RK_p = 1$ the amplifier transfer function $K_p = \infty$, which is evidence of its instability and capability for self-excitation. It is clear that using (5.11) we can obtain the threshold population difference corresponding to the stable mode boundary. However, this result will have a more universal character if we express the threshold condition through the resonator Q factor, which is one of the general characteristics of any oscillation system. The Q factor is accepted as being defined as

$$Q = \frac{\omega W_{\text{acc}}}{P_{\text{loss}}},$$

where W_{acc} is the energy accumulated in the resonator and P_{loss} is the power loss. Let the Q factor of the empty resonator be specified. Taking into account that

$$W_{\text{acc}} = \frac{\varepsilon_a}{2}\int_V |\dot{E}|^2 dV,$$

we obtain the power loss:

$$P_{loss} = \frac{\varepsilon_a \omega}{2Q} \int_V |\dot{E}|^2 dV.$$

Here, the integration must be calculated over the whole volume V of the resonator.

The output power consumed by the resonator from the active medium is expressed through the conductivity of the active medium:

$$P_a = \frac{\sigma_a}{2} \int_{V_a} |\dot{E}|^2 dV.$$

Now the integration is taken over the volume V_a occupied by the active medium.

The threshold population difference can be found from the condition $P_{loss} = P_a$. If, for simplicity, we consider that the active medium uniformly fills the whole volume of the resonator ($V_a = V$), the threshold condition takes the form

$$\frac{\omega \varepsilon_a}{Q} = \sigma_a.$$

To find the threshold population difference n_{21}^{thr} we use (5.5). At resonance $\xi = 0$ ($\omega = \omega_{12}$),

$$n_{21}^{thr} = \frac{3\varepsilon_a \hbar}{Qd^2 T_2}. \tag{5.15}$$

Let us estimate the value of the threshold population difference for ammonia. Substituting values of $\varepsilon_a = \varepsilon_0$, $d^2 = 1.2 \cdot 10^{-59}$ C^2 m^2, $T_2 = 6.5 \cdot 10^{-9}$s, and $Q = 10^4$, we obtain $n_{21}^{thr} \cong 4 \cdot 10^{18} m^{-3}$. As we can see, it is commensurable with the equilibrium value $n_{12}^e \cong 6.5 \cdot 10^{18}$ m^{-3}.

If this threshold was overcome, the regenerative amplifier would change into a generator. The first quantum generator was created using an ammonia molecular beam. For its excitation it is necessary to remove 0.1% of the particles approximately located on the lower level (see Sect. 2.7).

The power of the molecular generator is extremely low, but measurements have shown that the short-term relative instability of its frequency is less than 10^{-10}. The reason is that in complicated self-oscillating systems with a variety of resonances, in which the quantum generator is concerned, the frequency of continuous waves is determined by the resonance system with the larger Q factor.

In the radiofrequency range the Q factor of the amplification line is $Q_L \approx 10^8$ and the Q factor of the resonator is $Q \approx 10^4$. Therefore, the frequency stability of the molecular generator oscillations is determined by the frequency stability of the quantum transition. In the optical range, obtaining high stability of the oscillation frequency represents a complex enough problem. This is connected with the fact that in this range, $Q \gg Q_L$ and several proper frequencies of the resonator fall into the amplification band. Therefore, unavoidable random perturbations of the resonator (for instance, random variations in the length) lead to commensurable variations in the oscillation frequency. We shall return to this issue when we have equations at our disposal that describe the dynamics of the quantum generator.

5.4   Influence of the Saturation Effect on Quantum Amplifier Features

The equation system for a two-level quantum system is nonlinear. As shown in Sect. 2.12, this causes the saturation effect to arise, in which, with growth of the power flow density of the electromagnetic wave, the population difference aspires to zero. Of course, this is fully concerned with the active medium.

Let us examine the influence of saturation on the characteristics of the quantum amplifier by using the equation of the energy conservation law of the electromagnetic field—the Poynting theorem. For the active part of the power density, the differential form of this equation takes the form

$$\mathrm{div}\mathbf{I} = -P_{\mathrm{loss}},$$

where P_{loss} is the specific loss power in the medium. For a plain wave propagating in an active medium,

$$\frac{dI}{dz} = \alpha_p I. \tag{5.16}$$

In linear approximation the amplification factor α_p is proportional to the population difference n_{21}^{st}. System nonlinearity will be demonstrated in the fact that with growth of the power flow density, the population difference will decrease (see Sect. 2.12). For an active medium, by analogy with (2.41),

$$n_0 = \frac{n_{21}^{\mathrm{st}}}{1 + \dfrac{\left|\dot{E}\right|^2}{\left(1+\xi^2\right)E_{\mathrm{sat}}^2}}.$$

Therefore, Eq. (5.16) can be written as follows, taking into account the saturation effect for $\xi = 0$:

$$\frac{dI}{dz} = \frac{\alpha_0 I}{1 + I/I_{\mathrm{sat}}}. \tag{5.17}$$

Here, α_0 is the amplification factor in the linear approximation and I_{sat} is defined by (2.43).

The solution of Eq. (5.17) does not present great difficulties. If in the plane $z = 0$ the power flow is equal to $I(0)\Pi(0)$, the integration of (5.17) gives

$$\ln\left(\frac{I(L)}{I(0)}\right) + \frac{I(L) - I(0)}{I_{sat}} = \alpha_0 L,$$

which can be transformed into the equation for K_p determination:

$$\ln\left[K_p(I)\right] + \left[K_p(I) - 1\right]\frac{I(0)}{I_{sat}} = \alpha_0 L. \tag{5.18}$$

Figure 5.5a shows the function of the gain versus the ratio $I(0)/I_{sat}$ obtained on the basis of (5.18). The saturation effect decreases the gain from its maximal value $K_p(0) = K_0 = \exp(\alpha_0 L)$ to $K_p(\infty) = 1$.

Fig. 5.5 Function of the power transfer coefficient versus the normalized power flow. (**a**) Accurate function: *1* denotes $\alpha_0 L = 1.0$, *2* denotes $\alpha_0 L = 0.5$, and *3* denotes $\alpha_0 L = 0.25$. (**b**) Comparison of exact (*solid*) and approximated (*dotted*) functions $K_p(I)$ for $\alpha_0 L = 0.25$. The saturation effect equalizes the level populations and decreases the gain

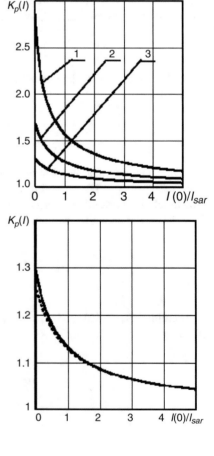

In the future, we shall need the function of the gain versus the power flow written in an explicit form to describe quantum generators. The approximated expression for $K_P(I)$ can be found under the condition of low amplification when $K_P \approx 1$. As we have already seen, this is true for gaseous media, at least. However, even if the amplification factor of the active medium $\alpha_0 > 1$, the value of $\alpha_0 L$ in quantum generators is less than the unit and $K_P(I) < K_0 \cong 1$. This is connected with the fact that harmonic oscillations with a high degree of coherence can be obtained only by using resonators with a high Q factor.

A small difference in $K_p(I)$ from the unit allows a change in the logarithm in (5.18) by the approximated expression $\ln(1 + \Delta K) \approx \Delta K$, where $\Delta K = 1 - K_p$. Solving the approximated equation, we obtain the gain in power:

$$K_p(I) = 1 + \frac{\alpha_0 L}{1 + I(0)/I_{\text{sat}}}. \tag{5.19}$$

Formula (5.19) approximates well the accurate function $K_P(I)$ and can be used in a wide range of values, $I(0)/I_{\text{sat}}$ (Fig. 5.5b).

5.5 Self-Excitation Conditions and the Power of Continuous Oscillations

Earlier, we discussed the issue of regenerative amplifier stability. If $RK_P > 1$, the oscillation amplitudes in the resonators, as a result of heat emission, begin to increase without limit. Nevertheless, during their growth, the saturation effect begins to play a larger role (see Sect. 2.12). The gain reduces and the oscillation amplitude finally achieves some stationary value. Thus, a mode of continuous periodic oscillations occurs in the resonator. Let us examine the features of this mode. For this, we use the Poynting theorem.

Let us present the electromagnetic field in the resonator in the form of a two-wave superposition propagating along the resonator longitudinal axis. Such a presentation of oscillation types in resonators is well known. We can use the waves of the wave guide, which, with a short-circuit from the two end segments, forms the resonator, as partial traveling waves. So, for example, for an open resonator we can use waves of the lens wave guide as partial waves. Such a presentation allows application of the method used earlier for description of a quantum amplifier.

For simplicity, we shall use partial waves in the form of plain electromagnetic waves. Let $I_1(z)$ be the power flow of the wave propagating in the direction of the z axis, and let $I_2(z)$ be the power flow of the wave propagating toward the z axis. Then, assuming that the saturation effect depends upon the total power, we write the following equation system:

$$\frac{dI_1}{dz} = \frac{\alpha_0 I_1}{b(\omega) + a(I_1 + I_2)};$$

$$\frac{dI_2}{dz} = -\frac{\alpha_0 I_2}{b(\omega) + a(I_1 + I_2)}. \tag{5.20}$$

Here, $a = I_{sat}^{-1}$, $b(\omega)$ considers the system frequency properties (see (2.13)).

Equation system (5.20) is the differential form of the energy conservation law in a quantum generator. Therefore, these equations are often called balance equations.[3]

The solution to system (5.20) must satisfy the boundary conditions at points $z = 0$ and $z = L$. We write them assuming that the reflection factors from the mirrors are known. In a quantum generator, reflection factor R from one of the mirrors (for instance, the left one) is usually equal to about 1. Then, at the point $z = 0$, $I_1(0) = I_2(0)$. At the point $z = L$, $I_2(L) = RI_1(L)$. At $R = 1$ the losses in radiation are absent and the Q factor of the resonator becomes infinite. In a real resonator, part of the energy always leaves the volume. In open resonators this is connected with electromagnetic wave diffraction on the mirror; in other types of resonators, part of the energy is absorbed by the walls. In both cases, losses can be taken into consideration by adding into the boundary conditions the energy loss part δ. Taking into account the additional losses, we can write the boundary conditions in the form

$$I_1(0) = I_2(0), I_2(L) = I_1(L)(R - \delta). \tag{5.21}$$

It follows from (5.9) that

$$\frac{dI_1}{I_1} = -\frac{dI_2}{I_2}.$$

This means that the product of the flow densities of the partial waves is constant at all points of the volume:

$$I_1(z)I_2(z) = C_1 \text{ or } I_2(z) = C_1/I_1(z). \tag{5.22}$$

Using the boundary conditions (5.21), we can write

$$C_1 = I_1(0)^2 = I_1(L)^2(R - \delta). \tag{5.23}$$

Substituting (5.22) in the first equation of the system (5.20), we obtain

---

[3]Do not confuse these equations with equations for level populations, which are often also called balance equations. Equation (5.9) is purely classical, and the equations for populations are quantum ones.

$$\frac{dI_1}{dz} = \frac{\alpha_0 I_1}{b + a(I_1 + C_1/I_1)}.$$

This is the differential equation with separable variables. Its solution, satisfying the boundary conditions for $z = 0$, takes the form

$$b \ln \left(\frac{I_1(z)}{I_1(0)}\right) + aI_1(z)\left(1 - \frac{I_1^2(0)}{I_1^2(z)}\right) = \alpha_0 z.$$

In the section $z = L$ we have

$$b \ln \left(\frac{I_1(L)}{I_1(0)}\right) + aI_1(L)\left(1 - \frac{I_1^2(0)}{I_1^2(L)}\right) = \alpha_0 L.$$

It follows from (5.23) that

$$\frac{I_1^2(0)}{I_1^2(L)} = R - \delta.$$

Therefore,

$$-\frac{b}{2} \ln (R - \delta) + aI_1(L)(1 - R + \delta) = \alpha_0 L. \qquad (5.24)$$

Solving (5.24), we obtain the power flow in the section $z = L$:

$$I_1(L) = \frac{2\alpha_0 L + b \ln (R - \delta)}{2(1 - R + \delta)} I_{\text{sat}}. \qquad (5.25)$$

The values $I_1(L) \geq 1$ have a physical sense only. This is possible if

$$2\alpha_0 L + b \ln (R - \delta) \geq 0.$$

From this inequality the threshold condition for self-excitation of the quantum generator follows, which simultaneously corresponds to the boundary of the stable mode of the regenerative amplification (5.14), $RK_p = 1$. At $b(\omega) = 1$ we have

$$\exp (\alpha_0 L)\sqrt{R - \delta} = 1. \qquad (5.26)$$

The first item in (5.26) is the gain in the linear approximation for the single pass of the resonator, and the second item is a "feedback factor."

The obtained value of the power flow density of the wave allows determination of the output power of the quantum generator:

Fig. 5.6 Function of quantum generator power versus the mirror reflection factor under conditions of resonance $b(\omega = \omega_{21}) = 1$ for $\alpha_0 L = 0.1$, cross-section $S_{\text{cross}} = 1 \text{ cm}^2$, and saturation parameter $I_{\text{sat}} = 0.1 \text{ W/cm}^2$: *1* denotes $\delta = 0$, *2* denotes $\delta = 0.002$, *3* denotes $\delta = 0.01$, and *4* denotes $\delta = 0.02$. The initial data correspond approximately to those from the first gas discharge laser using a mixture of neon and helium with an operating wavelength of 1.15 μm

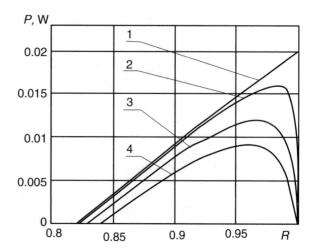

$$P_{\text{out}} = S_{\text{cross}} I_1(L)(1 - R) = S_{\text{cross}}(1 - R)\frac{2\alpha_0 L + b\ln(R - \delta)}{2(1 - R + \delta)} I_{\text{sat}}, \qquad (5.27)$$

where S_{cross} is the cross-sectional area of the resonator.

Figure 5.6 shows the output power function versus the reflection factor of the mirror. With R growth greater than its threshold value, according to (5.25) the power flow falling on the output mirror increases but its transmission simultaneously decreases $(1 - R)$. As a result, there is a value of the reflection factor at which the output power is maximal. This value corresponds to the optimal coupling of the generator with an external medium (a load), when the output power is equal to the power dissipated inside the generator.

On the basis of (5.27), one can find the power limit of the generator. It is clear that P_{out} will be greater if the losses in the resonator are connected only with radiation through the mirror. In this case in (5.27), $\delta = 0$ and it becomes a monotonically increasing function that has a maximum at a maximally allowed reflection factor value equal to 1 (see Fig. 5.6). Hence, the limit value of the output power is

$$P_{\text{out}}^{\text{lim}} = \alpha_0 I_{\text{sat}} V,$$

where $V = S_{\text{cross}} L$ is the active medium volume. Using (5.8) and (2.43), we have the remarkably brief result

$$P_{\text{out}}^{\text{lim}} = \frac{n_{21}^{\text{st}} \hbar \omega_{21}}{2T_1} V,$$

from which we can draw a series of useful conclusions. Firstly, the power limit does not depend upon the transition dipole moment, i.e., upon the probability of induced transitions. Secondly, $P_{\text{out}}^{\text{lim}}$ depends upon the energy transfer speed only from the pumping source to the operating transition. The higher this speed is, the more oscillation power will be required for population equalization of operating levels. We can give two examples confirming these conclusions.

The amplification of a laser using a neon and helium mixture at a wavelength of 3.39 μm with a length of 1 m is more than 100 ($\alpha_0 \approx 5$ m^{-1}). A laser using carbonic acid gas has an amplification of about 1.1 ($\alpha_0 \approx 0.1$ m^{-1}). However, the difference in relaxation time T_1 values is greater than one order, leading to the fact that the output power obtained from the volume unit of the carbonic acid laser is more than an order greater than the value obtained from the laser using a neon and helium mixture.

5.6 Steady-State Mode Equations for a Quantum Generator

The balance equations (5.20) describe the energy characteristics of the steady-state mode of a quantum generator. We can add a more visual and known character to this description.

As we have already mentioned, a resonator filled with an active medium is a regenerative amplifier, which in the threshold condition (5.15) overcomes self-excitation. Since the gain depends upon the power flow, the system aspires to some steady-state mode. We can find the conditions of existence and the steady-state parameters of the quantum generator. For this we use the Poynting theorem, which determines the energy balance of the electromagnetic field.

Let us examine the quantum generator model whose geometry is presented in Fig. 5.7. Here, all types of losses are reduced to flows of output power I_{out} and to

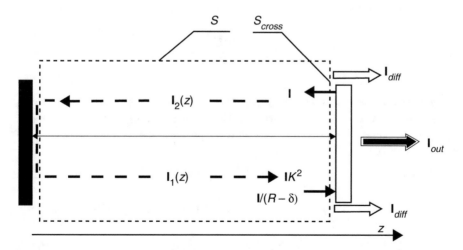

Fig. 5.7 Geometry of a quantum generator. The power flows, which are expressed through flow **I** coming from the right into volume V limited by cross-surface S_{cross}, are indicated. The power sum coming into the volume and the output power of the active medium are connected with energy W accumulated in it according to the Poynting theorem

diffraction losses I_{diff}. Therefore, we must add the power P_a generated by the active medium to the balance equation of the quantum generator. That is why, for our model, the Poynting theorem takes the form

$$\frac{dW_{\text{acc}}}{dt} = P_a - \oint_S I d\mathbf{S}. \tag{5.28}$$

Equation (5.28) does not contain some specific generator parameters. Therefore, it determines only the general regulations. This equation can be filled in a practical sense to introduce some heuristic field models into it.

First of all, we take into consideration that the resonator field represents the harmonic oscillation with a slowly changed amplitude. This will permit us to use relations that were obtained for harmonic oscillations, for calculation of quadratic quantities W_{acc}, P_a, $\mathbf{I}$ slowly changing in time. We shall also assume that the resonator fields take the form of plane waves.

Let us determine the Poynting vector through the closed surface that covers the internal cavity of the resonator (Fig. 5.7). The power coming into and going out from the allocated volume can be expressed through power flow $\mathbf{I}$, which comes in from the right:

$$\oint_S \mathbf{I} d\mathbf{S} = I\left(\frac{1}{R-\delta} - 1\right) S_{\text{cross}}, \tag{5.29}$$

where S_{cross} is the cross-sectional area.

At calculation of

$$P_a = \int_V \frac{\sigma_a |\dot{E}|^2}{2} dV,$$

we again use the model of plane partial waves. Since the field configuration in the resonator is close to the standing wave and its length is a multiple of $\lambda/2$, the value of P_a can be presented by the difference in wave power in the section $z = L$:

$$P_a = \left(K_p^2 - 1\right) I S_{\text{cross}}.$$

Together with (5.29), this gives the final form of the energy balance equation:

$$\frac{dW_{\text{acc}}}{dt} = \left(K_p^2 - \frac{1}{R-\delta}\right) I S_{\text{cross}}. \tag{5.30}$$

It follows from (5.30) that the system will be in equilibrium if

$$K_P(I) - \frac{1}{\sqrt{R - \delta}})I = 0. \tag{5.31}$$

According to (5.31), the quantum generator can be either in a quiescent state $I = 0$ or in a steady-state mode:

$$K_p(I) = \frac{1}{\sqrt{R - \delta}}. \tag{5.32}$$

The quiescent state cannot be realized physically, because equilibrium heat emission is always present. Therefore, it is more correct to speak about the damping oscillation state. Such a condition is possible if $\frac{dW_{acc}}{dt} < 0$. For this, it is necessary for the stability condition of the linear system to be satisfied:

$$K_p(0) < \frac{1}{\sqrt{R - \delta}},$$

which we have already discussed above (see (5.14)).

An equation similar to (5.32) is widely used in the oscillator theory for description of the steady-state mode. We use (5.18) and write

$$K_p(I) = 1 + \frac{\alpha_0 L}{b + 2aI}. \tag{5.33}$$

The multiplier 2 in the denominator is introduced because the saturation effect is determined by the total power of the partial waves.

Figure 5.8 shows the functions of the transfer coefficient (5.33) at $b = 1$ versus the normalized power flow for several population difference values. In the heat equilibrium state, $\alpha_0 < 0$. Transfer function $K_P < 1$. In the population inversion state, $\alpha_0 > 0$ and $K_p > 1$. The saturation effect equalizes the level populations and with $I(0)$ growth, $K_p(I) \to 1$.

Fig. 5.8 Function of the resonance transfer coefficient of the power versus the normalized power flow for $b(\omega = \omega_{21}) = 1$. The population difference corresponds to
(1) $\alpha_0 L = 0.05$,
(2) $\alpha_0 L = 0.02$,
(3) $\alpha_0 L = 0.01$, and
(4) $\alpha_0 L = -0.05$ (heat equilibrium)

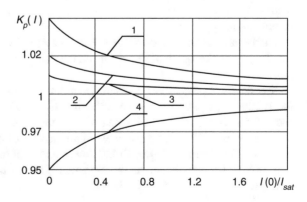

Fig. 5.9 Solution of the steady-state equation. At random deflection to point *1*, the energy going out to the resonator via the active medium will be greater than the losses on reflection, and the oscillations will increase, aspiring to return to point *0*. If the system is at point *2*, the direction of energy transfer changes to the opposite one. Therefore, point *0* is stable

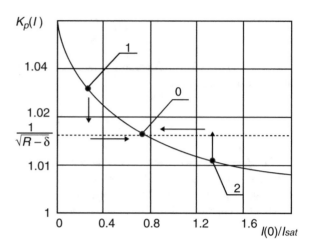

An illustration of a graphic solution to Eq. (5.32) is shown in Fig. 5.9. If the steady-state mode point exists, a question arises about the solution stability of Eq. (5.32). Addressing (5.30), it is easy to determine that at $K_p(I) > \frac{1}{\sqrt{R-\delta}}$ the energy transferring to the resonator from the active medium is greater than the power loss, so the oscillation power increases. If $K_p(0) < \frac{1}{\sqrt{R-\delta}}$, the oscillations' power decreases. Hence, the steady-state mode is stable. If we expand the function $K_p(I)$ into a power series in the vicinity of the steady-state point, we may strictly prove that $\frac{dW_{acc}}{dt} < 0$ if $\frac{dK_p}{dI} < 0$ at this point.

In the self-oscillator theory we can distinguish two modes of self-excitation—soft and rigid—in connection with the stability problem solution. The difference in these modes is manifested by the function of the steady-state value of the oscillation amplitude, for instance, versus the feedback value. The case considered by us concerns the soft mode of quantum generator self-excitation.

In the oscillator the rigid excitation mode arises if the gain function represents a nonmonotonous function of the amplitude. In our case, to obtain the rigid mode it is necessary that $K_p(I)$ changes as shown in Fig. 5.10. Without stopping on the way to realization of such a function, we would like to remind the reader about the main features of the rigid mode.

Firstly, at several values of the reflection factor there are two solutions to the steady-state mode equation. The points on the rising branch of the function $K_p(I)$ are unstable.

Secondly, the presence of two solution points essentially changes the function of the oscillation power versus the reflection factor. For reflection factor growth, oscillator self-excitation will happen only in the case where the line $(R - \delta)^{-0.5}$ occupies point 3. With that, the power will immediately take the value corresponding to point 3. With further R growth the power flow will increase. If we decrease the reflection factor, oscillation suppression will occur at point 1 and the oscillation power will decrease to zero by a jump.

Fig. 5.10 Rigid mode of self-excitation of a quantum generator. With a change in the mirror reflection factor, the line of feedback moves vertically. Point *1* is unstable, point *2* is stable, point *3* is the point of self-excitation with transition to point *4*, point *4* is the point of oscillation arising at *R* growth, and point *5* is the point of the oscillation break with a decrease in *R*

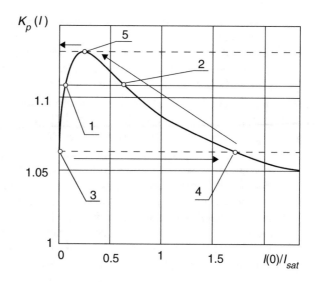

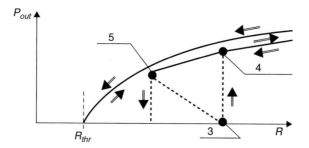

Fig. 5.11   Function of output power versus the mirror reflection factor in the soft and rigid modes of self-excitation. Functions of such types are typical for free-running oscillators; therefore, they require no specific explanation. In the rigid mode a hysteresis is observed. The numbering of the points on the graph corresponds exactly to the point plot in Fig. 5.10

Figure 5.11 shows the steady-state values of the normalized flow of the output power as a function of the mirror reflection factor. In the same manner we may draw the function of Π versus the population difference.

The peculiarities of the rigid mode will be demonstrated in the power time dependence just after the quantum generator is switched on. The action of the pumping source will lead to the fact that the population difference will be measured in time from the value $n_{21} = -n_{12}^e$ to n_{21}^{st}. If we consider the process as a diabatic one, we have the possibility to consequently pass all states following a population difference change in time. In this case the character of the process of transition into the steady-state mode will have the same features as the functions presented in Fig. 5.11.

In the rigid mode the moment when generation arises will be delayed with respect to the moment when the pumping source is switched on. This leads to more effective application of its energy because the occurrence of generation will not greatly disturb the population inversion.

The method of rigid mode realization will become clear with a glance at Fig. 5.8. Note that in the heat equilibrium state, the transfer function increases with power growth and $\frac{dK_p}{dI} > 0$, while in the presence of inversion it decreases and $\frac{dK_p}{dI} < 0$. Therefore, if we add a region with nonlinear absorption inside the resonator, then the through transfer coefficient $K_p(I) = K_p^{cond}(I)K_p^{abs}(I)$ will have a segment in which the amplification will increase (see Fig. 5.10). This solution was offered and implemented in 1964. Solutions of organic colorants or a glass containing cadmium selenides and sulfides are usually used as absorbing substances.

Instead of an absorbing medium we may use a nonlinear mirror whose reflection factor depends upon the power flow. To slow down the process of the steady-state oscillation transient at its initial stage, it is necessary to have a reflection increasing with I growth. An illustration of this idea is presented in Fig. 5.12. A semiconductor whose reflection factor from the surface increases with electron concentration growth in the conduction band (see Sect. 4.4) is the material to use for such a mirror. In turn, the concentration increases owing to interband transitions under the action of the incident radiation.

Both methods can be considered in control of the resonator Q factor, and for this reason they are often referred to as methods of Q factor modulation. It is known that diffraction losses in open resonators essentially depend upon the parallelism of the mirrors that form the resonator. The Q factor of the typical resonator decreases to a half with mirror rotation by 2 angular minutes. A resonator with a rotating reflector

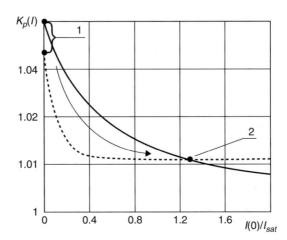

Fig. 5.12 Solution of the steady-state equation at application of a nonlinear mirror, exceeding the threshold in the moment when the generation arises (point *1*) and at steady-state (point *2*). The peculiarity of this mode consists in the fact that the exceeding of the threshold is small in the moment when the oscillation arises. As a result, the pumping source is able to provide a larger population difference value. After that, the reflection factor increases sharply and the generator transfers to steady-state (point 2)

was probably first used to obtain short and powerful light pulses. The power increase is so significant that this mode is called the giant-pulse generation mode. Such a name is quite correct as the pulse power achieves values measured in gigawatts. The pulse duration is reduced to values measured in picoseconds. A light pulse lasting 1 ps has a spatial dimension of only 0.3 mm!

The equations for the quantum generator investigated above are concerned exclusively with the steady-state mode. Nevertheless, the oscillation transient process has an important practical significance. We have already touched upon it during our discussion of the Q factor modulation process. We can try to describe the oscillation transient with the help of (5.30). Indeed, if we use (5.4) and (5.8) in (5.33), we can represent the right part of (5.30) as a function of n_{21}^{st}. The equation solution obtained by such an approach together with (5.3) gives a description of the transient. However, this model is still too rough. A strict description of the processes in a quantum generator can be found by solving the Maxwell equations together with the constitutive equations for the medium, (5.2) and (5.3).

5.7 Equations for Oscillations in the Resonator

To obtain oscillation equations for a quantum generator we formulate the mathematical task that should be solved for this.

The initial equation system is

$$\nabla \times \mathbf{H} = \frac{\partial \mathbf{D}}{\partial t} = \frac{\partial}{\partial t}(\varepsilon_a \mathbf{E} + \mathbf{P});$$

$$\nabla \times \mathbf{E} = -\frac{\partial \mathbf{B}}{\partial t} = -\mu_a \frac{\partial \mathbf{H}}{\partial t};$$

$$\frac{d^2 \mathbf{P}}{dt^2} + \frac{2}{T_2}\frac{d\mathbf{P}}{dt} + \omega_{12}^2 \mathbf{P} = -\frac{2\omega_{21}d^2}{3\hbar}n_{21}\mathbf{E};$$

$$\frac{dn_{21}}{dt} + \frac{n_{21} - n_{21}^{\bar{n}\grave{o}}}{T_1} = \frac{2}{\hbar\omega_{21}}\frac{d\mathbf{P}}{dt}\mathbf{E}.$$

This system is called semiclassical or quasiclassical in the literature. It should be solved in the specified initial conditions and in the boundary conditions.

It is impossible to obtain eigenfunctions and eigenvalues for this nonlinear boundary problem. Therefore, we need to use approximate methods of solution.

Let us examine the problem stated for us as the problem of resonator excitation by the polarization current:

$$\mathbf{j}_{\text{pol}} = \frac{d\mathbf{P}}{dt},$$

which is connected with the polarization of the active medium.

The schedule for solution of the problem of resonator excitation by an external current was well developed as far back as the middle of the twentieth century. Its sense consists in application of the real field expansion over the orthogonal eigenfunctions of the ideal resonator. In the literature it has been reported that there is a possibility of extending the understanding of the excitation problem. For this, it is sufficient to introduce currents that will imitate the difference between real and ideal resonators. If these differences are small, we can consider them small perturbations. Then, we may be limited only by the eigenfunction. Thus, for instance, we can take into account losses in the walls of a real resonator and determine its Q factor. This method for solving the linear tasks of resonator oscillations has been further developed, and at present it is considered a projection method for solution of electrodynamic problems.[4]

In solution of problems of radiation and excitation, we usually use vector $\mathbf{A}$ and scalar Φ potentials instead of fields $\mathbf{E}$ and $\mathbf{H}$. Their convenience consists in the fact that instead of six scalar equations, only four equations remain. It is known that only eddy fields satisfy the boundary conditions at the walls of the ideal resonator. Therefore, only the vector potential remains for us to express the electric and magnetic fields:

$$\mathbf{E} = -\frac{\partial \mathbf{A}}{\partial t}; \quad \mathbf{H} = \nabla \times \mathbf{A}.$$

As a result of transformation of the first Maxwell equation, we obtain the equation for the vector potential:

$$\nabla^2 \mathbf{A} = \varepsilon_a \mu_0 \frac{\partial^2 \mathbf{A}}{\partial t^2}.$$

The task of the eigenfunction and eigenvalue search can be solved by the method of variables separation.

Now we can present the solution in the product form,

$$\mathbf{A}(\mathbf{r}, t) = f(t)\mathbf{A}(\mathbf{r}).$$

The part depending upon time satisfies the equation

---

[4]We should pay attention to the fact that that this method has much more in common with matrix description in quantum mechanics, as well as with signal expansion on an orthogonal basis in the theory of signals.

$$d^2f/dt^2 = \omega^2 f,$$

and the vector potential distribution in the space is the solution to the equation

$$\nabla^2 \mathbf{A}(\mathbf{r}) + \left(\frac{\omega}{c}\right)^2 \mathbf{A}(\mathbf{r}) = 0,$$

with boundary conditions at the resonator walls,

$$[\mathbf{A}(\mathbf{r}) \times \mathbf{1}_n] = 0.$$

The solution gives eigenfunctions $\mathbf{A}_\nu(\mathbf{r})$ and eigenvalues ω_ν, with a sense of resonance frequencies. Eigenfunctions are orthogonal and form the orthonormalized basis:

$$\frac{1}{V} \int \mathbf{A}_\mu(\mathbf{r}) \mathbf{A}_\nu^*(\mathbf{r}) dV = \begin{vmatrix} 0 & \nu \neq \mu \\ 1 & \nu = \mu \end{vmatrix}. \tag{5.34}$$

The obtained functions are used for solving the problem of stimulated resonator oscillations. We write the initial heterogeneous equation for the vector potential as

$$\nabla^2 \mathbf{A} - \varepsilon_a \mu_0 \frac{\partial^2 \mathbf{A}}{\partial t^2} = -\mu_0 \mathbf{j}. \tag{5.35}$$

Let us represent the field in the resonator in the form of expansion over the basis:

$$\mathbf{A} = \sum_\mu c_\mu(t) \mathbf{A}_\mu(\mathbf{r}). \tag{5.36}$$

We substitute (5.36) in (5.35), multiply by $\mathbf{A}_\nu^*(\mathbf{r})$, and integrate over the volume. Taking into consideration (5.34), we obtain the equation for time-changing coefficients:

$$\frac{d^2 c_\nu}{dt^2} + \omega_\nu^2 c_\nu = \frac{1}{\varepsilon_a} j_\nu, \tag{5.37}$$

where

$$j_\nu = \frac{1}{V} \int_V \mathbf{j}(t, \mathbf{r}) \mathbf{A}_\nu^*(\mathbf{r}) dV.$$

This equation allows determination of all coefficients c_ν and, hence, fields in the resonator. Equation (5.37) coincides (in terms of its form) with the equation for an electric circuit containing an oscillating circuit without losses and an external source. If the current is the harmonic oscillation with frequency ω, we can use the complex amplitude approach. Then,

$$\dot{C}_\nu = \frac{1}{\varepsilon_a} \frac{j_\nu}{\omega_\nu^2 - \omega^2}.$$

We can see that at $\omega = \omega_\nu$ the oscillation amplitude becomes infinite. This is a result of complete absence of losses. However, in a real resonator, of course, losses are always present because a part of the energy leaves the volume. Let us modify (5.37) to take losses into account.

Let us examine the resonant cavity fully filled by a medium with conductivity σ. We assume that the current in the right part of Eq. (5.37) is equal to

$$\mathbf{j} = \sigma \mathbf{E} = -\sigma \frac{\partial \mathbf{A}}{\partial t}.$$

We expand the current density in a series of eigenfunctions of the resonator. The component of this current is

$$j_\nu = -\sigma \frac{dc_\nu}{dt} \frac{1}{V} \int_V \mathbf{A}_\nu \mathbf{A}_\nu^* dV.$$

The integration here is taken over all of the resonator volume. Therefore, Eq. (5.37) takes the form:

$$\frac{d^2 c_\nu}{dt^2} + \omega_\nu^2 c_\nu = -\frac{\sigma}{\varepsilon_a} \frac{dc_\nu}{dt}$$

or a universal form that does not depend upon the specific mechanism of losses and the resonator type:

$$\frac{d^2 c_\nu}{dt^2} + \frac{\omega_\nu}{Q_\nu} \frac{dc_\nu}{dt} + \omega_\nu^2 c_\nu = 0.$$

Here, Q_ν is the resonator Q factor corresponding to the ν-th type of oscillations.

Now we can write the stimulated oscillation equation with losses:

$$\frac{d^2 c_\nu}{dt^2} + \frac{\omega_\nu}{Q_\nu} \frac{dc_\nu}{dt} + \omega^2 c_\nu = \frac{1}{\varepsilon_a} j_\nu.$$

With harmonic influence,

$$\dot{C}_\nu = \frac{1}{\varepsilon_a} \frac{\dot{j}_\nu}{\omega_\nu^2 - \omega^2 + j\omega \frac{\omega_\nu}{Q_\nu}}.$$

The amplitudes of all natural oscillations are finite, and we can write the final solution in the form of a series of eigenfunctions. If Q factors of oscillation type $Q_\nu \gg 1$, that the one term of this series will dominate and the sum can be changed by the single term

$$\mathbf{A} = c_\nu(t) - \mathbf{A}_\nu(\mathbf{r}).$$

In the future we shall use exactly such a field presentation.

In the quantum generator the polarization of the operating transition and the polarization current $\mathbf{j}_{\mathrm{pol}} = \frac{\partial \mathbf{P}}{\partial t}$ act as the field source. We present polarization in the form of a series of eigenfunctions:

$$\mathbf{P} = \sum_\mu s_\mu(t)\mathbf{A}_\mu(\mathbf{r}).$$

Then,

$$\frac{d^2 c_\nu}{dt^2} + \frac{\omega_\nu}{Q_\nu}\frac{dc_\nu}{dt} + \omega^2 c_\nu = \frac{1}{\varepsilon_a}\frac{1}{V}\int_V \mathbf{A}_\nu \mathbf{A}_\nu^* dV \frac{ds_\nu}{dt}.$$

Here, integration is fulfilled over the volume occupied by the medium. For a completely filled resonator, because of the orthogonality condition (5.34), the integral in the right part becomes the unit and the equation takes a simpler form[5]:

$$\frac{d^2 c_\nu}{dt^2} + \frac{\omega_\nu}{Q_\nu}\frac{dc_\nu}{dt} + \omega^2 c_\nu = \frac{1}{\varepsilon_a}\frac{ds_\nu}{dt}.$$

Expansion of the polarization and the field into a series of eigenfunctions transforms the initial constitutive equations into equations with respect to coefficients c_ν and s_ν. From (2.24) we have

$$\frac{d^2 s_\nu}{dt^2} + \frac{2}{T_2}\frac{ds_\nu}{dt} + \omega_{21}^2 s_\nu = \frac{2\omega_{21} d^2 n_{21}}{3\hbar}\frac{dc_\nu}{dt}.$$

Substituting expansions $\mathbf{A} = \sum_\mu c_\mu(t)\mathbf{A}_\mu(\mathbf{r})$ and $\mathbf{P} = \sum_\mu s_\mu(t)\mathbf{A}_\mu(\mathbf{r})$ in the nonlinear equation for population difference (5.3), we find that

---

[5]Nonuniformity of volume filling can be taken into consideration by use of well-known mathematical formulas for integral estimation and introduction of a filling coefficient with an active medium.

$$\frac{dn_{21}}{dt} + \frac{n_{21} - n_{21}^{st}}{T_1} = -\frac{2}{\hbar\omega_{21}} \sum_{\nu} \frac{dc_\nu}{dt} A_\nu \sum_{\mu} \frac{ds_\mu}{dt} A_\mu.$$

With integration over the whole volume and taking into account the orthogonality condition (5.34), we obtain, for a single-mode generation regime and for homogeneous filling of the space,

$$\frac{dn_{21}}{dt} + \frac{n_{21} - n_{21}^{st}}{T_1} = -\frac{2}{\hbar\omega_{21}} \frac{dc_\nu}{dt} \frac{ds_\nu}{dt}.$$

Now we can write the equation system for stimulated oscillation in the resonator filled with an active medium in the final form:

$$
\begin{aligned}
\frac{d^2 c_\nu}{dt^2} + \frac{\omega_\nu}{Q_\nu} \frac{dc_\nu}{dt} + \omega_\nu^2 c_\nu &= \frac{1}{\varepsilon_a} \frac{ds_\nu}{dt}; \\
\frac{d^2 s_\nu}{dt^2} + \frac{2}{T_2} \frac{ds_\nu}{dt} + \omega_{21}^2 s_\nu &= \frac{2\omega_{21} d^2 n_{21}}{3\hbar} \frac{dc_\nu}{dt}; \\
\frac{dn_{21}}{dt} + \frac{n_{21} - n_{21}^{st}}{T_1} &= -\frac{2}{\hbar\omega_{21}} \frac{dc_\nu}{dt} \frac{ds_\nu}{dt}.
\end{aligned}
\tag{5.38}
$$

The equations for c_ν, s_ν, and n_{21} form a system of nonlinear equations whose solution is connected with definite difficulties. Therefore, we will discuss the quantum generator dynamics in the small-oscillation mode. In this case we may consider system (5.38) a linear one.

5.8   The Small-Oscillation Mode

The linear approximation of the small-oscillation mode allows us to examine the stability of the positive feedback systems and obtain useful information about the generator's behavior.

We replace system (5.38) with the linear approximation. We write the system of uniform linear equations of the quantum generator as

$$
\begin{aligned}
\frac{d^2 c_\nu}{dt^2} + \frac{\omega_\nu}{Q_\nu} \frac{dc_\nu}{dt} + \omega_\nu^2 c_\nu - \frac{1}{\varepsilon_a} \frac{ds_\nu}{dt} &= 0; \\
-\frac{2\omega_{21} d^2 n_{21}^{st}}{3\hbar} \frac{dc_\nu}{dt} + \frac{d^2 s_\nu}{dt^2} + \frac{\omega_{21}}{Q_L} \frac{ds_\nu}{dt} + \omega_{21}^2 s_\nu &= 0.
\end{aligned}
\tag{5.39}
$$

Here, n_{21}^{st} is the given value of the stationary inversion population difference; Q_L and Q_ν are the Q factors of the absorption line and the resonator, respectively; and ω_ν is the resonant frequency of the ν-th oscillation type.

Stimulated oscillations will exist if, in the right part of one of the equations—for example, the first one—a stimulated force will act. In this case the field of equilibrium heat emission acts as such a source. Within the limits of semiclassical approximation we cannot write it in an explicit form. Therefore, we cannot consequently and strictly discuss the issue of oscillation arising in a quantum generator. Nevertheless, in stability analysis the specific view of the stimulated force has no significance.

The linear system (5.39) may be written with respect to dimensionless time $\theta = \omega t$:

$$\frac{d^2 c_\nu}{d\theta^2} + \frac{\omega_\nu}{Q_\nu \omega_{21}} \frac{dc_\nu}{d\theta} + \frac{\omega_\nu^2}{\omega_{21}^2} c_\nu - \frac{1}{\varepsilon_a \omega_{21}} \frac{ds_\nu}{d\theta} = \frac{1}{\varepsilon_a \omega_{21}^2} j;$$

$$-\frac{2d^2 n_{21}^{st}}{3\hbar} \frac{dc_\nu}{d\theta} + \frac{d^2 s_\nu}{d\theta^2} + \frac{1}{Q_{\ddot{E}}} \frac{ds_\nu}{d\theta} + s_\nu = 0. \tag{5.40}$$

Using the Laplace transformation, we transfer from (5.40) to the system of linear algebraic equations with respect to images:

$$\left(p^2 + \frac{\omega_\nu}{Q_\nu \omega_{21}} p + \frac{\omega_\nu^2}{\omega_{21}^2} \right) \tilde{N}_\nu(p) - \frac{1}{\varepsilon_a \omega_{21}} p S_\nu(p) = \frac{1}{\varepsilon_a \omega_{21}^2} J(p);$$

$$-\frac{2d^2 n_{21}^{st}}{3\hbar} p \tilde{N}_\nu(p) + \left(p^2 + \frac{1}{Q_{\ddot{E}}} p + 1 \right) S_\nu(p) = 0.$$

Solving this system (for instance, using the Cramer rule) we obtain

$$C_\nu(p) = \frac{\left(p^2 + \frac{1}{Q_L} p + 1 \right) J(p)}{\varepsilon_a \omega_{21}^2 \left[\left(p^2 + \frac{1}{Q_L} p + 1 \right) \left(p^2 + \frac{\omega_\nu}{Q_\nu \omega_{21}} p + \frac{\omega_\nu^2}{\omega_{21}^2} \right) - \frac{1}{\omega_{21} \varepsilon_a} \frac{2d^2 n_{21}^{st}}{3\hbar} p^2 \right]}.$$

The multiplier before the current can be considered as the operator resistance of some two-terminal network:

$$Z_\nu(p) =$$

$$= \frac{\left(p^2 + \frac{1}{Q_L} p + 1 \right)}{\varepsilon_a \omega_{21}^2 \left[\left(p^2 + \frac{1}{Q_L} p + 1 \right) \left(p^2 + \frac{\omega_\nu}{Q_\nu \omega_{21}} p + \frac{\omega_\nu^2}{\omega_{21}^2} \right) - \frac{1}{\omega_{21} \varepsilon_a} \frac{2d^2 n_{21}^{st}}{3\hbar} p^2 \right]}. \tag{5.41}$$

From circuit theory fundamentals we know that the circuit is stable if its poles lie in the left half-plane of complex frequency p. Hence, for continuous oscillation arising in the quantum generator, the poles in (5.41) should be in the right half-plane. Figures 5.13, 5.14, 5.15, and 5.16 show the poles' diagrams of operator resistance

Fig. 5.13 Pole diagram of operator impedance (5.41) in the state of heat equilibrium; $n_{21}^e = -5 \cdot 10^{18}$ m^{-3}, the resonator Q factor is 10^4, and the line Q factor is 10^7. The resonator resonance frequency ω_c coincides with the frequency of quantum transition ω_{21}. The vertical displacement of poles *1* and *2* is the result of the coupling of the two oscillatory systems. A similar phenomenon is observed in an electric circuit with two similarly coupled oscillatory circuits

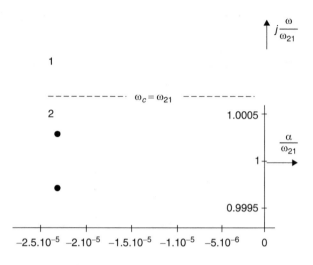

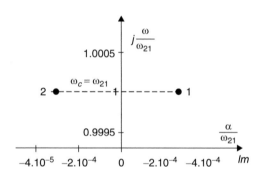

Fig. 5.14 Pole diagram of operator impedance (5.41) for a quantum generator using ammonia molecules with population inversion $n_{21}^{st} = 5 \cdot 10^{18}$ m^{-3}, a resonator Q factor of 10^4, and a line Q factor of 10^7. The resonance frequency ω_c of the resonator coincides with the frequency of the quantum transition ω_{21}. Pole *1* is in the right half-plane. It corresponds to time-increasing oscillations. The frequency of arising oscillations practically coincides with ω_{21}

$Z_\nu(p)$ for several specific cases. In the moment when the generator is switched on, $n_{21}^{st} = -n_{12}^e < 0$ and the poles are situated in the left half-plane (see Fig. 5.13). During the action of the pumping source they move and transfer into the right half-plane (see Fig. 5.14).

The position of the poles depends upon the mutual detuning of the resonator resonance frequency and the frequency of the operating transition. The character of the diagram variation is determined by the relative values of the resonator Q factor and the line Q factor.

The pole diagram $Z_\nu(p)$ for a molecular quantum generator in the centimeter range is shown in Fig. 5.15. The Q factor of the resonators used at these frequencies is of the order of 10^4, while Q_L is of the order of 10^7. At such Q factor ratios the

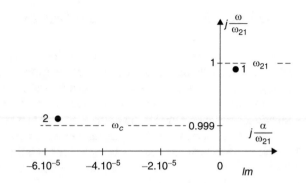

Fig. 5.15 Pole diagram of operator impedance (5.41) for a quantum generator using ammonia molecules at relative resonator detuning of 0.1%. The other parameters are the same as those in Fig. 5.14. It can be seen that the generation frequency practically coincides with the central frequency of the transition. However, the mutual influence of the two oscillatory systems displaces the poles vertically when approaching them

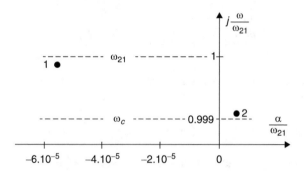

Fig. 5.16 Pole diagram of operator impedance (5.41) for a gas discharge optical quantum generator using a mixture of helium and neon. The resonator Q factor is 10^7 and the line Q factor is 10^4. The relative resonator detuning of 0.1% is practically directly transferred almost completely to the generation frequency

oscillation frequency is defined by the frequency of the operating transition ω_{21}, whose stability is extremely high. In the optical range the resonator Q factor is of the order of 10^7 and $Q_L \approx 10^4 - 10^5$. We can clearly see from Fig. 5.16 that for such Q factor ratios the oscillation frequency is defined by the resonant frequency of the resonator. Earlier, we could discuss the problem of the frequency stability of a quantum generator only on the basis of some heuristic considerations, but now this issue can be solved in the quantitative sense as well.

5.9   Abbreviated Equations for a Quantum Generator

The quantum generator represents a dynamic system in which oscillations are very close, in terms of their form, to harmonic ones. This is caused by the extremely high value of the Q factor. In the centimeter range the Q factor of the absorption line is of the order of 10^7, and in the optical range the resonator Q factor is of the same order. Therefore, without doubt, the approximated methods of analysis that have been well developed in oscillation theory can be used for solving problems regarding quantum generators. The basis of these methods consists in replacing the full equations with so-called abbreviated equations for slowly changing amplitudes. Let us complete the abbreviation procedure for the system (5.38).

We write the initial equation system as

$$
\begin{aligned}
\frac{d^2 c_\nu}{dt^2} + \frac{\omega_\nu}{Q_\nu}\frac{dc_\nu}{dt} + \omega_\nu^2 c_\nu &= \frac{1}{\varepsilon_a}\frac{ds_\nu}{dt}; \\
\frac{d^2 s_\nu}{dt^2} + \frac{2}{T_2}\frac{ds_\nu}{dt} + \omega_{21}^2 s_\nu &= \frac{2\omega_{21} d^2 n_{21}}{3\hbar}\frac{dc_\nu}{dt}; \\
\frac{dn_{21}}{dt} + \frac{n_{21} - n_{21}^{\text{st}}}{T_1} &= -\frac{2}{\hbar\omega_{21}}\frac{dc_\nu}{dt}\frac{ds_\nu}{dt}.
\end{aligned}
\tag{5.42}
$$

Let us present variables $c_\nu(t)$ and $s_\nu(t)$ in the forms of harmonic oscillations with their amplitude and phase slowly changing in time:

$$
c_\nu(t) = \operatorname{Re}\{\dot C_\nu \exp(j\omega t)\}; \quad s_\nu(t) = \operatorname{Re}\{\dot S_\nu \exp(j\omega t)\},
$$

where $\dot C_\nu(t)$ and $\dot S_\nu(t)$ are slowly changing complex amplitudes such as

$$
\begin{aligned}
\frac{dc_\nu}{dt} &\pounds \operatorname{Re}\{j\omega\dot C_\nu \exp(j\omega t)\}; \\
\frac{d^2 c_\nu}{dt^2} &\approx \operatorname{Re}\left\{\left[j2\omega\frac{d\dot C_\nu}{dt} - \omega^2\dot C_\nu\right]\exp(j\omega t)\right\}; \\
\frac{ds_\nu}{dt} &\approx \operatorname{Re}\{j\omega\dot S_\nu \exp(j\omega t)\}; \\
\frac{d^2 s_\nu}{dt^2} &\approx \operatorname{Re}\left\{\left[j2\omega\frac{d\dot S_\nu}{dt} - \omega^2\dot S_\nu\right]\exp(j\omega t)\right\}.
\end{aligned}
\tag{5.43}
$$

The question about the choice of frequency ω, which is included in these equations, can be solved in different ways for optical generators and masers. Therefore, we will complete the abbreviation procedure separately for two limiting cases.

As we already understood in Sect. 5.8, in the optical range the oscillation frequency is close to the resonator frequency; hence, frequency ω should be equal to resonator frequency ω_ν. After substantiation of (5.43) in (5.42), we obtain

$$\frac{d\dot{C}_\nu}{dt} + \frac{\omega_\nu}{2Q_\nu}\dot{C}_\nu = \frac{1}{2\varepsilon_a}\dot{S}_\nu;$$

$$\frac{d\dot{S}_\nu}{dt} + \frac{\omega_{21}}{2Q_L}(1+j\xi)\dot{S}_\nu = \frac{\omega_{21}d^2n_{21}}{3\hbar}\dot{C}_\nu; \qquad (5.44)$$

$$\frac{dn_{21}}{dt} + \frac{n_{21} - n_{21}^{st}}{T_1} = -\frac{\omega_{21}}{\hbar}\,\mathrm{Re}\left(\dot{C}_\nu\dot{S}_\nu^*\right).$$

Here, $\xi = Q_L\frac{\omega_\nu^2 - \omega_{21}^2}{\omega_{21}\omega_\nu}$ is the generalized resonator detuning with respect to the center of the spectral line.

This system can be solved by numerical methods for a wide enough range of included parameters. Nevertheless, under condition $Q_L \ll Q_\nu$ we may obtain, instead of (5.44), equations with regard to quadratic quantities.

In the optical range the time variation speeds $\dot{C}_\nu$ and $\dot{S}_\nu$ are defined by the constant $\tau = \frac{Q_\nu}{\omega_\nu} \gg T_2$. Therefore, in the second equation we can neglect the derivative in the left part. Then,

$$\dot{S}_\nu = \frac{2d^2n_{21}}{3\hbar}\frac{Q_L}{1+j\xi}\dot{C}_\nu$$

and after substitution of this expression in (5.44) and some simple transformations, we obtain the following system of equations:

$$\frac{dn_{21}}{dt} + \frac{n_{21} - n_{21}^{no}}{T_1} = -\frac{2\omega_{21}d^2n_{21}}{3\hbar^2}\frac{Q_L}{1+\xi^2}|\dot{C}_\nu|^2;$$

$$\frac{d|\dot{C}_\nu|^2}{dt} + \frac{\omega_\nu}{Q_\nu}|\dot{C}_\nu|^2 = \frac{2d^2n_{21}}{3\varepsilon_a\hbar}\frac{Q_L}{1+\xi^2}|\dot{C}_\nu|^2, \qquad (5.45)$$

which describes the time change of the energy in the resonator and the population difference (the system energy). With transition to energy variables, we of course lose all of the information about phase relations. The obtained equations are accepted as kinetic or velocity equations, since they define the speed of the energy exchange between the field in the resonator and the medium.

At zero detuning $\xi = 0$, system (5.45) becomes identical, in terms of its form, to the known equations of Statz and deMars, which were obtained at the earliest stage of quantum electronics in 1960. These equations are also called equations of population difference balance and equations of photon density in the resonator. They were obtained without the use of the Maxwell equations on the basis of the transition probability concept.

The probability method is simple and visual; however, it is absolutely impossible to determine, within its limits, the application boundaries of the balance equations. Now we can do that considering the transition from (5.42) to (5.44). For this transition it is necessary that $\tau \gg T_2$. Use of $\mathrm{Re}\left(\dot{C}_\nu\dot{S}_\nu^*\right)$ in the right part of the last equation is possible if $T_1 \gg \tau$. The conditions limiting the application area of the balance equation are discussed in detail in the literature.

In the microwave range the oscillation frequency is close to the frequency of the spectral line; hence, frequency ω should be equal to transition frequency ω_{21}. We again substitute (5.43) in (5.42) and obtain

$$
\begin{aligned}
\frac{d\dot{C}_\nu}{dt} + \frac{\omega_\nu}{2Q_\nu}(1 + j\xi)\dot{C}_\nu &= \frac{1}{2\varepsilon_a}\dot{S}_\nu; \\
\frac{d\dot{S}_\nu}{dt} + \frac{\omega_{21}}{2Q_L}\dot{S}_\nu &= \frac{\omega_{21}d^2n_{21}}{3\hbar}\dot{C}_\nu; \\
\frac{dn_{21}}{dt} + \frac{n_{21} - n_{21}^{st}}{T_1} &= -\frac{\omega_{21}}{\hbar}\,\mathrm{Re}\left(\dot{C}_\nu\dot{S}_\nu^*\right).
\end{aligned}
\tag{5.46}
$$

The kinetic equations for $\omega_{21} \cong \omega_\nu$ take the form

$$
\begin{aligned}
\frac{dn_{21}}{dt} + \frac{n_{21} - n_{21}^{st}}{T_1} &= -\frac{\omega_{21}\varepsilon_a}{\hbar Q_\nu}|\dot{C}_\nu|^2; \\
\frac{d|\dot{C}_\nu|^2}{dt} + \frac{\omega_{21}}{Q_L}|\dot{C}_\nu|^2 &= \frac{2d^2n_{21}}{3\varepsilon_a\hbar}\frac{Q_\nu}{1+\xi^2}|\dot{C}_\nu|^2.
\end{aligned}
\tag{5.47}
$$

Systems (5.45) and (5.47) are different in two aspects. Firstly, the right part of the first equation of system (5.47) does not depend upon the population difference. This means that the differential equation for population difference is a linear equation. Secondly, now the process dynamics is determined by the time of transverse relaxation T_2 $\left(\frac{\omega_{21}}{Q_L} = \frac{2}{T_2}\right)$. All of this provides a spare confirmation of the fact that if $Q_L \gg Q_\nu$ the processes in the quantum generator are defined by the properties of the spectral line (see Sect. 5.7).

The above-obtained equation systems allow description of the oscillation development process in the quantum generator.

5.10   Dynamics of a Single-Mode Quantum Generator

The generator behavior in the process of its operation is a reflection of the variety of properties and features of its structure. Therefore, the time variation of the generator's oscillation parameters represents a significant interest that is both practical and purely scientific. Now we are ready to discuss the dynamics of the quantum generator. Of course, the models we used above are not so detailed as to describe phenomena in real quantum generators. For details, the reader must refer to the specialist literature. However, they may be useful for those readers who decide to deal seriously with quantum generators. Our statement, as a result of simplifications, is close to description of traditional oscillators and other circuits with feedback. The possibility of a simplified description of a quantum generator once more emphasizes the fact that the term "quantum" is concerned exclusively with description of the medium properties only; the quantum generator itself is, as a whole, a classical device.

In order not to complicate the problem excessively, we shall concentrate on the single-mode regime analysis under conditions of homogeneous widening of the spectral line. We shall also assume that any external impacts on the generator are eliminated. Such a mode is called free-running generation.

For transient analysis in an optical quantum generator we use equation system (5.45). Preliminarily, we reduce it to more suitable form. For this, we consider the steady-state case ($d/dt = 0$). Then,

$$\frac{n_{21} - n_{21}^{st}}{T_1} = -\frac{2\omega_{21}d^2 n_{21}}{3\hbar^2} Q_L |\dot{C}_v|^2;$$

$$\frac{\omega_v}{Q_v} |\dot{C}_v|^2 = \frac{2d^2 n_{21}}{3\varepsilon_a \hbar} Q_L |\dot{C}_v|^2. \tag{5.48}$$

From the second equation of (5.48),

$$n_{21} = \frac{3\varepsilon_a \hbar \omega_v}{2d^2 Q_v Q_L} = N_0.$$

Substituting this result in the first equation of (5.48), we obtain

$$C_0^2 = \frac{\left(n_{21}^{st} - N_0\right)3\hbar^2}{T_1 2\omega_{21}d^2 N_0 Q_L} = \frac{(N^{st} - 1)3\hbar^2}{T_1 2\omega_{21}d^2 Q_L}; \quad N^{st} = \frac{n_{21}^{st}}{N_0}.$$

Fulfilling the normalization of (5.45) to N_0 and C_0, we have

$$\frac{dN}{dt} + \frac{N - N^{st}}{T_1} = \frac{N\varphi}{T_1}(1 - N^{st});$$

$$\frac{d\varphi}{dt} + \frac{\omega_v}{Q_v}\varphi = \frac{\omega_v}{Q_v}N\varphi. \tag{5.49}$$

Here, $N = \frac{n_{21}}{N_0}$ is the normalized population difference and $\varphi = \frac{|\dot{C}_v|^2}{C_0^2}$ is the normalized square of the vector potential amplitude.

The quantity N_0 is factually the threshold concentration difference, which we can make sure of by analyzing the second equation of the system (5.47). Really, for the oscillation amplitude to increase, the following is necessary:

$$\frac{\omega_v}{Q_v} \leq \frac{2d^2 n_{21}}{3\varepsilon_a \hbar} Q_L.$$

Hence,

$$(n_{21})_{\text{thr}} = \frac{3\varepsilon_a \hbar \omega_v}{2d^2 Q_v Q_L},$$

which is exactly equal to N_0.

System (5.49) has two independent solutions: $N = \varphi = 1$ (the mode of stationary generation) and $N = N^{\text{st}}$, $\varphi = 0$ (the quiescent state). It is important that any transient process leads the system to one of these states, depending on which state is stable in this case.

In spite of the essential simplification of the initial equations, system (5.49) still remains so difficult that it does not permit an analytical solution.[6] Approximate methods for analysis of quantum generator dynamics under additional limitations on parameters of kinetic equations can be found in the specialist literature. The transient can be divided into time segments, within whose limits the variable values have some restrictions. More curious readers can find these literature sources. We shall complete the discussion of the numerical solution.

Figure 5.17 shows the transient process at pulse pumping for 50 μs, which enables the threshold to be exceeded ten times. The longitudinal relaxation time $T_1 = 5$ μs, and the time of the oscillation transient is 0.1 μs. These initial data do not correspond to any specific laser; they are selected exclusively for consideration of the clearness of the result. Oscillations arise in the moment when the threshold is achieved. At first, the process runs very slowly and the population difference continues to increase. However, oscillations that have arisen gradually act more noticeably against the pumping source. As a result, instead of the threshold being exceeded ten times, the population difference achieves a maximal value of slightly more than three. The right slope of the first peak is fully connected with the damping of free-running oscillations in the resonator, as in this time interval the population difference is below the threshold level.

The phenomenon of laser radiation ripples is a well-known experimental fact. We can distinguish regular damping ripples and regular nondamping ripples.

Figure 5.18 shows the solution of the kinetic equation system with initial data close to the laser parameters on glass doped with neodymium. On the phase portrait we can clearly see that during the generation process the population difference

---

[6]It should be taken into account that, besides pure mathematical simplifications, our equations may contain a variety of simplifications of physical aspects that are not mentioned in the text. For instance, we privately assume that the medium is homogeneous in the space, but the saturation effect changes the concentration difference; hence, n_{21} is a function of the coordinates. Details can be found in the specialist literature.

Fig. 5.17 Results of numerical solution of kinetic equations: time domain plot (**a**) and phase portrait (**b**), showing the time change of N (*1*) and φ (*2*) in the quantum generator (**b**). The pumping pulse lasts for 5.0 µs. The initial equilibrium population difference is equal modulo to the threshold (the point $N = -1$). The longitudinal relaxation time $T_1 = 5$ µs and the oscillation transient time in the resonator is 0.1 µs. In the absence of generation, the pumping enables the threshold to be exceeded 10 times ($N^{st} = 10$). However, during generation the maximal value of N is only slightly greater than 3

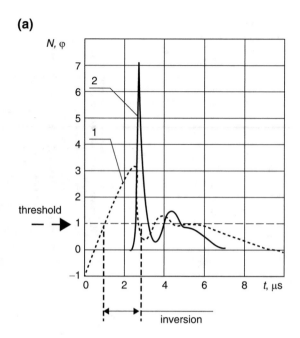

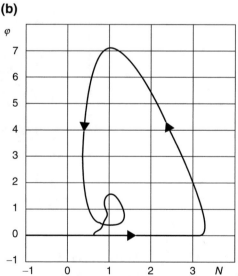

deflects absolutely insignificantly from the threshold value in spite of the fact that the pumping source is able to enable the threshold to be exceeded ten times.

Figures 5.17 and especially 5.18 show that in the free-running generation mode, the power of the pumping source is used extremely ineffectively. We have already discussed this problem in Sect. 5.6. The outlet consists in application of lasers with

Fig. 5.18 Mode with nondamping ripples with transition to the steady-state mode $\varphi = 1$ (**a**). The transverse relaxation time $T_1 = 1$ ms and the oscillations' damping time in the resonator is 0.1 μs. Pumping in the absence of generation provides $N^{st} = 10$. The maximal value $\varphi \approx 100$. The zero time moment corresponds to the point of achievement of the excitation threshold. It can be seen in the phase portrait that the relative deflection of the population difference from the threshold value is no greater than 10% (**b**)

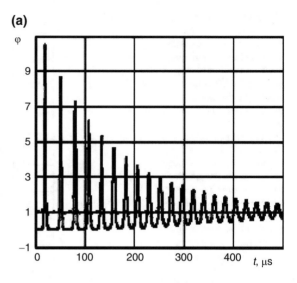

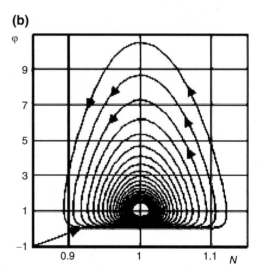

Q factor modulation. Kinetic equations allow calculation of the transients for such a generator. For this it is enough to replace the initial condition $N(0) = -\frac{n_{12}^e}{N_0}$ with the condition $N(0) = N^{st}$. The giant-pulse mode is shown in Fig. 5.19. The amplitude at the maximum is almost five orders greater than the steady-state value and about 10,000 times greater than the maximal value in Fig. 5.18.

Kinetic equations cannot give a solution in the form of nondamping ripples. Besides, irregular ripples are observed in experiments much more often. Earlier, it was shown that these phenomena are connected with external impacts on the generator. The discussion of this issue goes far beyond the scope of this book.

Fig. 5.19 Laser pulse with Q factor modulation (**a**). The transverse relaxation time $T_1 = 1$ ms, and the oscillation damping time in the resonator is 10^{-8} s. In the initial time moment the population difference exceeds the threshold value 10 times. The radiation $\varphi(t)$ [*solid curve*] go on maximum coincides with the moment of the threshold value (*1*) intersection with function $N(t)$ [*dashed curve*]. The slope (*2*) $\varphi(t)$ is the free-running damping oscillations in the resonator. This can be seen in the generator phase portrait (**b**)

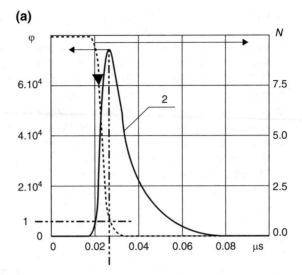

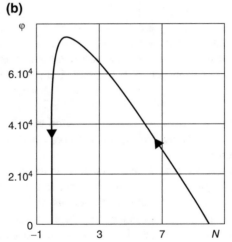

5.11 The Multimode Regime: Mode Synchronization

From electrodynamics and radio wave propagation we know well that the sizes of the open resonators should be large compared with the wavelength. For this reason a large number of oscillation/mode types with their own different frequencies fall into the frequency domain in which the self-excitation conditions are fulfilled. The frequency interval between modes is of the order of $\Delta f \approx c/2L$, where c is the light speed in the medium filling the volume, and it changes from approximately 10^8 Hz (in gas lasers with a length of about 1 m) to 10^{12} Hz (in semiconductor generators with a length of about 0.1 mm). In the optical range, with some exclusions, the width of the spectral line is greater or even much greater than the frequency interval between the proper frequencies. Thus, in accordance with the classification of the traditional oscillation theory, lasers are multimode generators.

In the process of oscillation development in multimode generators, oscillation competition occurs, which leads to domination by one of a variety of possible modes. However, in regard to mode competition, the phenomena in lasers differ from similar processes in electric circuits. Prior to this moment we diligently reduced the equations for a quantum generator to a form identical to that of the equations in circuit theory. Now we must, at last, pay attention to the differences.

The thing is that in traditional generators, mode interaction occurs in the restricted area of the space inside the lumped nonlinear element. This is true even for a multimode generator with a delay line in the feedback circuit, which structurally is very close to the laser. In quantum generators, mode interaction occurs in the nonlinear space with sizes in the hundreds of wavelengths. The result of mode competition can be found by integration over the whole volume of the active medium. The competition will essentially be weaker since the spatial field distribution of the generated modes is close to the orthogonal basis of the "cold" resonator (5.34). It is so much weaker that the laser is able to operate in the multimode regime. The separate oscillations of the quantum generator are weakly connected to each other, and the radiation represents a set of harmonic oscillations with random phases. Naturally, this promises nothing good. Remember the circuit and signal theory, where we studied the properties of a sum of sinusoids whose phases were distributed uniformly on a segment from $-\pi$ to π. To obtain single-mode (single-frequency) generation, as a rule, we need to provide specific measures for suppression of unwanted oscillations, which, for instance, could decimate the proper frequencies of the resonator.

Evidently, the mathematical aspects of the described phenomena represent a complicated problem. In the simplest case of a nonuniformly widened spectral line we may consider, with definite restrictions, that all modes are independent. However, even in this approximation the problem remains very bulky, and what can we say about an attempt to take into account intermode beats? Therefore, we shall limit our discussion to the very important aspect of multimode generation: synchronization of modes.

Multimode generation can be useful if the oscillations of all frequencies will be synchronized in phase. Really, the spectrum of a multimode generator represents a discrete set of almost equidistant harmonic components. Exactly the same set appears at expansion of periodic functions in the Fourier series, with the only difference being that each harmonic has a quite definite phase. Therefore, if we could impose the necessary phase on each oscillation, generation of a periodic pulse sequence would instead observe chaotic generation. The repetition frequency would become equal to the frequency of the intermode interval, and the pulse duration would be defined by the bandwidth of the spectral line. What could we expect in this case?

At mode synchronization the repetition frequency would be in the range from hundreds of megahertz to thousands of megahertz (terahertz units). However, a greater impression is provided by the possible values of the pulse duration. For a laser with an active element of glass doped with neodymium the spectral line width is of the order of about 10^{12} Hz, and this means that the pulse duration is several picoseconds. The organic colorants that are used in liquid lasers have the widest spectral line. In lasers using colorants we can obtain pulses with a duration of about 10^{-14} s. Such a light pulse occupies a segment with a length of only 3 μm in the space.

From the engineering point of view, the mode synchronization regime is provided in either an active way or a passive way. Active mode synchronization is provided by laser parameter modulation—for instance, a resonator Q factor with a frequency equal to the intermode interval. For such modulation, we use optical modulators. We can also modulate the pumping power. We know well that with any type of modulation, side components appear (being apart from it on the modulation frequency) near the carrier frequency. These side components contain information about the oscillation phase of the carrier frequency. Now the oscillations of each mode are developed in the presence of the external stimulated force. In oscillators this is the well-known phenomenon of frequency locking, and it is responsible for mode synchronization in a laser.

In the passive method of synchronization an additional cell with a nonlinear substance is placed inside the resonator. With suitable choices of the medium parameters and the cell position, a regime of mode synchronization occurs.[7]

It must be absolutely clear to us that even for experts in the field of quantum generation, mathematical description of the mode synchronization regime represents an extremely difficult problem, which is far from being finally resolved. For us, this problem is so complex that, as we have seen in this section, no one serious mathematical relation appears not to mention some equations.

5.12 Problems for Chapter 5

5.1. The following pumping source is provided:

---

[7]The idea of mode synchronization has found applications in multimode generators with a delay line on the surface acoustic waves. A detector and a modulator are included in the circuit. As a result, the device generates a periodic sequence of radio pulses. The pulse shape depends upon the apodization law of interdigital transducers (the frequency response of the filter). Generation of periodic pulses with an infill frequency of 100 MHz, duration of 50 ns, and period of 1.8 μs has been reported (H. Gilden, J.M. Reeder, M.F. Lewis, in Proc. IEEE US Symp. (1975), p. 251).

$$n_{\text{pump}}(t) = 2n_{12}^e \left(1 - 1,5 \exp\left(-\frac{1}{T_{\text{pump}}}\right)\right).$$

Determine the law of population difference variation of the two-level system.

5.2. Write the balance equations for the three-level system $W_1 = 0 < W_2 < W_3$. The noncoherent pumping source acts on the transition $1 \leftrightarrow 3$. Find the conditions at which inversion arises on the transition $1 \leftrightarrow 2$.

   Assistance: For noncoherent pumping emission this problem can be solved using transition probability language. The probability of stimulated transitions is proportional to the spectral density of the emission power. In the optical range ($\hbar\omega \gg k_B T$) we can neglect the up-transitions caused by contacts with the thermostat.

5.3. Transitions between levels of a superfine structure that arise from interactions of electron and nuclear spins are used in a hydrogen generator. The transition frequency $f_{21} = 1.420$ GHz. The transitions are connected with magnetic dipole interaction. After averaging over the orientations we obtain the equation system for **M**, which looks like (2.24) and (2.25) in form. At resonance, $\dot{\mathbf{M}} = -j\frac{\mu^2 T_2}{3\hbar}\mu_0 n_{12}\dot{\mathbf{H}}$, where $\mu = \frac{\gamma\hbar}{2} = \frac{eg}{2m_e} = g\mu_B$ is the magnetic dipole moment of the transition.

   The inverse population difference is 10^{18} m^{-3}, $T_2 = 1$ ms. Determine the threshold value of the resonator Q factor that gives the possible existence of nondamping oscillations exceeding the threshold.

5.4. Derive the formula for the amplification factor of an H_{10}-type wave in a rectangular metal wave guide filled with a paramagnetic substance.

   Specify: the population difference n_{21}^{st}, the electron paramagnetic resonance (EPR) frequency f_H, the line width Δf, the wave guide section, the relative dielectric permeability of the crystal lattice ε, and the filling coefficient of the wave guide cross-section k_{fil}.

   Assistance: Partial filling can be taken into account in the appropriate decreases in the volume parameters of the medium. For simplicity, assume that the magnetic susceptibility is scalar.

5.5. In a metal wave guide, some paramagnetic crystal doped with ions with spin 1/2 is placed. The EPR frequency is 10 GHz, the line bandwidth is 30 MHz, the population difference is $n_{21} = 10^{23}$ m^{-3}, and the damping in the wave guide is 0.01 m^{-1}. Determine the specific amplification if the substance filling coefficient is 0.2.

   Instruction: Use the result of task 5.4.

5.6. Calculate the output power of a gas discharge laser using a CO_2–N_2–He mixture. The specific amplification in power is $\alpha_0 = 0.1$ m^{-1}, the time of the longitudinal relaxation $T_1 = 0.1$ μs, the population difference $n_{21} = 10^{18}$ m^{-3}, the length of the gas discharge tube is 1 m and its diameter is 5 cm, the reflection factor of the output mirror is 90%, and the diffraction losses of the resonator are 0.1%.

   Instruction: Use Eq. (5.15).

5.7. The resonator of a gas discharge laser is formed by mirrors with $R = 1$ and $R = 0.95$. The dipole moment of the transition is 1.4 debyes and the relaxation time $T_2 = 1$ ns. The wavelength is 0.6328 μm and corresponds to the center of the amplification line. Determine the threshold population difference if the laser length is 1 m.

Instruction: Use Eq. (5.16). The relative dielectric permeability may be assumed to equal 1.

5.8. Using Eq. (5.17), derive a formula for the output power of a quantum generator with a symmetric configuration.

5.9. On the basis of the abbreviated equation system (5.30), obtain the self-excitation condition of a quantum generator.

Instruction: This task should be solved using linear approximation assuming that $n_{21} = n_{21}^{st}$.

5.10. Using the result of the previous task, determine the generation frequency under the condition that $\omega_{21} \neq \omega_v$.

5.11. Conduct an investigation of quantum generator dynamics, solving equation system (5.35) by using numerical methods on a computer.

Chapter 6
Nonlinear Interaction of Electromagnetic Waves with a Substance

6.1 Introduction: Methods for Analysis of Nonlinear Interaction of Electromagnetic Waves with a Medium

In analyzing systems with both continuous and discrete energy spectra, we have made sure that our equations describing polarization **P**, magnetization **M**, and current density **J** as substance reactions in electric fields **E** and magnetic fields **H** are nonlinear ones. This confirms that nonlinearity is peculiar to the substance. Moreover, we can affirm that all real physical systems are nonlinear. Nonlinearity provides the stability of the world surrounding us.

The nonlinear character of a medium's constitutive equations radically changes the process of electromagnetic wave propagation. Firstly, nonlinearity gives birth to electromagnetic waves on the harmonics of the initial wave. Secondly, the spatial wave configuration changes in a nonlinear medium. Lastly, the origination of different instabilities and hystereses is possible both in time and in space.

Nonlinear interactions of electromagnetic waves with a medium have been studied for a long time. In the first half of the twentieth century, investigations of nonlinear interactions of electromagnetic fields with electron–ion plasma were started. In particular, they were initiated by the discovery of features of radio wave propagation in the ionosphere. However, research in the field of nonlinear electrodynamics was most intensively begun when the first optical quantum generator appeared.[1] This is connected to the fact that for certain observation of nonlinear phenomena the size of the interaction area should be of the order of hundreds or

---

[1]Sergey Vavilov (1891–1951) conducted purposeful research on nonlinear optical effects prior to the appearance of lasers. In 1950, in a book titled *Light Microstructure*, he described optical nonlinearity and, probably for the first time, used the term "nonlinear optics." Vavilov founded numerous physical schools. In 1934, under his supervision, Pavel Cherenkov discovered luminescence of pure liquids under the action of radioactive substance radiation—the Cherenkov–Vavilov effect.

© Springer Nature Switzerland AG 2020
V. V. Shtykov, S. M. Smolskiy, *Introduction to Quantum Electronics and Nonlinear Optics*, https://doi.org/10.1007/978-3-030-37614-7_6

thousands of wavelengths. Investigations of nonlinear phenomena in light propagation led to the appearance of a new scientific–engineering direction: nonlinear optics.

During the study of any physical phenomena, two approaches are possible: the molecular–kinetic (microscopic) one and the phenomenological (macroscopic) one.

Using the microscopic approach and assuming a definite mechanism of the physical phenomenon at interaction of a field with a substance, it is possible to discover peculiarities of specific interactions and to calculate the quantitative parameters and characteristics describing the interaction results.

The macroscopic approach is expedient to use for revealing general regulations peculiar to different mechanisms in field interaction with a substance.

Of course, phenomenological macroscopic approaches can be used to study the same circle of problems together, supplementing each other.

In this chapter, we consider some nonlinear effects and examples of their practical applications. Statements will be based on the phenomenological description of radiation interaction with a substance.

6.2 Phenomenological Description of Nonlinear Effects

In Sect. 2.5 we determined the connection between the polarization vector and the intensity of the harmonic electric field in linear approximation. In the general case, this connection can be represented in the form of the linear equation

$$\dot{\mathbf{P}} = \dot{\chi}(\omega)\dot{\mathbf{E}}. \tag{6.1}$$

This equation for each component of vector $\dot{\mathbf{P}}$ is similar to the relations of circuit theory. So, for instance,

$$\dot{P}_x = \dot{\chi}_{xx}\dot{E}_x + \dot{\chi}_{xy}\dot{E}_y + \dot{\chi}_{xz}\dot{E}_z$$

is a linear function of all components of vector $\mathbf{E}$.

Under the arbitrary law of field time variation we can transfer from the frequency domain to presentation in the time domain through the convolution integral

$$\mathbf{P}(t) = \int_0^t \chi(\tau)\mathbf{E}(t - \tau)d\tau. \tag{6.2}$$

Here, the impulse characteristic $\chi(t)$ of the system is used, which is the Fourier transformation of the frequency response $\dot{\chi}(\omega)$.

In strong fields the linear connection between $\mathbf{P}$ and $\mathbf{E}$ may be disturbed. For a two-level system this is manifested in the form of a saturation effect (see Sect. 2.12).

If we need to take nonlinearity into account, the linear Eq. (6.1) transfers to the more complicated equation

$$\mathbf{P} = \widehat{P}(\mathbf{E}), \tag{6.3}$$

where $\widehat{P}(\mathbf{E})$ is some functional (or operator) of the electric field vector. It is impossible to accurately solve the problem on nonlinear interaction of the field and the medium. Nevertheless, as in the theory of electric circuits, we may try to find relations that will approximate the system properties with the desired degree of accuracy. The "volumetric nonlinearity" in most cases is small; therefore, it is quite enough to have the power approximation:

$$\mathbf{P}(t) = \mathbf{P}_L(t) + \int_0^t d\tau_1 \int_0^{t_1} \chi(\tau_1, \tau_2) \mathbf{E}(t - \tau_1) \mathbf{E}(t - \tau_1 - \tau_2) d\tau_2$$

$$+ \int_0^t d\tau_1 \int_0^{t_1} d\tau_2 \int_0^{t_2} \chi(\tau_1, \tau_2, \tau_3) \mathbf{E}(t - \tau_1) \mathbf{E}(t - \tau_1 - \tau_2) \mathbf{E}(t - \tau_1 - \tau_2 - \tau_3) d\tau_3 + \dots.$$

This equation is the formal generalization of (6.2) to the nonlinear case. Using the Fourier transformation and transferring to polarization writing through the coordinate components of the vectors,[2] we have

$$\dot{P}_i(\omega_p) = \chi_{ij}(\omega_p)\dot{E}_j(\omega_p) + \chi_{ijk}(\omega_p, \omega_q, \omega_r)\dot{E}_j(\omega_p)\dot{E}_k(\omega_r)$$
$$+ \chi_{ijkl}(\omega_p, \omega_q, \omega_r, \omega_s)\dot{E}_j(\omega_q)\dot{E}_k(\omega_r)\dot{E}_l(\omega_s) + \dots \tag{6.4}$$

The theory of combination frequencies in electric circuits corresponds to Eq. (6.4). The spectral component of polarization on frequency ω_p is formed owing to the field on frequency ω_p and field combination of the second, third, and higher orders. Depending on the ratio between the frequency values in (6.4), we may meet either the fields $\dot{E}(\omega)$ themselves or the complex conjugated fields $\dot{E}(-\omega) = \dot{E}^*(\omega)$.

It should be taken into consideration that nonlinear items in polarization can be connected not only with peculiarities of electron behavior in a strong electric field but also with the action of a magnetic field, as well as with the interaction of electromagnetic waves with other forms of motion: acoustic, spin, plasma, and other types of waves. A complete consecutive phenomenological description of phenomena in different media can be produced on the basis of thermodynamics.

---

[2]Each coordinate component $P_i(E_x, E_y, E_z)$ can be expanded into the Taylor power series as a function of several variables. According to the accepted rule, we assume in (6.4) that summing is provided over the repeated indices.

6.3   Classification of Nonlinear Effects

As we can see, (6.4) contains an infinite number of items. However, we can specify a value of the field intensity at which this infinite series can be replaced by a finite sum. It is usually enough to keep items till the third order. In this connection, we may consider quadratic and cubic nonlinearities.

Quadratic nonlinearity can be written in the form

$$\dot{\mathbf{P}}_{\text{quad}}(\omega_3) = \chi(\omega_3, \omega_1, \omega_2)\dot{\mathbf{E}}(\omega_1)\dot{\mathbf{E}}(\omega_2). \tag{6.5}$$

The quadratic properties of the medium are described by the constitutive tensor of the third rank. Such a tensor in the general case contains 27 elements.

Quadratic nonlinearity is responsible for generation of the second harmonic:

$$\dot{\mathbf{P}}_{\text{quad}}(2\omega) = \chi(2\omega, \omega, \omega)\dot{\mathbf{E}}(\omega)\dot{\mathbf{E}}(\omega),$$

optical detection:

$$\dot{\mathbf{P}}_{\text{quad}}(0) = \chi(0, \omega, -\omega)\dot{\mathbf{E}}(\omega)\dot{\mathbf{E}}^*(\omega),$$

and light modulation:

$$\dot{\mathbf{P}}_{\text{quad}}(\omega) = \chi(\omega, 0, \omega)\dot{\mathbf{E}}(0)\dot{\mathbf{E}}(\omega).$$

Light modulation can be considered as a variation in the dielectric permeability under the action of a permanent field:

$$\varepsilon(\omega) = \varepsilon(\omega) + \chi(\omega, 0, \omega)\dot{\mathbf{E}}(0).$$

Variations in the optical properties of crystals under the action of a permanent electric field were observed and studied for the first time in 1893 by Friedrich Pockels.[3] Now this electro-optical effect is called the Pockels effect.

The item proportional to the third power of the electric field intensity gives

$$\dot{\mathbf{P}}_{\text{cub}}(\omega_4) = \chi(\omega_4, \omega_1, \omega_2, \omega_3)\dot{\mathbf{E}}(\omega_1)\dot{\mathbf{E}}(\omega_2)\dot{\mathbf{E}}(\omega_3). \tag{6.6}$$

---

[3]Friedrich Pockels (1865–1913) was a German physicist. He obtained a doctorate from the University of Göttingen in 1888, and from 1900 to 1913 he was a professor of theoretical physics at the University of Heidelberg. In 1893 he discovered that a steady electric field applied to certain birefringent materials causes the refraction index to vary approximately in proportion to the strength of the field. This phenomenon is now called the Pockels effect.

Cubic nonlinearity is described by the constitutive tensor of the fourth rank, which has 81 elements in total.

Thanks to cubic nonlinearity, we can observe generation of the third harmonic:

$$\dot{\mathbf{P}}_{\text{cub}}(3\omega) = \chi(3\omega, \omega, \omega, \omega)\dot{\mathbf{E}}(\omega)\dot{\mathbf{E}}(\omega)\dot{\mathbf{E}}(\omega), \tag{6.6}$$

and quadratic (over-field) detection in the presence of a permanent field (for instance, photoconductivity):

$$\dot{\mathbf{P}}_{\text{cub}}(0) = \chi(0, \omega, \omega, 0)\dot{\mathbf{E}}(\omega)\dot{\mathbf{E}}^*(\omega)\dot{\mathbf{E}}(0) = \chi(0, \omega, \omega, 0)\left|\dot{\mathbf{E}}(\omega)\right|^2\dot{\mathbf{E}}(0). \tag{6.6}$$

If $\omega_1 = \omega_2 = \omega_4 = \omega$ and $\omega_3 = -\omega$, we deal with the self-impact phenomenon:

$$\dot{\mathbf{P}}_{\text{cub}}(\omega) = \chi(\omega, \omega, -\omega, \omega)\dot{\mathbf{E}}(\omega)\dot{\mathbf{E}}^*(\omega)\dot{\mathbf{E}}(\omega) = \chi(\omega, \omega, -\omega, \omega)\left|\dot{\mathbf{E}}(\omega)\right|^2\dot{\mathbf{E}}(\omega)$$

as we can consider it a manifestation of the dependence of the dielectric permeability of the medium upon the wave intensity:

$$\varepsilon(\omega) = \varepsilon(\omega) + \chi(\omega, -\omega, \omega, \omega)\left|\dot{\mathbf{E}}, (\omega)\right|^2.$$

If one field is permanent, the cubic nonlinearity

$$\dot{\mathbf{P}}_{\text{cub}}(\omega) = \chi(\omega, 0, 0, \omega)\dot{\mathbf{E}}^2(0)\dot{\mathbf{E}}^2(\omega)$$

leads to the rise of stimulated anisotropy in an optically isotropic medium.

The phenomenon of the stimulated birefringent property was discovered and studied in 1875 by John Kerr.[4]

It is clear that the problem of nonlinear interaction represents a bulky enough task. Fortunately, as for linear susceptibility, the internal symmetry peculiar to crystals reduces the number of independent elements of the constitutive tensors, and many of them become zero (see Sect. 2.10).

Since quadratic effects are described by the third-rank tensor, we can realistically state that these effects are forbidden in media with an inversion center.

Really, the transformation matrix corresponding to the symmetry center takes the form

---

[4]John Kerr (1824–1907) was a Scottish physicist and a pioneer in the field of electro-optics. He is best known for the discovery (in 1875) of what is now called the Kerr effect, in which a birefringent property arises in optically isotropic substances under the action of static fields. He also demonstrated (in 1876) a similar phenomenon with magnetic fields, which is now called the magneto-optical Kerr effect. He became a member of the Royal Society in 1890. Light from lasers allows achievement of the effect using the light's own electric field, which is called the AC Kerr effect.

$$\mathbf{A} = \begin{vmatrix} -1 & 0 & 0 \\ 0 & -1 & 0 \\ 0 & 0 & -1 \end{vmatrix}.$$

Repeating the same actions as those in Sect. 2.12, we obtain

$$\chi_{imn} = a_{ij}\chi_{ijk}a_{mj}a_{kn}.$$

Because the transformation matrix is diagonal, for any pair of elements,

$$\chi_{ijk} = -\chi_{ijk}.$$

This means that all elements of the third-rank tensor are equal to zero. Hence, physical effects described by the third-rank tensor cannot exist in structures with inversion centers.

The absence of an inversion center is a necessary condition for crystals to have piezoelectric properties. Therefore, we can contend that piezoelectrics have quadratic nonlinearity ($P_{NL} \sim E^2$). However, we must note that the absence of a symmetry center (inversion) in itself is a necessary but not at all sufficient condition for the existence of quadratic (over-field) effects.

From (6.5) it is clear that the tensor of quadratic susceptibility is symmetric over the last pair of indices, i.e., $\chi_{ijk} = \chi_{ikj}$. However, the interchangeable properties of the indices of tensor elements require proof in the general case.

Instead of the tensor χ_{ijk}, we often use the equivalent tensor $\mathbf{d}$, whose elements are

$$d_{ijk} = \frac{\chi_{ijk}}{2}.$$

The multiplier 1/2 is introduced for convenience of description of the generation of the second harmonic.[5]

The symmetry of tensor χ_{ijk} allows application instead of triple-index writing of the reduced indexation. Since replacement of jk with kj does not change the values of the tensor components, the double index jk can be changed by the only index, K.

Values of K for different combinations of jk are presented in Table 6.1.

It is also suitable to completely change the axis letter indices with figures, assuming that for the first index, $x = 1$, $y = 2$, and $z = 3$. Then, the expression $\chi_{x4} = \chi_{14}$ will be true.

---

[5]Sometimes this relation can be written in the form $d_{ijk} = \frac{\chi_{ijk}}{2\varepsilon_0}$.

Conversion of the indexation system is introduced by the following rule:

$$d_{iK} = \begin{cases} d_{ijk} & jk \leftrightarrow K = 1,2,3 \\ 2d_{ijk} & jk \leftrightarrow K = 4,5,6 \end{cases},$$

which corresponds to the accepted rules of conversion:[6]

$$(EE)_K = E_j E_k \;\; jk \leftrightarrow K = 1,2,3,4,5,6.$$

Now we express polarization on the summed frequency in the matrix form using reduced writing for tensor **d**:

$$\begin{matrix} \dot{P}_1 \\ \dot{P}_2 \\ \dot{P}_3 \end{matrix} = \begin{vmatrix} d_{11} & d_{12} & d_{13} & d_{14} & d_{15} & d_{16} \\ d_{21} & d_{22} & d_{23} & d_{24} & d_{25} & d_{26} \\ d_{31} & d_{32} & d_{33} & d_{34} & d_{35} & d_{36} \end{vmatrix} \times \begin{vmatrix} \dot{E}_1 \dot{E}_1 \\ \dot{E}_2 \dot{E}_2 \\ \dot{E}_3 \dot{E}_2 \\ \dot{E}_2 \dot{E}_3 \\ \dot{E}_1 \dot{E}_3 \\ \dot{E}_1 \dot{E}_2 \end{vmatrix}$$

or in the vector form:

$$\mathbf{\dot{P}}_{\text{quad}} = \mathbf{d}\dot{\mathbf{E}}_1\dot{\mathbf{E}}_2.$$

The number of components d_{iK} different from zero for any material is defined by the point group to which the material can be attributed. A view of tensors for different classes of crystals is shown in the Appendix 7. More detailed descriptions of the crystallographic properties of substances can be found in the literature.

The value of quadratic susceptibility depends upon the values of all three frequencies; hence, it is different for different types of nonlinear field transformation. The numerical value changes from 10^{-23} to 10^{-20} F/V.

As for cubic nonlinearity, the Neumann principle does not apply any restrictions. Thus, cubic effects are possible even in isotropic media, and the description is simpler than for quadratic ones. We therefore start with consideration of nonlinear properties of the medium from cubic nonlinear effects.

---

[6]In the literature we can encounter the transformation at which the multiplier 2 is attributed to the field E product.

6.4   Cubic Nonlinear Effects

In the previous section we clarified that in an isotropic medium, only cubic effects can exist. The constitutive equations of the two-level system obtained in Chap. 2 are valid for an isotropic medium. Therefore, our theory can answer the question of how cubic nonlinearity arises at the macroscopic level, and it allows numerical estimation of the medium parameters starting from the fundamental constants of a substance.

Let us consider the influence of a harmonic electric field on a substance:

$$\mathbf{E} = \text{Re}\left[\dot{\mathbf{E}}\exp\left(j\omega t\right)\right] \tag{6.7}$$

with complex amplitude $\dot{\mathbf{E}}$ and a frequency close to the resonance frequency ω_{21} of some electric dipole transition between nondegenerated energy levels 1 and 2.

The response of a substance to this impact—polarization $\mathbf{P}$—obeys Eq. (2.24). If the field is strong, we must take into consideration the population difference variation, which is described by Eq. (2.25). As a result, the saturation effect of the two-level system arises, which we discussed in Sect. 2.12. Let us now examine this phenomenon from the position of the power approximation.

It is known that a harmonic impact on any nonlinear system in the steady-state mode causes a response consisting of a set of harmonics. In our case, the substance nonlinearity becomes apparent in the fact that polarization besides the fundamental harmonic contains the third harmonic as well as odd harmonics of a higher order:

$$\mathbf{P} = \text{Re}\left\{\dot{\mathbf{P}}(\omega)e^{j\omega t}\right\} + \dot{\mathbf{P}}(3\omega)e^{j3\omega t} + \ldots\right\}. \tag{6.8}$$

The population difference besides the direct current (DC) component contains the second harmonic and even harmonics of a higher order. We write it in the following form:

$$n_{12} = n(0) + \text{Re}\left[\dot{n}(2\omega)e^{j2\omega t} + \dot{n}(4\omega)e^{j4\omega t} + \ldots\right]. \tag{6.9}$$

Indeed, it follows from (2.25) that n_{12} cannot contain the first harmonic of the input impact as the product of the periodic time function with a zero mean and $\mathbf{E} \sim \exp\left(j\omega t\right)$ is included in the right part of the equation. Since the right part of (2.24) contains the product of the function $\mathbf{E} \sim \exp\left(j\omega t\right)$ and the function n_{12}, which does not contain an item of type $\exp(j\omega t)$, the second harmonic cannot be introduced in the polarization structure. Continuing considerations in the same manner, we can transfer to Eqs. (6.8) and (6.9).

In this case, what is interesting for us is the complex amplitude of polarization on the frequency of the fundamental harmonic of $\dot{\mathbf{P}}(\omega)$. We can see from (2.24) that a contribution to the fundamental harmonic is made by the DC component of population difference $n(0)$. To obtain the expression for a complex amplitude of polarization, we repeat the transformations that were done in Sect. 2.12. We use the solution of the linearized Eq. (2.27) obtained in Sect. 2.6, replacing n_{12}^e with $n(0)$ in it, and obtain

$$\dot{P}(\omega) = \dot{\chi}\dot{E} = -j\frac{d^2n(0)}{3\hbar}\frac{T_2}{1+j\xi}\dot{E}, \tag{6.10}$$

where $\xi(\omega) = (\omega - \omega_{12})T_2$ is the generalized detuning. As usual, in (6.10) we assume that the absorption line width is much less than the resonance frequency: $1/T_2 \ll \omega_{21}$.

The difference of the mode under consideration consists in the fact that the deflection of $n(0)$ from n_{12}^e is taken into account because of transmission of part of the field energy to the substance. The DC component of population difference $n(0)$ was obtained in Sect. 2.12. In accepted terms,

$$n(0) = \frac{n_{12}^e}{1 + \frac{E}{E_{sat}^2\left(1+\xi^2\right)}}. \tag{6.11}$$

In (6.11), $E^2 = \left|\dot{E}\right|^2$ is the square of the electric field amplitude characterizing the emission intensity and

$$E_{sat}^2 = \frac{3\hbar^2}{T_1T_2d^2} \tag{6.12}$$

is the square of the saturation field amplitude (see Sect. 2.12).

Let us return to Eq. (6.10) for polarization. We substitute (7.11) in it and expand the fraction into a series of $\mathring{A}/E_{sat}$ powers:

$$\dot{P}(\omega) \approx -j\frac{\chi_0}{1+j\xi}\dot{E} + j\frac{\chi_0}{(1+j\xi)\left(1+\xi^2\right)}\frac{E^2}{E_{sat}^2}\dot{E} + \ldots, \tag{6.13}$$

where $\chi_0 = \frac{d^2n_{12}^e T_2}{3\hbar}$.

Introducing the linear part $\dot{\chi}_1$ (see (2.30) in Sect. 2.7) and the nonlinear part $\dot{\chi}_3$ of the complex susceptibility of a substance,[7] we can rewrite (6.13) in the form

$$\dot{P}(\omega) = \dot{\chi}_1(\omega)\dot{E} + \dot{\chi}_3(\omega)E^2\dot{E}. \tag{6.14}$$

From (6.13) and (6.14) we have

---

[7]In an isotropic medium, cubic susceptibility can be considered a scalar quantity.

$$\dot{\chi}_3(\omega) = j \frac{\chi_0}{(1 + j\xi)(1 + \xi^2)} \cdot \frac{1}{E_{\text{sat}}^2} = -\frac{\dot{\chi}_1}{E_{\text{sat}}^2 (1 + \xi^2)}. \tag{6.15}$$

Thus, the saturation effect leads to appearance of a cubic item in the expression for the polarization amplitude, and it is concerned with the class of cubic nonlinear effects.

The cubic susceptibility has real and imaginary parts. From (6.15) we obtain

$$\chi_3'(\omega) = \frac{\chi_0}{E_{\text{sat}}^2} \frac{\xi}{\left(1 + \xi^2\right)^2} \quad \text{and} \quad \chi_3''(\omega) = \frac{\chi_0}{E_{\text{sat}}^2} \frac{1}{\left(1 + \xi^2\right)^2}. \tag{6.16}$$

The internal nonlinearity that is peculiar to any two-level substance particles becomes apparent for $\text{Å} \sim E_{\text{sat}}$ or at the emission power flow density $I \sim I_{\text{sat}}$. In Sect. 2.12 an estimate of this value was made for the resonance transition in ammonia on the frequency $f_{12} = 23.866$ GHz, which gave the value $I_{\text{sat}} \cong 6.7$ mW/cm^2. In other cases the substance nonlinearity is demonstrated at essentially high emission intensity. So, for transition between electron levels of Cr^{3+} ions in a ruby at a resonance wavelength of $\lambda_{21} = 0.69$ μm we have $T_1 \approx 3$ ms, $T_2 \approx 7$ ps, and $d \approx 1.5 \cdot 10^{-32}$ C · m, which according to (6.12) gives $I_{\text{sat}} \cong 1$ kW/cm^2 ($E_{\text{sat}} \approx 10^5$ V/m). Radiation of such intensity can be obtained using pulsed solid-state lasers. Assuming that almost all active particles in the crystal are located at the lower level (i.e., the population difference is equal to the chromium ion concentration $2 \cdot 10^{25}$ m^{-3}) we obtain $|\dot{\chi}_3|_{\text{max}} \cong 10^{-16} \cdot E_{\text{sat}}^{-2} \cong 10^{-26} \frac{\text{F·m}}{V^2}$. Far from the resonance, the nonlinearity is much weaker. So, for carbon bisulfide at a wavelength of $\lambda_{21} = 0.69$ μm, $|\dot{\chi}_3|_{\text{max}} \cong 4 \cdot 10^{-31} \frac{\text{F·m}}{V^2}$.

In Sect. 6.3 we noted that cubic nonlinearity may be considered as a self-impact phenomenon, which is demonstrated in the dependence of medium properties upon radiation intensity. According to (6.14), the absolute dielectric permeability can be presented in the form

$$\dot{\varepsilon}_a\left(\omega, E^2\right) = \varepsilon' \varepsilon_0 + \dot{\chi}_1(\omega) \left(1 - \frac{E^2}{E_{\text{sat}}^2 \left(1 + \xi^2\right)}\right),$$

where $\varepsilon' \varepsilon_0$ is its linear part describing other mechanisms of field interaction with a medium.

How does the saturation effect manifest itself at powerful emission propagation in a medium of resonance two-level particles? To answer this question, we write the complex propagation constant of the wave in the medium as

$$\dot{\gamma} = \frac{\omega}{c} n \cdot \sqrt{1 + \frac{\dot{\chi}_1(\omega)}{\varepsilon'\varepsilon_0}\left(1 - \frac{E^2}{E_{sat}^2\left(1 + \xi^2\right)}\right)}^{-1/2}.$$

Here, $n = \sqrt{\varepsilon'}$ is the medium refraction index. At $\left|\frac{\dot{\chi}_1(\omega)}{\varepsilon'\varepsilon_0}\right| \ll 1$ the phase constant is

$$\beta\left(\omega, E^2\right) \cong \frac{\omega}{c} n \cdot \left[1 + \frac{\chi_1'(\omega)}{2\varepsilon'\varepsilon_0}\left(1 - \frac{E^2}{E_{sat}^2\left(1 + \xi^2\right)}\right)\right], \tag{6.17}$$

and the damping factor is

$$\alpha\left(\omega, E^2\right) \cong \alpha\left(1 - \frac{E^2}{E_{sat}^2\left(1 + \xi^2\right)}\right), \tag{6.18}$$

where α is the damping factor in the linear approximation. According to (6.17) and (6.18) the propagation constant depends upon E^2 or power flow density I. We should take into account that not just the saturation effect can be the reason for n dependence upon radiation intensity. There is a series of physical mechanisms of a nonresonance character that lead to the same result.

6.5 Nonlinear Absorption

To designate through α_0 the linear part of damping factor α_p independent of E^2, taking account of (6.18) under resonance conditions, we have

$$\alpha_p \cong \alpha_0\left(1 - \frac{E^2}{E_{sat}^2}\right). \tag{6.19}$$

Reduction of the damping factor with radiation intensity growth or, in other words, saturation of absorption causes equalization of the level population that leads to a decrease in the response of the substance in the electric field.[8]

Nevertheless, nonlinear phenomena do not always lead to absorption decreasing under the action of a powerful emission. There are physical mechanisms with which the absorption does not decrease but rises with E^2 growth. This occurs in so-called photochrome substances.

---

[8]This question was discussed in Sect. 5.3 in conformity with the quantum amplifier.

Nonlinear absorption leads to disturbance of the usual exponential law of wave damping in the medium. Really, the intensity of the plane wave traveling along the z axis obeys the equation

$$\frac{dI}{dz} = -\alpha_p I,$$

which has the exponential solution $I(z) = I(0) \exp(-\alpha_p z)$ only in the case that the damping factor does not depend on E^2.

In the general case, (6.19) should be presented in the form

$$\alpha_p \approx \alpha_0 (1 - aI),$$

in which constant a considers the cubic nonlinearity, which can be both positive and negative in quantity.

Taking into account the self-impact effect, we have

$$\frac{dI}{dz} = -\alpha_0 (1 - aI)I. \tag{6.20}$$

Equation (6.20) is a equation with separable variables to which a solution can easily be found using the method of direct integration:

$$I(z) = I(0) \frac{\exp(-\alpha_0 z)}{1 - aI(0)[1 - \exp(-\alpha_0 z)]}. \tag{6.21}$$

The transparency coefficient for the layer of finite length z is

$$K(z) = \frac{I(z)}{I(0)} = \frac{\exp(-\alpha_0 z)}{1 - b + b \exp(-\alpha_0 z)}. \tag{6.22}$$

Figure 6.1 shows the function $K(z)$ for different values of the product $b = aI(0)$. If $a > 0$, the transparency coefficient increases with the growth of the power flow density $I(0)$. That is why this phenomenon is called self-blooming.

Self-bloomed substances under radiation action have found applications in laser engineering. They play the role of the gate modulating the resonator Q factor. Usually we can use a bath with a saturated colorant. Application of the cell provides a rigid mode of self-excitation (see Sect. 5.6).

At $a < 0$ the absorption in the medium increases with growth of the power flow density and the transparency coefficient (6.22) decreases. This phenomenon is used in optical limiters of light intensity (for example, in "chameleon" glasses).

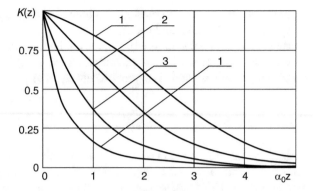

Fig. 6.1 Variation of power flow density with distance in the absorbing medium: *1* denotes $b = 0.9$, *2* denotes $b = 0.7$, *3* denotes $b = 0.0$, *4* denotes $b = -2.0$. If $b > 0$, at high wave intensities the power flow depends weakly upon the distance. This phenomenon is called self-blooming. In the case that $b < 0$, cubic nonlinearity is manifested in the fact that at high values of I_0 the power flow decreases fast

Table 6.1 Conformities for reduced indexation

ik	xx	yy	zz	yz	xz	xy
K	1	2	3	4	5	6

6.6   The Kerr Optical Effect

In the previous section we considered the function of the damping factor versus the wave intensity. However, not only the imaginary part but also the real part of the complex susceptibility $\dot{\chi}$ is a function of E^2. Therefore, according to (6.17), we write the refraction index of the medium as

$$n(\omega, E^2) \cong n \cdot \left[1 + \frac{\chi_0 \xi}{2\varepsilon' \varepsilon_0 (1 + \xi^2)} \left(1 - \frac{E^2}{E_{\text{sat}}^2 (1 + \xi^2)} \right) \right]. \qquad (6.23)$$

The linear part of the refraction index does not represent a special interest; therefore, we can write (6.23) in the form

$$n \cong n_1 + n_3 E^2, \qquad (6.24)$$

where n_1 is a linear part of the refraction index and

$$n_3 = - \frac{\chi_0 \xi}{2n\varepsilon_0 E_{\text{sat}}^2 (1 + \xi^2)^2}.$$

The quadratic function of the substance refraction index versus the amplitude of the field intensity in the wave is called the Kerr optical effect. This effect gives birth to a series of phenomena that are very interesting and important for practicing wave propagation in a medium with cubic nonlinearity.

Let us estimate the last item value in (6.24), i.e., the nonlinear component of the refraction index arising as a result of the saturation effect. At $\xi = \pm 1/\sqrt{3}$, n_3 achieves its maximum, whose modulus is approximately equal to

$$n_{max} \cong \frac{\chi_0}{6n\varepsilon_0 E_{sat}^2}.$$

The numerical value of this maximum for a ruby ($T_2 = 7$ ps, $d = 1.5 \cdot 10^{-31}$ $C \cdot m$, $n = 1.77$, $E_{sat} \approx 10^5$ V/m) at a wavelength of $\lambda_{21} = 0.69$ µm is $n_{max} \approx 10^{-16}$. Therefore, the nonlinear part of the refraction index, even under conditions of a transition to deep saturation, is a value much less than 1.

The value χ_3' is usually indicated in the literature. To use (6.16) we have

$$n_3 = \frac{\chi_3'}{2n\varepsilon_0}. \tag{6.25}$$

Far from the resonance, the cubic effect is essentially weaker. So, in liquid carbon bisulfide ($n = 1.5$) at a wavelength of 1 µm, $\chi_3' \approx 2 \cdot 10^{-31}$ F/m^2 and $n_3 \approx 10^{-20}$. But, in spite of the apparent insignificance of the effect, the light wave passing the distance of 10,000 wavelengths in the medium still "feels" such a small variation in the refraction index.

6.7 The Self-Focusing Phenomenon

When electromagnetic wave beams propagate in a nonlinear medium, whose properties are described by (6.24), field distribution variations happen in the cross-section. Sometimes these variations have the character of spatial instabilities. We consider the following physical situation as an example. Let a heterogeneous beam (in the cross-section) with the complex amplitude

$$\dot{E}(x, y, z) = \dot{U}(x, y, z) \exp(-j\beta z) \tag{6.26}$$

fall along the normal from a vacuum on the medium boundary in which the Kerr effect is observed. In (6.26), $\beta = \frac{\omega}{c}$, $\dot{U}(x, y, z)$ is the slowly changing coordinate function on the wavelength.

We assume that at the point $z = 0$, this beam has a parabolic profile (Fig. 6.2a):

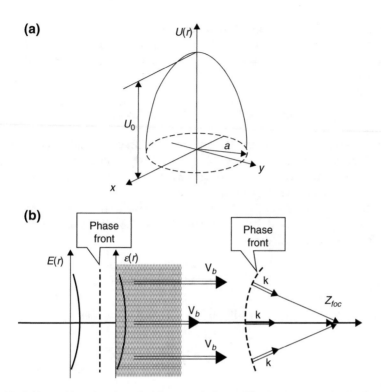

Fig. 6.2 Self-focusing of a beam with a parabolic profile (**a**) in a medium with cubic nonlinearity (**b**). A heterogeneous beam in the cross-section gives birth to irregularity of the refraction index. If the refraction index in the beam center is larger than in its periphery, the beam velocity of the central beam part V_b is larger than that for the peripheral parts. As a result, the phase front is bent. Curvature of the wave front is equivalent to the action of a focusing lens

$$|U(x,y)|^2 = U_0^2\left(1 - \frac{r^2}{a^2}\right),$$

where $r^2 = x^2 + y^2$.

What will happen with this beam in the nonlinear medium during its propagation along the z axis? To give an irrefragable answer, we have to solve the nonlinear Helmholtz equation, written taking into consideration (6.24):

$$\nabla^2\dot{E} + \beta^2\left(n_1 + n_3 E^2\right)\dot{E} = 0, \tag{6.27}$$

which the complex amplitude of the beam (6.26) obeys. The solution to (6.27) can be found in the literature. We shall try to represent a qualitative picture of the processes that occur and to obtain some quantitative estimations.

If $n_3 > 0$, the refraction index, under the action of radiation, turns out to be nearer the beam axis than its periphery. In other words, the refraction index profile, like the beam profile itself, is parabolic. The wave phase delay in the medium will depend on r in the same manner. Therefore, at distance z from the boundary, which is so small that the beam profile has no time to change essentially, its complex amplitude can be written as

$$\dot{U}(x,y,z) = U_0\sqrt{1 - r^2/a^2}\exp\left[-j\beta n_3 U_0^2\left(1 - r^2/a^2\right)z\right]. \qquad (6.28)$$

The presence of the exponential multiplier in (6.28) depending upon r is evidence of the fact that the beam phase front is not already a plain one (Fig. 6.2b). The nonlinear medium acts as a focusing lens, which allows expectation of beam contraction to a point at some distance z_{foc}.

To estimate this distance, we find the curvature radius of the beam phase front R (z) at point z. The phase difference of the wave on the beam axis and at distance r from the axis (see Fig. 6.2b) is

$$\beta\left(R - \sqrt{R^2 - r^2}\right) \cong \beta r^2/2R.$$

Comparing this expression with (6.28), it is easy to see that

$$R(z) = \frac{a^2}{U_0^2 z n_3}.$$

During beam advancement deep into the nonlinear medium, the curvature radius of the phase front decreases. Simultaneously, the beam radius decreases and the field amplitude grows. This, to a greater extent, bends the front and intensifies the beam compression process. We can estimate the distance at which its "collapse" occurs, roughly assuming $z_{\text{foc}} \cong R(z_{\text{foc}})$, from which

$$z_{\text{foc}} \cong \frac{a}{U_0\sqrt{n_3}}. \qquad (6.29)$$

We would assume that self-focusing of the beam with a profile close to a parabolic one takes place at an indefinitely small amplitude and at indefinitely weak nonlinearity of the substance. However, this is wrong.

The thing is that we have completely forgotten about the existence of diffraction, which contradicts self-focusing. It is known that the angle of diffraction divergence of the beam with radius a is equal to $\theta \sim \lambda/a$. At distance z the beam radius will increase by $\Delta a = \theta z \cong z\lambda/a$. If $\Delta a \leq a$ is fulfilled for $z \leq a^2/\lambda = L_{Fr}$, ($L_{Fr}$ is the Fresnel diffraction length), the diffraction role is still comparably small. Nevertheless, for $z > L_{Fr}$ the beam strongly widens and the medium's ability to self-focus is not enough to overcome the diffraction blooming of the beam. It follows from this that self-focusing occurs if $z_{\text{foc}} < L_{Fr}$. Substituting (6.29) in this inequality and taking into account that the power transported by the beam is $P = \frac{\pi U_0^2}{4Z_{\text{med}}}a^2$, where Z_{med} is the characteristic impedance of the medium, we obtain the following condition of self-focusing existence in the nonlinear medium

$$P \geq \frac{\pi}{4Z_{\text{med}}} \frac{\lambda^2}{n_3} .$$ (6.30)

Let us estimate the radiation power level necessary for self-focusing in liquid carbon bisulfide ($n_1 = 1.5$) at a wavelength of 1 μm. In accordance with (6.30) and for $n_3 \approx 10^{-20}$ we have $P \approx 10^6$ W, which is fully achievable with the help of modern pulsed lasers. For a ruby near the frequency of resonance absorption, P is only around $10^2 - 10^3$ W.

Initially, the results of many experiments with powerful laser radiation—for instance, involving destruction of transparent materials—did not coincide with the results of theoretical calculations. It turned out that the self-focusing phenomenon decreases the threshold power densities for many nonlinear optical effects and even essentially changes the phenomenon pattern.

In practice, self-focusing is sometimes a harmful phenomenon. Nonlinear optical devices often operate at power levels commensurable with (6.30). If there is the slightest irregularity in the input beam profile, the self-focusing "amplifies" it and finally the beam can be broken into separate threads, destroying the device operation. It is said that in this case, transverse instability of the wave occurs. Moreover, a high concentration of the wave power as a result of self-focusing may lead to overheating and destruction of the substance in so-called optical disruption.

6.8 Optical Bistability

The ability of a substance to change its refraction index under light action allows creation of a very interesting device, which can be considered as one of the most significant elements of optical computers.

Let us examine a Fabry–Pérot resonator, consisting of two semitransparent mirrors located opposite each another and filled by a substance with refraction index n.

Let a wave fall on one of the mirrors, and let the power flow density of this wave be I_1. Partially, this wave will be reflected from the mirror, giving birth to wave I_3, and partially it will pass through the resonator, having excited oscillations in it. This can be considered as superposition of two waves with intensities I_{i1} and I_{i2} propagating in opposite directions (Fig. 6.3a).

The transfer function in the power of the resonator is equal to the ratio of intensities of the passed wave Π_2 to the incident (see also Sect. 5.3):

Fig. 6.3 **a** A Fabry–Pérot resonator. **b** Function of the transfer coefficient in power versus the phase shift on the resonator length. The simplest model can be built with replacement of the open resonator modes by plain waves. Energy relations allow us to obtain the wave intensity inside the resonator, and phase relations allow us to obtain the resonance frequencies and the transparence factor

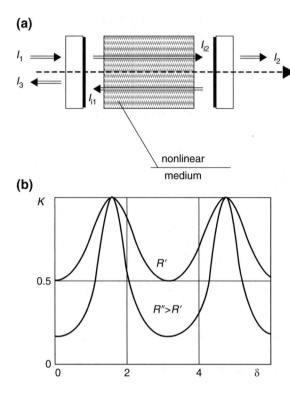

$$K = \frac{I_1}{I_2} = \frac{1}{1 + F \sin^2 \delta}, \tag{6.31}$$

where $\delta = \frac{2\pi n L}{\lambda}$ is the phase shift for resonator length L, $F = \frac{4R}{(1-R)^2}$ is the parameter characterizing its Q factor, and R is the reflection factor from the mirrors.

In accordance with (6.31), the frequency response of the resonator has a view of maximums periodically located on the frequency axis (Fig. 6.3b), and in each maximum the total phase shift δ of the wave is a multiple of π. Each such point corresponds to the natural resonator frequency.

We assume that the substance filling the resonator has cubic nonlinearity of type (6.24). Let the phase shift δ_0 be equal to $\frac{2\pi n_1 L}{\lambda}$ (see Fig. 6.3b) at a light intensity close to zero. What will occur with I_1 growth?

The growth in incident light intensity at $\chi'_3 > 0$ leads to a refraction index increase and to growth of δ, i.e., approaching the resonance. This will lead to a total intensity $I_i = I_{i1} + I_{i2}$ increase, which to a greater extent will increase δ and, relatively, transfer function K.

The process will progress in an avalanche-like manner. As a result, at small growth of the incident light intensity a jump-like increase of the passed light intensity will be observed.

Let us try to make more exact the character of function I_2 versus I_1. Since $I_2 = I_{i1} - I_{i2}$, and $I_{i2} = RI_{i1}$, the intensity of the passed light can be easily connected with the oscillation intensity in the resonator: $I_2 = I_1 \frac{1-R}{1+R}$. Taking (6.31) into consideration, we obtain the following equation system for finding I_2 as the function of I_1:

$$K = \frac{1}{1 + F \sin^2(\delta_0 + aI_i)},$$ (6.32)

$$K = \frac{1 - R}{1 + R} \frac{I_i}{I_1},$$ (6.33)

where $a = \frac{4\pi n_3 Z_{med} L}{\lambda}$.

We demonstrate its solution by the graphic method. In Fig. 6.4a the curve corresponds to (6.32) and the straight-line family corresponds to Eq. (6.33) for different values of the incident light intensity. With the growth of I_1 an increase in T is observed, which leads to deflection of the function $I_2(I_1)$ from the linear one (Fig. 6.4b).

With a definite combination of δ_0 and F in some region of I_1 values, a situation is possible where equation system (6.32) and (6.33) has three solutions, with high, intermediate, and low transmission. Two of them are stable, and appropriate segments of the function I_2 versus I_1 are shown by the solid curve in Fig. 6.4b. The third one—the intermediate solution—is unstable (it is shown by the dotted segment in Fig. 6.4b). From the figure we see that function $I_2(I_1)$ is really strongly nonlinear. For different combinations of the initial detuning δ_0 and the resonator Q factor, this function can have a hysteresis or may not have one. In the presence of a hysteresis

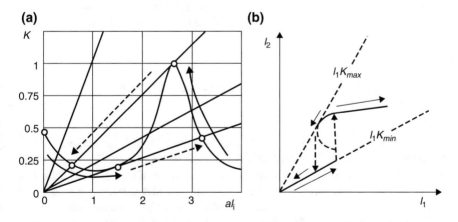

Fig. 6.4 The phenomenon of optical bistability. **a** Graphic solution of the equation system. **b** Nonlinear function of output beam intensity versus input beam intensity. The resonator is filled with a nonlinear medium with cubic nonlinearity. Field I_i changes its dielectric properties and thus tunes the resonator. Solution of the equation system determines the connection between output and input flows. Function $I_2(I_1)$ has a hysteresis, because of which we can create a bistable element

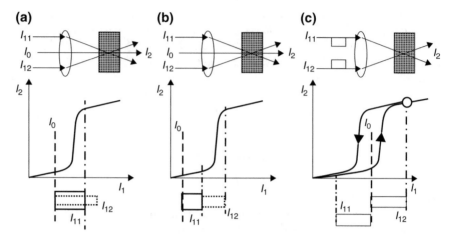

Fig. 6.5 Optical elements of digital logic. **a** OR element. **b** AND element. **c** Trigger. These elements form the basis of an optical computer that can perform up to 10^{13}–10^{15} operations per second, with the number of parallel channels being about 10^6

our device can be in two stable states with low and high transmission; therefore, the phenomenon under consideration is known as optical bistability.

On the basis of bistable resonators we can create different digital logic elements in which 0 and 1 signals are transmitted not in the form of high and low voltage levels but in the form of high and low light intensities. Examples of such elements are shown in Fig. 6.5. These elements form the basis of an optical computer able to perform up to 10^{13}–10^{15} operations per second for a number of parallel channels of the order of 10^6. Bistable elements are preformed on the basis of superlattices— periodic heterostructures (GaAs/Ga$_x$Al$_{1-x}$As). They ensure a switching time of less than 10^{-9} s at a specific switching energy of the order of 10^{-12} J/μm^2.

6.9 Generation of the Third Harmonic: Multiphoton Processes

The above-considered nonlinear effects were connected with the response of the medium arising on the frequency of the field influencing the substance. However, polarization (7.45) besides the fundamental harmonic contains also the third harmonic and other odd harmonics. These harmonics, under definite conditions, can appear as sources exciting electromagnetic waves on frequencies 3ω, 5ω, etc.

Let us examine in detail the generation of the third harmonic when the frequency of the third harmonic 3ω is equal to the resonance frequency ω_{21} for some transition. First of all, we shall find the local connection of the complex amplitude of the third harmonic of polarization $\mathbf{P}(3\omega)$ with the complex amplitude of the field generating it. For this we use equation system (2.24) and (2.25).

In accordance with (2.25), the population difference contains components with the complex amplitude $n_{12}(2\omega)$, due to the presence of the product $\frac{\partial P}{\partial t}E$ in the right part of (2.25). This alternate part of the population difference is the reason for the appearance of polarization $\dot{P}(3\omega)$ on frequency 3ω in accordance with (2.24).

First of all, we find the complex amplitude of polarization on the fundamental frequency, substituting $\omega_{21} = 3\omega$ in (2.28):

$$\dot{P}(\omega) = \frac{d^2}{4\hbar\omega}n_{12}^e\dot{E}(\omega). \qquad (6.34)$$

The next stage is determination of the second harmonic amplitude of the population difference. Substitution of (6.8) and (6.9) in (2.25) gives

$$\left(j2\omega + T_1^{-1} \right)n_{12}(2\omega) = -j\frac{1}{3\hbar}\dot{P}(\omega)\dot{E}(\omega).$$

Taking into consideration that $\omega \gg T_1^{-1}$ and using formula (6.34), we have

$$n_{12}(2\omega) = -\frac{\dot{P}(\omega)\dot{E}(\omega)}{6\hbar\omega} = -\frac{d^2 n_{12}^e}{24\hbar^2\omega^2}\dot{E}(\omega)\dot{E}(\omega). \qquad (6.35)$$

Addressing (2.24), we obtain the complex amplitude of the third harmonic of polarization. Under the condition of resonance $\omega_{21} = 3\omega$,

$$\dot{P}(3\omega) = -j\frac{d^2 T_2}{6\hbar}n_{12}(2\omega)\dot{E}(\omega).$$

Finally, after substitution of (6.35), we obtain

$$\dot{P}(3\omega) = \dot{\chi}_3(3\omega)\dot{E}(\omega)\dot{E}(\omega)\dot{E}(\omega), \qquad (6.36)$$

where

$$\dot{\chi}_3(3\omega) = j\frac{d^4 T_2 n_{12}^e}{144\hbar^3\omega^2} = j\frac{\chi_0}{E_{sut}^2}\frac{1}{16T_1 T_2\omega^2} \qquad (6.37)$$

is the nonlinear susceptibility at generation of the third harmonic. Fulfilling all calculations for $3\omega \neq \omega_{12}$, we may state that the nonlinear susceptibility on the frequency of the third harmonic sharply decreases with detuning growth as the nonlinear susceptibility on the fundamental frequency (6.16). Substituting in (6.37) the above-mentioned parameters of the ruby, we have $|\chi_3(3\omega)| \approx 3 \cdot 10^{-44} \frac{F\cdot m}{V^2}$.

The polarization current can set up a correspondence with the nonlinear polarization on the third harmonic. This current excites the electromagnetic field changing with frequency 3ω. To calculate it we need to solve the Maxwell equation system together with nonlinear constitutive equations. Of course, it is impossible to obtain the exact solution to this problem. Therefore, we use approximate methods based on the smallness of the local nonlinearity.

In the simplest case it is assumed that the field of the initial wave with frequency ω is specified and, hence, the nonlinear polarization is specified at all points of the interaction area. Such a solution method is called a specified field method. We use it for analysis of the generation of the third field harmonic.

Let us assume that on the nonlinear medium boundary with the coordinate $z = 0$, a plain electromagnetic wave falls whose amplitude is specified:

$$\dot{E}(\omega) = \dot{E}_1 \exp\left(-j\beta_1 z\right).$$

We remember that phase constant $\beta_1 = n_1\omega/c$, where n_1 is the medium refraction index of the fundamental frequency. The field of this wave gives birth to the polarization wave on the frequency of the third harmonic with the complex amplitude

$$\dot{P}(3\omega) = \dot{P}_3 \exp\left(-j3\beta_1 z\right), \tag{6.38}$$

which plays the role of stimulated force in the right part of the equation

$$\nabla^2 \dot{E}(3\omega) + \beta_3^2 \dot{E}(3\omega) = -9\omega^2 \mu_0 \dot{P}(3\omega). \tag{6.39}$$

This equation is still complicated enough. Therefore, we use the weakness of the local nonlinearity and shall find the solution in the plain wave form

$$\dot{E}(3\omega) = \dot{E}_3 \exp\left(-j\beta_3 z\right) \tag{6.40}$$

with the complex amplitude $\dot{E}_3$ slowly changing along the z axis. In (6.40), $\beta_3 = n_3 3\omega/c$ is the phase constant of the frequency of the third harmonic.

Let us find the equation for $\dot{E}_3(z)$. We substitute (6.38) and (6.40) into the wave equation (6.39). First of all, let us examine the first item in (6.39):

$$\nabla^2 \dot{E}(3\omega) = \frac{\partial^2 \dot{E}(3\omega)}{\partial z^2} = \left\{ \frac{\partial^2 \dot{E}_3}{\partial z^2} - 2j\beta_3 \frac{\partial \dot{E}_3}{\partial z} - \beta_3^2 \dot{E}_3 \right\} \cdot \exp\left(-j\beta_3 z\right).$$

The condition of amplitude variation slowness compared with the fast oscillating exponential multiplier means that $\partial \dot{E}_3/\partial z \ll \beta_3 \dot{E}_3$. This allows us to neglect the second z derivative. Then we obtain the following abbreviated equation for the complex amplitude of the third harmonic:

$$\frac{\partial \dot{E}_3}{\partial z} = -j\frac{3\pi}{\lambda \varepsilon_0 n_3} \dot{P}_3 \exp\left(-j\Delta\beta z\right),$$ (6.41)

where $\Delta\beta = 3\beta_1 - \beta_3$.

Equation (6.41) is solved by simple integration. Its solution under the boundary condition $\dot{E}_3(0) = 0$ takes the form

$$\dot{E}_3 = \frac{3\pi}{\lambda \varepsilon_0 n_3} \dot{P}_3 \frac{\exp\left(-j\Delta\beta z\right) - 1}{\Delta\beta}.$$

Hence, the modulus of the field complex amplitude on the frequency of the third harmonic is

$$\dot{E}_3 = \frac{3\pi}{\lambda \varepsilon_0 n_3} \mid \dot{P}_3 \mid \frac{\sin\left(0.5\Delta\beta z\right)}{0.5\Delta\beta}.$$ (6.42)

Let us discuss this result. First of all, we note that in condensed media, frequency dispersion occurs, so in the optical range, n_1, as a rule, differs from n_3. For this reason, $\Delta\beta \neq 0$ and the third harmonic amplitude depends upon the coordinate according to the sine law. With $\Delta\beta$ growth, the maximal possible value of E_3 becomes smaller and smaller (Fig. 6.6). This is connected with the fact that the polarization wave turns out to be nonsynchronized with the electromagnetic wave excited by it and alternately sometimes returns the energy to the third harmonic of the field and sometimes takes it back. At fulfillment of the so-called condition of phase synchronism only or phase matching,

$$3\beta_1 = \beta_3,$$ (6.43)

the polarization wave at any point of the space returns the energy to the field and the third harmonic amplitude grows linearly with the coordinate.

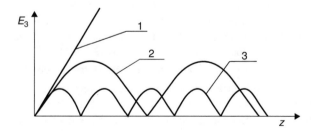

Fig. 6.6 Function of the field third harmonic amplitude versus the longitudinal coordinate. Under conditions of phase matching, $\Delta\beta = 0$ (denoted by *1*), and third harmonic waves radiated by the secondary sources $\dot{P}_3 \exp\left(-j3\beta_1 z\right)$ are in-phase combined at the observation point. An infinite amplitude increase is impossible because of exhaustion of the first harmonic of the initial wave. If the conditions for phase matching are not fulfilled, the function of the third harmonic versus z becomes nonmonotonic (*2* denotes $\Delta\beta_1 \neq 0$ and *3* denotes $\Delta\beta_2 > \Delta\beta_1$)

An impression can be formed that at fulfillment of (6.43) with an increase in the thickness of the nonlinear substance z layer, the third harmonic amplitude achieves an arbitrarily large value. Of course, this is wrong. The result (6.42) is obtained on the assumption that the fundamental frequency amplitude—and, hence, the polarization amplitude as well—do not depend on the coordinate.

As a matter of fact, during excitation, the third harmonic takes the energy exactly from the first harmonic of the field and the amplitude of the last gradually decreases. The process of wave exhaustion limits the maximal possible value of the third harmonic amplitude. In our solution we can take it into consideration, replacing everywhere above the phase constant β_1 with the complex propagation constant

$$\dot{\gamma}_1 = \beta_1 - j\alpha_1.$$

Here, α_1 is the damping coefficient considering the energy transfer to the third harmonic. Then, under the condition of phase matching,

$$\dot{E}_3 = \frac{3\pi}{\lambda\varepsilon_0 n_3}\,\dot{P}_3\,\frac{1 - \exp(-3\alpha_1 z)}{3\alpha_1}.$$

The limiting amplitude value is

$$\dot{E}_3 = \frac{\pi}{\lambda\alpha_1\varepsilon_0 n_3}\,|\,\dot{P}_3\,|\,.$$

Of course, the numerical value of α_1 is unknown and for its determination we should include some additional considerations. A self-consistent problem solution can be obtained using the coupled-wave method, which is discussed in Sect. 6.12.

To satisfy the synchronism condition (6.43) in the optical range we need to use the specific approach based on application of anisotropic crystals. It will be described below in the example of the second harmonic generation.

Let us estimate the effectiveness of the third harmonic generation at a wavelength of $\lambda_3 = 0.69$ μm in a ruby crystal with length $z = 1$ cm for $E_1 = 3 \cdot 10^6$ V/m ($I_1 = 1$ MW/cm^2). Assuming that $n_1 = 1.74$ and $n_3 \cong 1.76$, we obtain that the parameter characterizing the loss value in the amplitude due to nonfulfillment of (6.41) is $\Delta\beta \approx 5900$. Then, in accordance with (6.42) and (6.36) for $|\chi_3(3\omega)| \approx 3 \cdot 10^{-44}\,\frac{F \cdot m}{V^2}$, we have $E_3 \sim 10^{-11}$ V/m. As we can see, the effectiveness of the third harmonic generation in the ruby even at resonance is very small.

Now we shall look briefly at the process of the third harmonic generation not from the wave but from the corpuscular point of view. This consists in absorption by a substance particle of three photons with the same energy $\omega_1\hbar = \omega_{12}\hbar/3$, its transition from the lower energy level to the upper one (Fig. 6.7, left panel), and the further reverse transition accompanied by radiation of the one photon with energy:

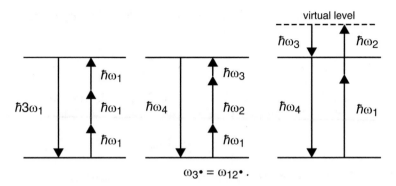

$$\omega_3{}^\bullet = \omega_{12}{}^\bullet.$$

Fig. 6.7 Multiphoton processes in a two-level system. *Left:* Absorption of three photons and emission of the photon on the triple frequency. *Middle:* Absorption of three photons of different frequencies and emission of the photon on the summed frequency. *Right:* Absorption of two photons and emission of a pair of photons through the virtual level under condition $\omega_1 + \omega_2 = \omega_3 + \omega_4$. These processes demonstrate the energy conservation law. For reliable observation of multiphoton processes the condition of phase matching should be fulfilled, which is a consequence of the impulse conservation law

$$\omega_3 \hbar = \omega_{12} \hbar.$$

As for any particle interaction, in this process the laws that must be satisfied are the energy conservation law:

$$3\omega_1 \hbar = \omega_3 \hbar$$

and the impulse conservation law:

$$3\hbar\beta_1 = \hbar\beta_3,$$

which is equivalent to the condition of phase synchronism (6.43).

From the common considerations it is clear that besides the three-photon absorption on the one frequency ω_1, absorption of three photons of three different frequencies ω_1, ω_2, and ω_3 may occur (Fig. 6.7, middle panel), as well as radiation of one photon on frequency ω_4. The main point is that the energy conservation law should be satisfied:

$$\omega_1 + \omega_2 + \omega_3 = \omega_4.$$

The influence on the nonlinear medium of the waves with frequencies ω_i and with the phase constant β_i, $i = 1, 2, 3, 4$,

$$\dot{\mathbf{E}}(\omega_i) = \dot{\mathbf{E}}_{\mathbf{i}} \exp\left(-j\beta_i z\right), \qquad (6.44)$$

leads, as a result of three-photon absorption, to polarization appearance on the combined frequency ω_4 with the amplitude

$$\dot{\mathbf{P}}(\omega_4 = \omega_1 + \omega_2 + \omega_3) = \chi(\omega_4, \omega_1, \omega_2, \omega_3)\dot{\mathbf{E}}(\omega_1)\dot{\mathbf{E}}(\omega_2)\dot{\mathbf{E}}(\omega_3). \qquad (6.45)$$

This polarization effectively excites the wave $\dot{\mathbf{E}}(\omega_4)$ in the case that the impulse conservation law or the condition of phase synchronism is satisfied:

$$\beta_1 + \beta_2 + \beta_3 = \beta_4.$$

There may be a process of two-photon absorption on frequencies ω_1 and ω_2 and radiation of two photons on frequencies ω_3 and ω_4 (Fig. 6.7, right panel). The energy conservation law in this case takes the form

$$\omega_1 + \omega_2 = \omega_4 + \omega_3. \qquad (6.46)$$

The process of the birth of the "new" photon pair has a spontaneous character. It means that the radiation of frequencies ω_3 and ω_4 is noncoherent. The process becomes stimulated if there are stimulated fields in the interaction area. Under definite conditions these fields will be amplified, and to introduce feedback, the appearance of generation is possible.

From the point of view of the wave approach, this phenomenon is observed in a cubic nonlinear medium when three waves $\dot{\mathbf{E}}(\omega_1), \dot{\mathbf{E}}(\omega_2)$, and $\dot{\mathbf{E}}(\omega_3)$ propagate in this medium, inducing polarization on the combined frequency $\omega_1 + \omega_2 + \omega_3 = \omega_4$ with the amplitude

$$\begin{aligned}\dot{\mathbf{P}}(\omega_4 = \omega_1 + \omega_2 - \omega_3) &= \dot{\chi}(\omega_4, \omega_1, \omega_2, \omega_3)\dot{\mathbf{E}}(\omega_1)\dot{\mathbf{E}}(\omega_2)\dot{\mathbf{E}}(-\omega_3) \\ &= \chi(\omega_4, \omega_1, \omega_2, \omega_3)\dot{\mathbf{E}}(\omega_1)\dot{\mathbf{E}}(\omega_2)\dot{\mathbf{E}}^{*}(\omega_3),\end{aligned} \qquad (6.47)$$

which excites wave $\dot{\mathbf{E}}(\omega_4)$. Simultaneously in this process, wave $\dot{\mathbf{E}}(\omega_3)$ is amplified by energy from waves $\dot{\mathbf{E}}(\omega_1)$ and $\dot{\mathbf{E}}(\omega_2)$. To make sure of it, we need to form a system of coupled equations of the (6.41) kind for all four waves.

Four waves participate in all above-mentioned nonlinear transformations. For this reason, such processes are called four-wave interactions.

It should be remembered that χ_3 is the tensor of the fourth rank. Therefore, nonlinear phenomena in the general case require sufficiently complicated mathematics for their description. Moreover, nonlinear susceptibilities for each specific case of nonlinear transformation may have absolutely different values even for the same material. Everything depends upon the nonlinearity mechanism that predominates in each specific case.

6.10 Phase Conjugation

Earlier we considered collinear wave interaction in a nonlinear medium. A series of interesting and practically useful phenomena arises at noncollinear wave propagation. One such effect will be discussed in this section. It is called phase conjugation (PC), and it has already found practical applications.

The PC phenomenon has much more in common with holography. Therefore, for better understanding of the sense of this phenomenon, we offer some information about holography.

The holographic method for recording and reproduction of wave fields was devised in 1948 by Dennis Gabor.[9]

Let us examine the process of hologram recording and reproduction in the very simplest case, when the signal wave is plain with wave vector $\mathbf{k}_3$:

$$\dot{\mathbf{E}}_3 = \mathbf{E}_3 \exp\left(-j\mathbf{k}_3\mathbf{r}\right). \tag{6.48}$$

Let us have at our disposal a photoemulsion that is capable, after development, of changing its refraction index proportionally to the light intensity. If at hologram recording on the photoemulsion together with the signal wave, the reference plain wave falls, which is coherent to the signal wave

$$\dot{\mathbf{E}}_1 = \mathbf{E}_1 \exp\left(-j\mathbf{k}_1\mathbf{r}\right)$$

with wave vector $\mathbf{k}_1$, in the photoemulsion layer an interference pattern is formed (Fig. 6.8a) with the period Λ depending upon the angle between $\mathbf{k}_1$ and $\mathbf{k}_3$. After development, the plain array of the refraction index arises in the layer with the same period Λ. We can attribute wave vector $\mathbf{k}_\Lambda$, directed along the normal to its plane and equal to $2\pi/2\pi/\Lambda$, to the periodic array (as a wave). In this case the period and orientation of the holographic array can be easily determined from the vector relation

$$\mathbf{k}_3 = \mathbf{k}_1 + \mathbf{k}_\Lambda. \tag{6.49}$$

The hologram reproduction consists in the fact that only one wave $\dot{\mathbf{E}}_1$ dispersed on the hologram via the diffraction grating (Fig. 6.8b) falls on its surface. Since the Bragg diffraction condition coincides with (6.49) and hence is satisfied automatically, the reference wave reflecting from the array with high effectiveness[10] gives

---

[9]Dennis Gabor (1900–1979) was a Hungarian–British electric engineer and a physicist, most notable for inventing holography (1948), for which he received the 1971 Nobel Prize in Physics. He worked in Germany, the UK, and the USA. He created the general theory of holography. In 1956 he made the first holographic microscope. He worked in electronics, information theory, and communications theory.

[10]The reference wave passes through the hologram; such a hologram is called the transmission hologram. Together with the transmission hologram, we may record the reflecting hologram.

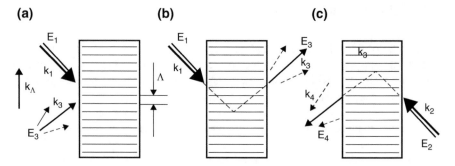

Fig. 6.8 Geometry of processes of recording (**a**) and reproduction (**b**) of holograms (**b, c**). In the photoemulsion layer an interference pattern is formed with the period Λ. After development a plain lattice of the refraction index arises with the same period. At reproduction the wave diffracts on this lattice and recovers the signal wave

birth to the initial signal wave that is required to be obtained at the reproduction stage.

In a real situation the signal wave outgoing from the object is not a plain one. But any wave beam can be presented in the form of superposition of plain waves with different complex amplitudes and directions of wave vectors. Each of these partial waves interfering with the reference wave at the recording stage creates its own "partial" array in the emulsion layer and, accordingly, is reproduced at diffraction on this array of the reference wave. Therefore, holographic recording of any arbitrarily complicated object is possible (see the dotted arrows in Fig. 6.8). The reconstructed field in the hologram is perceived by an observer as an imaginary image of the real object. If the observation angle is changed, the object aspect will be changed.

Now let us imagine that on the hologram at reproduction it is not the E_1 wave that falls but the wave $\dot{E}_2$ ($k_2 = -k_1$) that is directed exactly toward it. Instead of the signal wave, the inverse wave E_4 with respect to it arises with wave vector $k_4 = -k_3$ (Fig. 6.8c) as a result of Bragg diffraction. If the initial signal beam has a complicated profile and wave front, each of the partial plain waves that are its components undergo an inversion, i.e., a change of the wave vector sign to the opposite at hologram reproduction by the inverse wave (with respect to the reference wave) $\dot{E}_2$. This is equivalent to a sign change of the partial wave phases in beam $\dot{E}_4$ with respect to their values in beam E_3, while their amplitude remain unchanged.

From the mathematical point of view, such a transformation means that the wave amplitude $\dot{E}(\omega_4 = \omega)$ is a complex conjugate with regard to the amplitude of the signal wave. This is the PC phenomenon. Really, the complex conjugate amplitude (6.48) is

$$\dot{E}_4 = \dot{E}_3^* = E_3 \exp\left(jk_3 r\right),$$

and this is nothing other than the inverse wave.

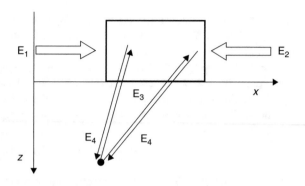

Fig. 6.9 Four-wave mixing. Two contradirectional reference waves (E_1 and E_2) and the signal wave (E_3), which is subject to inversion, fall at once into a nonlinear medium. As a result of interaction, the wave $E_4 \sim E_1 E_2 E_3^*$ is formed, which is exactly inverse with respect to the signal wave

Thus, if, for instance, the signal wave is divergent, the inverse wave is convergent. If the signal wave on some segment of the wave front has a phase delay, the inverse wave with the same profile on this segment has a phase advance of exactly the same value. These unique properties of the inverse wave would find wide applications if the inverse wave appeared on the mirror output without delay after application of the signal wave.

It turns out that if the photoemulsion is changed to a substance with cubic nonlinearity, the holographic array of the refraction index owing to (6.21) will appear simultaneously with the start of the interference pattern. Therefore, in contrast to the usual holography, the reproduction stage can be combined in time with the recording stage. It means that the hologram arising in the nonlinear medium can change in time, tracking the variation of the signal beam. That is why this type of holography is sometimes called dynamic holography.

Let us imagine that on the layer of a nonlinear substance, three coherent waves fall: the signal wave E_3 and two pumping waves $\dot{E}_1$ and $\dot{E}_2$, for which $\mathbf{k}_2 = -\mathbf{k}_1$ (Fig. 6.9). Two processes run simultaneously:

1. Wave $\dot{E}_1$ interferes with wave $\dot{E}_3$, and the arising light pattern gives birth to an array of the refraction index on which the diffraction of the pumping wave E_2 immediately leads to the rise of the inverse wave E_4.
2. The pumping wave $\dot{E}_2$ interferes with the signal wave $\dot{E}_3$, and this results in a hologram appearance on which the second pumping wave E_1 diffracts and gives birth to the inverse wave $\dot{E}_4$.

As a result of wave interaction in the nonlinear medium, as a response to the signal wave $\dot{E}_3$ over time, defined by the inertia properties of physical processes, the wave $\dot{E}_4$ appears with phase conjugation. This phenomenon is known as four-wave mixing (FWM).

We can approach the PC phenomenon in a cubic nonlinear medium from other positions. These are all reasons to consider it as a specific case of wave generation with the combined frequency $\omega_4 = \omega_1 + \omega_2 - \omega_3$, which we discussed at the end of the previous section.

Indeed, in this case, $\omega_4 = \omega_1 = \omega_2 = \omega_3 = \omega$, and in full accordance with the energy conservation law (6.46) the frequency of the inverse wave ω_4 is also equal to ω.

As for the condition of phase synchronism $\mathbf{k}_1 + \mathbf{k}_2 = \mathbf{k}_3 + \mathbf{k}_4$, it is fulfilled automatically at any incident angle θ since $\mathbf{k}_1 = -\mathbf{k}_2$ and $\mathbf{k}_3 = -\mathbf{k}_4$.

According to such an approach, the signal wave and two pumping waves induce in the medium a polarization wave with wave vector $\mathbf{k}_4$ and the complex amplitude

$$\dot{\mathbf{P}}(\omega_4 = \omega) = \chi_3(\omega, \omega, \omega, -\omega)\dot{\mathbf{E}}_1(\omega)\dot{\mathbf{E}}_2(\omega)\dot{\mathbf{E}}_3^*(\omega), \tag{6.50}$$

which is a source effectively exciting the inverse wave

$$\dot{\mathbf{E}}_4 = \dot{\mathbf{E}}_3 \exp\left(\, j\mathbf{k}_3\mathbf{r}\right).$$

The equation for the slowly changing amplitude of the inverse wave is obtained in the same manner as Eq. (6.41). For $\mathbf{k}_1 = -\mathbf{k}_2 = k\mathbf{1}_{xx}$, $-\mathbf{k}_3 = \mathbf{k}_4 = k\mathbf{1}_z$ (for the coordinate system, see Fig. 6.9), assuming that the conditions of the phase matching are fulfilled, we obtain (by using (6.50))

$$\frac{\partial \mathbf{E}_4}{\partial z} = -j\frac{\pi\chi_3}{\lambda\varepsilon_0 n}\dot{\mathbf{E}}_1\dot{\mathbf{E}}_2\dot{\mathbf{E}}_3^*. \tag{6.51}$$

Wave $\dot{\mathbf{E}}_4$ arises in the nonlinear layer rather than falling on it from outside. The solution to (6.51) with layer thickness L gives its amplitude on the output of the PC mirror:

$$\dot{\mathbf{E}}_4 = -j\frac{\pi\chi_3 L}{\lambda\varepsilon_0 n}\dot{\mathbf{E}}_1\dot{\mathbf{E}}_2\dot{\mathbf{E}}_3^*. \tag{6.52}$$

Using (6.52) we estimate the value of the reflection factor from the PC mirror using FWM at $\lambda \approx 1$ µm on a silicon plate with thickness $L = 10$ mm. If we consider that the power flow density in the pumping beams $I_{1,2} = 1$ kW/sm^2 (this corresponds to $E_{1,2} = 10^5$ V/m), for $\chi_3 \approx 10^{-27}$ $\frac{\text{F·m}}{\text{V}^2}$ we have

$$R = \left|\frac{E_4}{E_3}\right| \cong 0.05.$$

This value is close to those observed experimentally.

From (6.52) it follows that with layer thickness growth we can obtain an arbitrarily large reflection factor. This result is the consequence of application of specified field approximation.

If wave amplitudes $\dot{\mathbf{E}}_1$, $\dot{\mathbf{E}}_2$, and $\dot{\mathbf{E}}_3$ are not considered as constants and we solve the equation system for coupled waves, we can make sure the value of R reaches some limiting value $R_{\max}$ with layer thickness growth. This value depends upon both

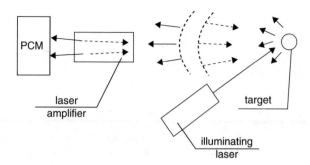

Fig. 6.10 Beam pointed at a target using a phase conjugation mirror (PCM). Radiation from a low-power illumination laser reflected from the target passes to a powerful laser amplifier, is amplified in it, and passes to the PCM. The beam reflected from it again passes through the amplifier and, at its output, turns out to be exactly phase inverse relative to the input wave. As a result, the powerful output beam focuses exactly at the point where the target is situated

the pumping wave amplitudes and the signal wave amplitude. Under definite conditions the reflection factor may be much greater than 1 owing to a power transfer from the pumping waves to the inverse wave.

To conclude, we examine some variants of practical application of the PC phenomenon.

Let us imagine that the problem lies in accurate pointing of a powerful laser beam toward the target. Such a difficult engineering problem can be solved using a PC mirror (Fig. 6.10).

Radiation from a low-power illumination laser reflected from the target reaches a powerful laser amplifier, is amplified in it, and passes on to the PC mirror. The beam reflecting from it passes through the amplifier and, owing to the above-mentioned property of compensation, the wave phase front on the amplifier output is exactly phase conjugated with the input wave. Hence, according to all optical laws, the powerful output beam should be focused exactly at the point where the target is situated.

Let us examine one more example. In powerful solid-body lasers there is a problem of distortion of the phase front arising from nonuniform heating of the active crystal. If one mirror in the laser resonator is changed to a PC mirror, all phase distortions will be automatically compensated for and the output beam will have a minimal angular divergence close to the diffraction limit (Fig. 6.11).

Besides the examples described above, the PC phenomenon has found wide applications in devices for optical signal processing. However, further study of it is outside the scope of our lecture course.

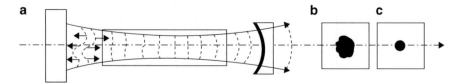

Fig. 6.11 Phase distortion compensation of a beam wave front. **a** Laser circuit. **b, c** Light spot in a far region using the usual mirror and a phase conjugation mirror. In powerful solid-state lasers there is a problem of wave front distortions arising from nonuniform heating of the active crystal. If one mirror in the laser resonator is replaced with a phase conjugation mirror, all phase distortions are automatically compensated for

6.11 Quadratic Nonlinear Effects: Generation of the Second Harmonic

As the first example of quadratic effect application, we examine generation of the second harmonic. For description of its process we use specified field approximation.

We assume that the fundamental harmonic field takes the form

$$\dot{\mathbf{E}}(\omega) = \dot{\mathbf{E}}_1 \exp\left(-j\beta_1 z\right),$$

where $\dot{\mathbf{E}}_1$ is the complex amplitude of the fundamental field harmonic at the point $z = 0$ and β_1 is the phase constant of frequency ω_1.

If the field of the fundamental harmonic is specified, the polarization on the second harmonic is

$$\dot{\mathbf{P}}(2\omega) = d(2\omega)\dot{\mathbf{E}}_1\dot{\mathbf{E}}_1 \exp\left(-j2\beta_1 z\right).$$

By analogy with (6.41), we write the equation for the complex amplitude of the second harmonic as

$$\frac{\partial \dot{\mathbf{E}}_2}{\partial z} = -j\frac{2\pi}{\lambda\varepsilon_0 n_2}d(2\omega)\dot{\mathbf{E}}_1\dot{\mathbf{E}}_1 \exp\left(-j\Delta\beta z\right),$$

where $\Delta\beta = 2\beta_1 - \beta_2$. For an interaction area with length L the solution to this equation is

$$\dot{\mathbf{E}}_2 = \frac{2\pi d(2\omega)}{\lambda\varepsilon_0 n_2}\frac{\exp\left(-j\Delta\beta L\right) - 1}{\Delta\beta}\dot{\mathbf{E}}_1\dot{\mathbf{E}}_1.$$

After simple transformations we obtain

$$\dot{E}_2 = -j\frac{2\pi L d(2\omega)}{\lambda \varepsilon_0 n_2} \exp\left(-j\Delta\beta\frac{L}{2}\right) \frac{\sin\left(\Delta\beta\frac{L}{2}\right)}{\Delta\beta\frac{L}{2}} \dot{E}_1 \dot{E}_1.$$

In practice, the harmonic generation is observed at interaction of wave beams of a limited section. If the cross-sizes of the beam are much greater than the wavelength, the real field can be replaced by the field of the plain wave.[11] In this approximation the power of the second harmonic transmitted by the wave through cross-section S_{cross} is

$$P_2 = \frac{8\pi^2 Z_0 d^2(2\omega)}{\varepsilon_0^2 n_1^2 n_2 S_{\text{cross}}} L^2 \left(\frac{\sin\left(\Delta\beta\frac{L}{2}\right)}{\Delta\beta\frac{L}{2}}\right)^2 P_1^2, \tag{6.53}$$

where Z_0 is the characteristic impedance of a vacuum. The transfer coefficient of the first harmonic into the second one is

$$K = \frac{P_2}{P_1} = \frac{8\pi^2 Z_0 d^2(2\omega)}{\varepsilon_0^2 n_1^2 n_2 S_{\text{cross}}} L^2 \left(\frac{\sin\left(\Delta\beta\frac{L}{2}\right)}{\Delta\beta\frac{L}{2}}\right)^2 P_1.$$

At a fixed value of the first harmonic power the transfer coefficient increases with reduction of beam section S_{cross}. Therefore, in experiments on generation of the second harmonic, wave beam focusing is used. Accurate analysis shows that there are optimal relations between the nonlinear crystal length L_{opt} and distance z_0, in which the cross-sectional area is twice as large ($L_{\text{opt}} \approx 5.7 z_0$). The transformation effectiveness essentially increases at transition to wave interaction on the optical wave guide, whose cross-section has an order of λ. At high power densities we should consider the self-focusing phenomenon.

From (6.53) it follows that the condition of the phase matching

$$\Delta\beta = 2\beta_1 - \beta_2 = 0 \tag{6.54}$$

is a prerequisite for effective generation of the second harmonic. The condition (6.54) can be written as

$$2\frac{\omega}{v_{\text{ph}}(\omega)} = \frac{2\omega}{v_{\text{ph}}(2\omega)} \quad \text{or} \quad v_{\text{ph}}(2\omega) = v_{\text{ph}}(\omega).$$

Thus, the phase matching requires equality of the phase velocities of the interacting waves. Therefore, we sometimes talk about matching of the wave velocities.

---

[11] For a more accurate description of processes in nonlinear media it is necessary to use the theory of wave beams.

Fig. 6.12 Functions of refraction indices of ordinary n_o and extraordinary n_e waves propagating in a potassium dihydrophosphate crystal versus the wavelength. The refraction index of the extraordinary wave depends upon its propagation direction with respect to the optical axis and changes from n_e to n_0. It allows provision of phase matching. Point 1 correspond to the first harmonic, and points 2 and 3 correspond to the second harmonic

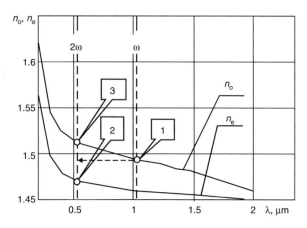

The condition of the phase synchronism (6.54) imposes the following requirement on the medium properties:

$$n(2\omega) = n(\omega). \tag{6.55}$$

Because of the natural frequency dispersion, this condition is impossible to fulfill on one wave type. The outlet consists in application of the anisotropic properties of crystals. We shall show how can it be done in the example of generation of the second harmonic in a potassium dihydrophosphate (KDP) crystal (KH_2PO_4).

KDP is a type of single-axis crystal in which two types of waves—ordinary and extraordinary—may be propagated. The functions of refraction indices versus the wavelengths of ordinary (o) and extraordinary (e) waves propagating in the KDP crystal are presented in Fig. 6.12. We see that the refraction index of the second harmonic is higher than that of the first one, but n_o is always greater than n_e.

An attempt can be made to fulfill the synchronism condition by using the fact that in single-axis crystals the refraction index of the extraordinary wave depends upon angle θ between the wave normal and the optical axis of the crystal. This function is shown in Fig. 6.13. In single-axis crystals there is a direction (more exactly, a cone of directions) of interacting wave propagation at which the refraction index of the ordinary wave of the first harmonic $n_o(\omega)$ is equal to the refraction index of the second harmonic $n_e(2\omega)$.

Let us assume that a neodymium laser is the radiation source. At a wavelength of $\lambda = 1.06$ μm the KDP refraction index of the ordinary wave $n_o(\omega) = 1.494$, while for the extraordinary wave, $n_e(\omega) = 1.460$. For the wave of the second harmonic, ($\lambda_2 = 0.56$ μm)$n_o(2\omega) = 1.512$ and $n_e(2\omega) = 1.470$. With variation of θ, $n_e(2\omega)$ varies from 1.512 to 1.470 and for $\theta = 60°$,

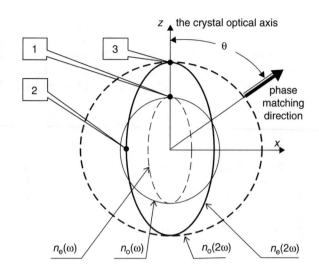

Fig. 6.13 Phase matching at generation of the second harmonic in a nonlinear single-axis crystal. In some direction of the wave vectors the refraction index of the ordinary wave of the first harmonic is equal to the refraction index of the extraordinary wave of the second harmonic. The directions of the phase matching form a cone with the vertex angle 2θ. Point 1 correspond to the first harmonic, and points 2 and 3 correspond to the second harmonic

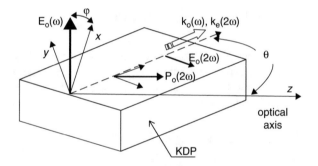

Fig. 6.14 Generation of the second harmonic in a potassium dihydrophosphate (KDP) crystal. The direction of phase synchronism for conversion of $\lambda = 1.06$ μm into $\lambda = 0.53$ μm is $\theta = 60°$. Vector $\dot{\mathbf{E}}_o(\omega)$ of the initial ordinary wave is perpendicular to the z axis. In this case, $E_z = 0$ and $\dot{\mathbf{P}}(2\omega) = d_{36}E_xE_y\mathbf{1}_z$. Nonlinear polarization is directed along the z axis and achieves its maximum at $\varphi = 45°$. Projection $\dot{\mathbf{P}}(2\omega)$ perpendicular to wave vector $\mathbf{k}_e(2\omega)$ excites the extraordinary wave of the second harmonic $\dot{\mathbf{E}}_e(2\omega)$

$$n_e(2\omega) = n_o(\omega) \tag{6.56}$$

Fulfillment of (6.56) is a necessary but insufficient condition for second harmonic generation. The thing is that the main wave should be an ordinary wave with vector $\dot{\mathbf{E}}(\omega)$ perpendicular to the z axis. The wave of the second harmonic should be an extraordinary wave with vector $\mathbf{E}(2\omega)$ perpendicular to $\dot{\mathbf{E}}(\omega)$. Nevertheless, it will appear only at appropriate orientation of the nonlinear polarization vector $\dot{\mathbf{P}}(2\omega)$. The orientation of vector $\mathbf{P}(2\omega)$ completely depends upon the properties of tensor $\mathbf{d}$. For KDP, intensive generation of the second harmonic becomes possible if vector $\mathbf{E}(\omega)$ is oriented at an angle of $45°$ with respect to crystallographic axes x and y. A sketch of a frequency doubler (optical second harmonic generator (OSHG)) on a KDP crystal is presented in Fig. 6.14.

The problem of phase matching at interaction of three waves—for instance, at generation of the third harmonic—can be solved in a similar manner.

It should be noted that together with crystals in which $n_o > n_e$ (which are called optically negative crystals), there are optically positive crystals in which $n_o < n_e$. The condition of the phase synchronism (6.56) for them should be

$$n_e(\omega) = n_o(2\omega).$$

The necessity of the phase matching condition reduces the number of point groups of non-center-symmetric crystals that allow effective provision of second harmonic generation. As shown in the literature, negative crystals in classes **422** and **622**, as well as positive crystals in classes **4mm** and **6mm**, are not at all suitable for generation of the second harmonic. Positive crystals in classes **4, 6, 422**, and **622** are practically useless for generation.

6.12   The Linear Electro-optical Effect

In Sect. 6.3 we described the electro-optical effect in a general manner. In the modern interpretation it can be attributed to quadratic effects. The effect consists in appearance of quadratic polarization on the frequency of the initial field in the presence of a permanent electric field:

$$\dot{\mathbf{P}}_{quad}(\omega) = \chi(\omega, 0, \omega)\mathbf{E}(0)\dot{\mathbf{E}}(\omega). \tag{6.57}$$

It can be considered as a linear variation of relative dielectric permeability under the action of the permanent field:

$$\varepsilon(\omega) = 1 + \frac{\chi(\omega)}{\varepsilon_0} + \frac{\chi(\omega, 0, \omega)}{\varepsilon_0}\dot{\mathbf{E}}(0). \tag{6.58}$$

If we use (6.57) together with the Maxwell equations, we may obtain a solution to the problem of the joint impact of two fields in the same manner as we did in Sects. 6.10 and 6.12.

Historically, however, it was traditional to consider linear electro-optical effects by using optical methods—namely, induced anisotropy. We can do so until the sizes of the interaction area are small in comparison with the wavelength of the modulating field.

Let us examine the Pockels effect following the generally accepted presentation method. First, we offer some information about anisotropic medium optics.

To describe the optical properties of crystals it is convenient to use the ellipsoid of the refraction indices or the optical indicatrix, which characterizes the dependence of the refraction index of the substance upon the wave vector direction. In an arbitrarily

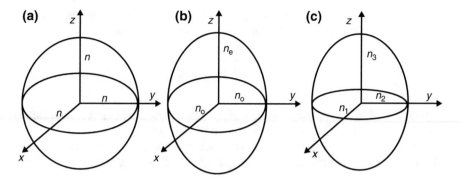

Fig. 6.15 Optical indicatrixes for the main types of crystals. **a** Isotropic dielectrics. **b** Single-axis crystals. **c** Double-axis crystals. The indicatrix for isotropic dielectrics is a sphere. The indicatrix for single-axis crystals is an ellipsoid of revolution that is elongated (negative crystals) or flattened (positive crystals). In the general case, all three refraction coefficients have different values

directed coordinate system the optical indicatrix (or the index ellipsoid) is described by the ellipsoid equation:

$$\frac{x^2}{n_{11}^2} + \frac{y^2}{n_{22}^2} + \frac{z^2}{n_{33}^2} + \frac{2yz}{n_{23}^2} + \frac{2xz}{n_{13}^2} + \frac{2xy}{n_{12}} = 1, \qquad (6.59)$$

where $n_{ij} = \sqrt{\varepsilon_{ij}}$ is the tensor (more exactly, the tensor-like quantity) of the refraction index. Here, we use an already known indexation of the second-rank tensor's elements.

Using the tensor of polarization constants $a_{ij} = \frac{1}{\varepsilon_{ij}}$, we can write the equation of the indicatrix (6.59) in the form:

$$a_{11}x^2 + a_{22}y^2 + a_{33}z^2 + 2a_{23}yz + 2a_{13}xz + 2a_{12}xy^2 = 1. \qquad (6.60)$$

If the axes x, y, and z are directed along the main axes of the ellipsoid corresponding to refraction indices n_{11}, n_{22}, and n_{33}, Eq. (6.60) can be transformed into

$$a_{11}x^2 + a_{22}y^2 + a_{33}z^2 = 1. \qquad (6.61)$$

In optically homogeneous dielectrics (liquids, gases, and crystals with cubic symmetry) the refraction indices in all directions are the same and the optical indicatrix is degenerated into a sphere (Fig. 6.15a). Equation (6.61) takes the form

$$a_{11}\left(x^2 + y^2 + z^2\right) = 1. \qquad (6.62)$$

In so-called single-axis crystals the optical indicatrix represents the rotation ellipsoid (Fig. 6.15b). For such crystals it is accepted to extract two main values of the refraction index of ordinary n_o and extraordinary n_e beams. If, for instance, $a_{11} = a_{11} = 1/n_o^2$ and $a_{33} = 1/n_e^2$ instead of (6.61) we shall have

$$\frac{(x^2 + y^2)}{n_o^2} + \frac{z^2}{n_e^2} = 1. \tag{6.63}$$

Crystals of low symmetry are optically double axis. Their optical indicatrixes (Fig. 6.15c) are described in the general case by Eq. (6.61).

Electric field E changes the indicatrix. Indicatrix variations under the action of electric forces with accuracy up to the quadratic member can be presented in the form

$$\Delta a_{ij} = \frac{1}{\varepsilon_{ij}} - \frac{1}{\varepsilon_{ij}(0)} = r_{ijk}E_k + h_{ijkl}E_kE_l. \tag{6.64}$$

The first item describes the so-called linear Pockels effect (variation of the refraction index under the action of an electric field), and the second item describes the Kerr quadratic affect, which we discussed earlier.

Let us look in more detail at the Pockels effect:

$$\Delta a_{ij} = r_{ijk}E_k,$$

which is of great interest in optical electronics. Although this effect is called a linear one, as we said above, it is related to the quadratic nonlinearity of the medium. We shall determine this connection digressing from the tensor character of susceptibility. With this assumption,

$$\Delta a = \frac{1}{\varepsilon} - \frac{1}{\varepsilon(E = 0)} = \frac{1}{\varepsilon(E = 0) + \frac{\chi(\omega, 0, \omega)}{\varepsilon_0}E} - \frac{1}{\varepsilon(E = 0)} \cong -\frac{\chi(\omega, 0, \omega)}{\varepsilon_0 n^4}E.$$

Hence,

$$r = -\frac{\chi(\omega, 0, \omega)}{\varepsilon_0 n^4} = -\frac{2d}{\varepsilon_0 n^4}.$$

If we take into consideration the tensor character of coefficients r and d, we have

$$r_{Ki} = -Cd_{iK}, \tag{6.65}$$

where $K = jk$ is the doubled coefficient whose values are presented in Table 6.1; C is some constant.

Equation (6.65) allows us to obtain from the matrix of nonlinear susceptibility coefficients the matrix of electro-optical coefficients by means of mutual rearrangement of columns and rows, i.e., by transposition.

Let us obtain, for instance, the matrix of electro-optical coefficients for the crystal KH_2PO_4 (KDP), widely used in optical electronics. Since the symmetry of this crystal is $\overline{4}2m$, for this crystal, only two coefficients are nonzero: $d_{14} = d_{25}$ and d_{36}. On the basis of (6.65) the electro-optical coefficients, which are nonzero, are

$$r_{41} = r_{yzx} = r_{52} = r_{xzy} \text{ and } r_{63} = r_{xyz}.$$

The electro-optical coefficient matrix itself takes the form

$$\mathbf{r} = \begin{vmatrix} 0 & 0 & 0 \\ 0 & 0 & 0 \\ 0 & 0 & 0 \\ r_{41} & 0 & 0 \\ 0 & r_{41} & 0 \\ 0 & 0 & r_{63} \end{vmatrix}. \tag{6.66}$$

Knowing the matrix $\mathbf{r}$, we determine the variation of the optical indicatrix under the action of an electric field. For a single-axis KDP crystal, according to (6.66), the optical indicatrix (6.59) under the influence of an arbitrarily directed electric field $\mathbf{E}(0) = E_x\mathbf{1}_x + E_y\mathbf{1}_y + E_z\mathbf{1}_z$ takes the view

$$\frac{(x^2 + y^2)}{n_o^2} + \frac{z^2}{n_e^2} + 2r_{41}E_x + 2r_{41}E_y + 2r_{63}E_z = 1. \tag{6.67}$$

According to (6.67), under the influence of an arbitrarily directed electric field on a single-axis crystal, additional items appear in the equation of the optical indicatrix, indicating the orientation variation of the main axis of the ellipsoid.

Equation (6.67) may again be reduced to the form (6.63), i.e., to find the new direction of the main axes. For simplicity we shall consider that the electric field is directed along the z axis. Then Eq. (6.67) takes a simpler form:

$$\frac{(x^2 + y^2)}{n_o^2} + \frac{z^2}{n_e^2} + 2r_{63}E_z = 1. \tag{6.68}$$

The action of the electric field performs a rotation of the coordinate system by $45°$ around the z axis (Fig. 6.16). Really, substituting

$$x = x' \cos(45°) - y' \sin(45°) = \frac{1}{\sqrt{2}}(x' - y')$$

and

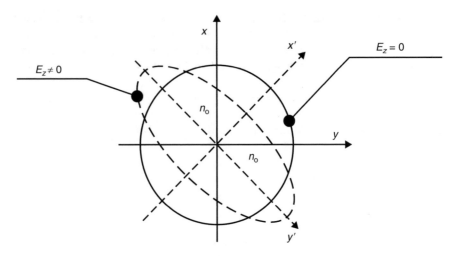

Fig. 6.16 Section variations in the optical indicatrix of a single-axis potassium dihydrophosphate crystal under the action of an electric field. A permanent electric field acting in the optical axis direction transforms the crystal from uniaxial to biaxial. This phenomenon is a linear electro-optical effect

$$y = x' \cos(45°) + y' \sin(45°) = \frac{1}{\sqrt{2}}(x' + y')$$

into (6.68) we obtain

$$\left(\frac{1}{n_o^2} + r_{63}E_z\right)x'^2 + \left(\frac{1}{n_o^2} - r_{63}E_z\right)y'^2 + \frac{z^2}{n_e^2} = 1. \tag{6.69}$$

According to (6.69), the refraction coefficients along the main axis are

$$n_{x'} = \frac{n_o}{\sqrt{1 + r_{63}E_z n_o^2}} \approx n_o - 0,5r_{63}E_z n_o^3. \tag{6.70}$$

Thus, under the action of an electric field, a single-axis crystal becomes a double-axis one.

Crystals without an inversion center have a piezoelectric effect. Therefore, together with the primary electro-optical effect, a secondary one is observed, connected with mechanical deformations under the action of an electric field. The cutoff frequency of the primary effect is a value of about 10^{13} s^{-1} (the oscillation frequency of the crystal lattice). For the secondary effect the cutoff frequency depends on the crystal size and is of the order of 10^6 s^{-1}. Thus, on a high frequency of about 10^9 Hz the first effect dominates.

6.13   The Electro-optical Amplitude Modulator

The Pockels effect is widely used for development of optical modulators and deflectors—devices for beam deflection.

Since the linear electro-optical effect consists in variation of the refraction index, an electro-optical phase modulator would be more natural. However, all photoreceivers are, fundamentally, quadratic detectors. Implementation of a coherent photoreceiver (an optical superheterodyne) is possible, but it presents definite difficulties. Therefore, in most cases, amplitude modulation in the optical range is used. In optical amplitude modulators the natural phase modulation is transformed into amplitude modulation.[12] We will examine further the principle of construction of such modulators.

Let us assume that in a KDP crystal an electromagnetic wave that is polarized on the x axis propagates along the z axis, i.e.,

$$\dot{\mathbf{E}} = E \exp\left(-j\frac{\omega n_x}{v_{ph}} Z\right) \mathbf{1}_x.$$

We shall consider that the optical indicatrix of this crystal is described by Eq. (6.69) and the main half-axes of the ellipsoid are directed along x', y', and z (see Fig. 6.16) and are equal to $n_{x'}$, $n_{y'}$, and n_z, respectively.

We expand wave $E = E\mathbf{1}_x$ to two orthogonal waves $E_{x'}$ and $E_{y'}$. Then, using Eq. (6.70) we obtain

$$
\begin{aligned}
\dot{E}_{x'} &= \frac{E}{\sqrt{2}} \exp\left[-j\frac{\omega}{v_{ph}}\left(n_0 - 0,5r_{63}E_z n_0^3\right)z\right]; \\
\dot{E}_{y'} &= \frac{E}{\sqrt{2}} \exp\left[-j\frac{\omega}{v_{ph}}\left(n_0 + 0,5r_{63}E_z n_0^3\right)z\right].
\end{aligned}
\tag{6.71}
$$

At wave propagation in the crystal with length L, a phase difference arises between orthogonal waves $\dot{E}_{x'}$ and $\dot{E}_{y'}$, according to (6.71):

$$\Delta\varphi = \frac{2\pi}{\lambda} r_{63} n_0^3 E_z L. \tag{6.72}$$

If the voltage

$$U = E_z L \tag{6.73}$$

---

[12]The amplitude modulation can be obtained by directly using, as an example, the displacement of the fundamental absorption edge in semiconductors under the action of an electric field (see Sect. 4.6).

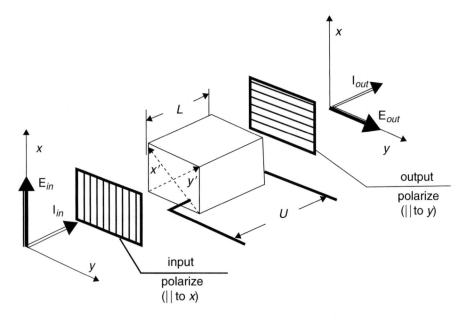

Fig. 6.17 An electro-optical amplitude modulator with a longitudinal field. A linear polarized wave falls from the left to the crystal edge. At U variation the phase difference between projections of the electric field $E_{x'}$ and $E_{y'}$ changes. This leads to wave polarization type variation. A crossed input polarizer passes through the horizontal field component

is such that $\Delta\varphi = \frac{\pi}{2}$, the linear polarization will transfer to the circular one. At

$$\Delta\varphi = \pi, \tag{6.74}$$

the polarization again becomes linear but is oriented already along the y axis. Using this property, we may create an optical amplitude modulator.

Let us examine one of the possible circuits of amplitude in the optical modulator presented in Fig. 6.17. The modulator consists of a KDP crystal placed between two crossed polaroids. The main indicatrix axes for $U = 0$ are oriented along axes x, y, and z. At zero voltage the modulator is nontransparent. The maximal transparency is achieved when the condition in (6.74) is fulfilled, to which the voltage corresponds:

$$U_{\lambda/2} = \frac{\lambda}{2r_{63}n_0^3}. \tag{6.75}$$

The voltage $U_{\lambda/2}$ that is necessary for ensuring the propagation difference of $\lambda/2$ is called the half-wave voltage. It is clear that it is preferable to use a crystal for which the half-wave voltage is lower under otherwise equivalent conditions.

The transfer coefficient of the electro-optical modulator

$$K = \frac{P_{\text{out}}}{P_{\text{in}}}$$

can be found on the basis of (6.72) and (6.73). If we take into consideration (6.75), then

$$K = \sin^3\left(\frac{U}{U_{\lambda/2}}\right).$$

Figure 6.18 (left panel) shows the modulation characteristics of the modulator. To decrease nonlinear distortions it is necessary to displace the operating point in the middle of the linear part of this characteristic. For this, the bias voltage needs to be equal to $0.5U_{\lambda/2}$. Looking at Table 6.2, it is easy to understand that this involves some difficulties. The outlet involves application of a wave with circular polarization (Fig. 6.18, right panel). To convert linear polarization into circular polarization, in optics they use anisotropic plates. The thickness is selected to ensure a phase difference of 90° between ordinary and extraordinary waves. Such plates are called quarter-wave plates.

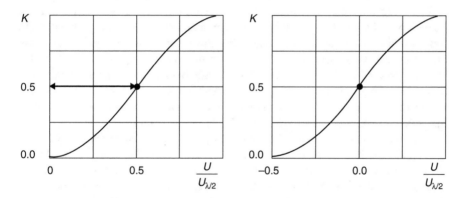

Fig. 6.18 Transmission coefficient of an electro-optical modulator with linear polarization at input (*left*) and circular polarization at input (*right*). For a wave with linear polarization, bias voltage $U_0 = 0.5\ U_{\lambda/2} \sim 1$ kV is required. In application of a wave with circular polarization, bias voltage is not required

Table 6.2 Characteristics of some crystals used in electro-optical modulators

Crystal	Crystal abbreviation	T_{Curie} (K)	$U_{\lambda/2}$ (kV)	$\Delta\lambda$ (μm)
KH_2PO_4	KDP	123	7.4	0.2–1.2
$LiNbO_3$	LN	1470	3.5	0.35–4.5
$LiTaO_3$	LT	890	2.5	0.35–4.5
Ba, $NaNb_5O_{15}$	BNN	833	1.8	0.4–4.5
KD_2PO_4	DKDP	222	3.4	0.2–1.6
$(Ba, Sr)Nb_2O_6$	BSN	330	0.45	0.4–4.5

A high degree of solubility in water is a disadvantage of KDP and ammonium dihydrogen phosphate (ADP) crystals, which were used in the first sample modulators. The results obtained sparked intensive searches for other materials with the Pockels effect. Table 6.2 lists the values of $U_{\lambda/2}$ for some crystals used in modulators in the optical range ($\lambda = 0.63$ μm).

The materials listed in Table 6.2 can be used in ferroelectrics. Therefore, the temperature of the phase transition T_{Curie} is indicated for them. The last column indicates the transparency range of the materials.

The electro-optical effect essentially increases near T_{Curie}. This explains the high value of the electro-optical constant of a barium–sodium–niobate (BSN) crystal. However, this crystal is of little use, because of its own optical properties. The most suitable crystal, on the basis of its set of properties, is a monocrystal of lithium niobate, and they are grown on an industrial scale.

In addition, we note that the above-considered modulator scheme in which a voltage is applied along the wave propagation direction (the so-called longitudinal modulator scheme) has several disadvantages. The main one is the necessity to use specific electrodes that are transparent to the modulating radiation.

In this regard, a transverse modulator scheme is preferable, in which a modulating field is applied at a right angle to the propagation direction of the optical radiation. In this case the electrodes do not contradict the passage of the light. The phase difference can be increased with application of a longer crystal, because in this case it is proportional to the product of the electric field intensity and the length. The half-wave voltage decreases with plate thickness b reduction because in a transverse modulator scheme, $U = Eb$. In particular, a large voltage reduction is achieved using integrated optical wave guides ($b \approx \lambda \approx 1$ μm, $U_{\lambda/2} \leq 10$ V).

The small inertia property of the electro-optical effect allows creation of high-speed light modulators with a cutoff frequency of the order of gigahertz. In the above-described configuration, limitations in frequency are connected with the crystal size, which should be less than half the wavelength of the modulating frequency. This limit can be overcome in traveling-wave modulators under the condition of phase matching ($v(\Omega_{mod}) = v(\omega_{carr})$). Coupled-wave equations are better to use for description of such modulators.

6.14 Coupling Mode Equations

Above, we confirmed that the specified field approximation does not give an accurate description of nonlinear wave interaction. This is due to the fact that harmonic waves of combined frequencies, in which the problem solution is presented, interact with each other though the mechanism of nonlinear polarization. Thus, waves in a nonlinear medium turn out to be coupled. This coupling can be described by some equation system. We should not think that equations of coupled waves (or modes or types of waves or oscillations) are the exclusive privilege of nonlinear optics. The phenomenon of wave coupling arising on spatial irregularities is well known in electrodynamics. For instance, this coupling in wave guides leads to energy exchange of the main wave type into higher types, which is demonstrated in the damping increase.

In the general case, the equation system for coupled waves has a complicated enough view. Therefore, we shall limit ourselves to the case of three-wave interaction in a medium with quadratic nonlinearity.[13]

So, let the field in a quadratic nonlinear medium be the sum of three plain waves with frequencies ω_1, ω_2, and ω_3 propagating along the z axis:

$$\dot{\mathbf{E}} = \sum_{i=1}^{3} \mathrm{Re}\left[\dot{\mathbf{E}}_i \exp\left(\omega_i t - j\beta_i z\right)\right],$$

where $\dot{\mathbf{E}}_i$ is the slow functions of z and $\beta_i = \frac{\omega_i n_i}{c}$ is a phase constant.

These waves give birth in the medium to nonlinear polarization waves on combined frequencies, whose slowly changing complex amplitudes are

$$\dot{\mathbf{P}}_i = d\dot{\mathbf{E}}_j \dot{\mathbf{E}}_k.$$

Acting as in Sect. 6.10, we obtain an equation system for slowly changing amplitudes:

$$\frac{\partial \dot{\mathbf{E}}_i}{\partial z} = -j \frac{\omega_i^2 \mu_0}{2\beta_i} d\dot{\mathbf{E}}_j \dot{\mathbf{E}}_k \exp\left(-j\Delta\beta z\right), \tag{6.76}$$

where $\Delta\beta = \beta_j + \beta_k - \beta_i$.

For distinctness, we examine the generation of the summed frequency $\omega_3 = \omega_1 + \omega_2$. In this case the system from (6.76) takes the following form:

$$\frac{\partial \dot{\mathbf{E}}_1}{\partial z} = -j \frac{\omega_1^2 \mu_0}{2\beta_1} d\dot{\mathbf{E}}_3 \dot{\mathbf{E}}_2^* \exp\left(-j\Delta\beta z\right);$$

$$\frac{\partial \dot{\mathbf{E}}_2}{\partial z} = -j \frac{\omega_2^2 \mu_0}{2\beta_2} d\dot{\mathbf{E}}_3 \dot{\mathbf{E}}_1^* \exp\left(-j\Delta\beta z\right); \tag{6.77}$$

$$\frac{\partial \dot{\mathbf{E}}_3}{\partial z} = -j \frac{\omega_3^2 \mu_0}{2\beta_3} d\dot{\mathbf{E}}_1 \dot{\mathbf{E}}_2 \exp\left(j\Delta\beta z\right).$$

Here, $\Delta\beta = \beta_3 - \beta_1 - \beta_2$.

The system of equations in (6.77) describing the interaction of three coupled waves is an unsolvable problem in elementary functions in the general case.

---

[13]We understand, of course, that for quadratic nonlinearity also, strictly speaking, the number of waves is indefinite because waves of combined frequencies give birth to polarization on new frequencies. However, these additional items are small quantities of a higher order and we **should** (!) omit them.

Nevertheless, in solving the specific problem, some of the interacting waves are initial. These waves can be considered with some reliability as being specified and practically independent of the z coordinate. This reminds us of the known specified field approximation. Really, this is not entirely so. In solution of some problems the system of three equations can be reduced to a system of two equations without losing the interaction of the two that remain.

Thus, for instance, in the solution to the second harmonic generation problem, $\omega_1 = \omega_2 = \omega$ and $\omega_3 = 2\omega$. Hence, waves $\mathbf{E}_1$ and $\mathbf{E}_2$ become indistinguishable. Therefore, the equation of coupled waves for this process can be written as

$$\frac{\partial \dot{\mathbf{E}}_1}{\partial z} = -j\frac{\omega}{2n_1} Z_0 d\dot{\mathbf{E}}_2 \dot{\mathbf{E}}_1^* \exp(-j\Delta\beta z);$$
$$\frac{\partial \dot{\mathbf{E}}_2}{\partial z} = -j\frac{2\omega}{2n_2} Z_0 d\dot{\mathbf{E}}_1 \dot{\mathbf{E}}_1 \exp(j\Delta\beta z). \tag{6.78}$$

Here, $\dot{\mathbf{E}}_1$ and $\dot{\mathbf{E}}_2$ are the complex amplitudes of the first and second harmonic, respectively, and β_1 and β_2 are the phase constants of the waves, $\Delta\beta = \beta_2 - 2\beta_1$. It is necessary to solve system (6.78) under boundary conditions in the section $z = 0$: $\dot{\mathbf{E}}_1 = \dot{\mathbf{E}}_{10}$ and $\dot{\mathbf{E}}_2 = 0$. As a result, we obtain the solution describing the energy exchange from the first harmonic wave to the second harmonic wave. The system can be made more exact if we add in equation items considering natural wave damping.

6.15   Parametric Amplification: The Parametric Generator

We use (6.77) to solve the problem of parametric amplification of waves in a medium with quadratic nonlinearity. This process has practical significance because it allows development of a parametric tunable optical generator.

Let us examine the parametric processes that can arise in a nonlinear medium, and discuss issues of its practical application.

Let an intensive field on the summed frequency be specified. We shall call it the pumping field, $\dot{\mathbf{E}}_{\text{pump}} = \dot{\mathbf{E}}_3$. Then (6.77) will reduce to the two-equation system

$$\frac{\partial \dot{\mathbf{E}}_1}{\partial z} + \alpha \dot{\mathbf{E}}_1 = -j\frac{\omega_1}{2n_1} Z_0 d\dot{\mathbf{E}}_{\text{pump}} \dot{\mathbf{E}}_2^* \exp(-j\Delta\beta z);$$
$$\frac{\partial \dot{\mathbf{E}}_2}{\partial z} + \alpha \dot{\mathbf{E}}_2 = -j\frac{\omega_2}{2n_2} Z_0 d\dot{\mathbf{E}}_{\text{pump}} \dot{\mathbf{E}}_1^* \exp(-j\Delta\beta z). \tag{6.79}$$

For greater reliability of the solution result, in the left parts of the equations, items are added considering wave damping. For simplicity, we neglect tensor character $\mathbf{d}$, replacing it with scalar coefficient d.

In (6.79) we can see that the action of pumping can be considered as a process of the medium parameters' variation with frequency $\omega_{\text{pump}} = \omega_1 + \omega_2$. Hence, in this case, we deal with the parametric process. We encountered similar processes when we studied the theory of electric circuits.

Without stopping at the solution procedure (which is well known to readers) we write the result obtained under the following conditions at the point $z = 0$: $\dot{E}_1 = \dot{E}_{10}$ and $\dot{E}_2 = 0$.

$$
\begin{aligned}
\dot{E}_1 &= \dot{E}_{10} \left[\cosh(gz) + j\frac{\Delta\beta}{2g} \sinh(gz) \right] \exp(-\alpha z) \exp\left(-j\frac{\Delta\beta}{2} z \right); \\
\dot{E}_2 &= -j\frac{\omega_2}{2gn_2} Z_0 d\dot{E}_{\text{pump}} \dot{E}_{10} \sinh(gz) \exp(-\alpha z) \exp\left(-j\frac{\Delta\beta}{2} z \right),
\end{aligned}
\tag{6.80}
$$

where $g = \frac{1}{2}\sqrt{\frac{\omega_1\omega_2}{n_1 n_2} Z_0^2 \left| d\dot{E}_{\text{pump}} \right|^2 - \Delta\beta^2}$ is the parameter characterizing the specific parametric amplification.

Under conditions of phase synchronism, $\Delta\beta = 0$ and the solution to (6.80) takes the simpler form

$$
\begin{aligned}
\dot{E}_1 &= E_{10} \cosh(gz) \exp(-\alpha z); \\
E_2 &= -j\sqrt{\frac{\omega_2 n_1}{\omega_1 n_2}} E_{10} \sinh(gz) \exp(-\alpha z).
\end{aligned}
\tag{6.81}
$$

If $g = \frac{1}{2}\sqrt{\frac{\omega_1\omega_2}{n_1 n_2}} Z_0 |d\dot{E}_{\text{pump}}| > \alpha$, the initial wave with frequency ω_1 will be amplified. We can reasonably call this wave the "signal" wave. Incidentally, with amplification, additional radiation on frequency ω_2 arises.[14] According to the accepted tradition, the wave with frequency ω_2 is called the "idler."

Let us estimate the value of the specific gain. One of the suitable crystals for a parametric amplifier is lithium niobate (LiNbO$_3$). Let $f_1 \approx f_2 \approx 3 \cdot 10^{14}$Hz and $f_{\text{pump}} \approx 6 \cdot 10^{14}$Hz. Lithium niobate has $d \approx 5 \cdot 10^{-23}$ F/V, $n_1 \approx n_2 \approx 2.2$. For pumping power density $I_{\text{pump}} \approx 5 \cdot 10^{10}$ W/m^2 (the radiation of a solid-body laser) the field intensity is $E_{\text{pump}} \approx 4 \cdot 10^6$ V/m. Substitution gives a g value near 1/cm. We cannot expect monocrystals of a large length. Therefore, even at a large pumping power density the gain is small. This explains why parametric processes are used mainly for creation of optical generators.

If a nonlinear crystal is placed into a resonator tuned to frequency ω_1 or ω_2 (or both right away), under definite pumping power density the parametric gain will compensate for the losses. This opens the way to new methods of optical generator creation. The practical significance of such a generator consists in the fact that it transforms coherent radiation of a fixed frequency into coherent radiation with a frequency coinciding with the natural frequency of the resonator. Since quantum

---

[14]Compare the obtained results with those in Sect. 6.10.

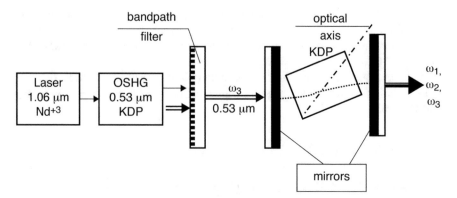

Fig. 6.19 Structural circuit of an optical parametric generator implemented by Sergei Akhmanov and Rem Khokhlov in 1962. The pumping source is a laser on glass doped with neodymium, doubling the frequency. The pumping radiation wavelength is 0.53 μm. The filter passes the radiation of the second harmonic. A parametric process occurs in a potassium dihydrophosphate (KDP) crystal placed in a resonator. The resonator mirrors are transparent for pumping radiation ω_3 and reflect radiation with frequencies ω_1 and ω_2. The parametric generation frequencies ω_1 and ω_2 depend upon the angle between the optical axis of the crystal and the resonator axis. Generation takes place on frequencies at which the phase-matching conditions are fulfilled. *OSHG* optical second harmonic generator

transitions are not used in a parametric generator, it may be tuned to a wide frequency range.

The parametric generator was created by Rem Khokhlov[15] and colleagues in 1962. Their experimental bench structure is presented in Fig. 6.19. The KDP crystal pumping was provided by the field of the second harmonic of a laser on neodymium glass radiation, which operated in a giant-pulse mode. Oscillation frequency tuning was provided by rotation of the crystal.

The frequency tuning of parametric generators is directly connected with the conditions of the impulse conservation $n_3\omega_3 = n_1\omega_1 + n_2\omega_2$ and energy conservation $_3 = \omega_1 + \omega_2$ of photons. Refraction indices in crystals depend upon the crystal orientation, the temperature, the electric field, and the pressure. Since both conditions should be fulfilled simultaneously, by changing the value of any external impacts, we may vary frequencies ω_1 and ω_2. We would like to note that we are speaking here about frequency tuning of the parametric gain rather than resonator tuning.

---

[15]Rem V. Khokhlov (1926–1977) was a Soviet physicist, one of the founders of nonlinear optics, an academician, and a rector of Lomonosov Moscow State University (1973–1977). He devised a method for analysis of nonlinear electronic devices and, on the basis of it, solved a series of radiophysical problems. He developed methods for asymptotic solution of wave equations based on the general theory of wave processes. He utilized these methods in nonlinear acoustics and created the theoretical fundamentals of nonlinear optics.

Let us estimate the value of the threshold pumping power. From (6.81) it follows that the gain in the crystal with length L with absence of loss is

$$K = \cosh(gL).$$

If the resonator is formed by mirrors with reflection factor R and the diffraction losses are negligible, the self-excitation condition (see Fig. 5.3) can be written as

$$\cosh(gL) \quad \text{or} \quad gL \geq \operatorname{arcosh}\left(\frac{1}{R}\right) = -\ln\left(\frac{R}{1 + \sqrt{1 - R^2}}\right).$$

If $R \approx 1$, then

$$g_{\text{thr}} = -\frac{1}{L}\ln\left(\frac{R}{1 + \sqrt{1 - R^2}}\right) \approx \frac{1 - R}{L}.$$

After replacement of the pumping electric field intensity with the power flow I_{pump}, we obtain

$$g_{\text{thr}}^2 = \frac{\omega_1 \omega_2 Z_0^3 d^2}{2 n_1 n_2 n_i} I_{\text{thr}} = \frac{(1 - R)^2}{L^2}.$$

Hence, the threshold pumping power density is

$$I_{\text{thr}} = \frac{(1 - R)^2 2 n_1 n_2 n_i}{L^2 \omega_1 \omega_2 Z_0^3 d^2}.$$

For a parametric generator using a lithium niobate crystal with a length of 1 cm and mirrors with a 98% reflection factor, substitution of parameters gives $I_{\text{thr}} = 2.24$ kW/cm^2. Such a power density value can be obtained even in the continuous mode. This once more confirms the attractiveness of the parametric generator as a source of coherent optical radiation.

6.16   Problems for Chapter 6

6.1. At a wavelength of 0.63 μm the half-wave voltage for a KDP crystal is 3.4 kV and the refraction index $n = 1.5$. Determine the numerical value of the electro-optical coefficient r_{63}.

6.2. A GaAs crystal $(\overline{4}3m)$ is used in a phase modulator at a wavelength of 10.6 μm. A permanent electric field acts in the direction [001]. The wave propagates in the direction [110]. The electro-optical coefficient $r_{41} = 1.1 \cdot 10^{-12}$ m/V, the refraction index is $n = 3.06$, the plate thickness is 5 mm, and its length is 2 cm. Determine the voltage at which the additional phase shift will be equal to $90°$.

6.3. The faces of the phase modulator crystal described in the previous task reflect radiation and therefore form a Fabry–Pérot resonator. Determine the function of the transfer factor of the device versus the control voltage. How will this device behave if we use the voltage that arises on the output of the photoreceiver registering the passed radiation as the control voltage?

6.4. Under radiation action, direct transitions occur from the valence band to the conduction band in an ideal intrinsic semiconductor. The absorption factor is described by the formula $\alpha_p(\omega) \approx A\sqrt{\hbar\omega - W_G}$. In each absorption act, an electron–hole pair is born. As a result, the semiconductor conductivity changes. This phenomenon is called inherent photoconductance. Present this effect in the form

$$\mathbf{J}_{\text{cub}}(0) = \sigma_3(0, \omega, -\omega, 0)\left|\dot{\mathbf{E}}(\omega)\right|^2\dot{\mathbf{E}}(0)$$

and express the nonlinear conductance through α_p.

Instruction: Study Chap. 4 attentively. Neglect there (where it is required) the difference of the temperature from zero. The effective masses of carriers are considered equal to the electron mass. The recombination time of pairs T_{rec} and the mobility of carriers μ are specified.

6.5. On the basis of the previous task solution, estimate the value of the cubic conductance of a semiconductor.

The constant $A = 10^6 \frac{1}{m\sqrt{eV}}$, and $\hbar\omega - W_G = 0.1$ eV.

6.6. For convenience of registration of infrared radiation with a wavelength of 10.6 μm (from a CO_2 laser), it is transformed into radiation with a wavelength of 0.96 μm in a proustite (Ag_3AsS_3) crystal. A neodymium laser with a wavelength of 1.06 μm is used as the local oscillator in the frequency converter. The power density of the local oscillator is 1 MW/m^2. The refraction index of the medium is $n_1 = n_2 = n_3 = 2.6$, and $d_{\text{eff}} = 3 \cdot 10^{-22}$ F/V. Determine the conversion coefficient for a crystal length of 1 cm.

Note: It is possible to obtain a conversion coefficient close to 1 on a lithium iodate crystal for radiation with a wavelength of 3.39 μm from a ruby laser.

6.7. Generation of the second harmonic with a wavelength of 1 μm happens in a planar optical wave guide. For simplicity, assume that the wave guide is implemented using an isotropic material with a refraction index $n_{\text{int}} = 2.3$ and is placed in a space with $n_{\text{ext}} = 2.0$. The first harmonic propagates on the main wave type, while the second propagates on a higher type. Give a graphic illustration of the phase-matching condition in the absence of frequency dispersion of the refraction index.

Instruction: Try to investigate this problem using numerical simulation on a computer.

6.8. Radiation of a quantum generator on carbonic acid gas contains harmonic oscillations with frequencies that differ about by 55 GHz. At wave propagation in a medium with quadratic nonlinearity, polarization arises on the frequency difference, which excites waves of the difference frequency in a metal wave guide. Determine the size of the wide wave guide wall a at which the phase-matching conditions are fulfilled, which is necessary for excitation of the H_{10} wave. The monocrystal GaAs is used as the nonlinear medium. The refraction index at a wavelength of 10 μm is 3.06, and the relative dielectric permeability on a frequency of 55 GHz ($\lambda \approx 5.4$ mm) is 11.2. The wave guide is fully filled with GaAs.

6.9. Derive a formula for the conversion coefficient in power under the conditions of the experiment described in the previous task. Neglect wave damping. Assume that the narrow wave guide wall b is equal to 0.5 a. What is the power on the device output with a length of 1 cm if $d_{\mathrm{eff}} = 12 \cdot 10^{-22}$ F/V and the power of each of the beams falling onto the front face of the crystal is 1 kW?

Chapter 7
Some Types of Quantum Generators and Amplifiers

7.1 Introduction: Classification of Quantum Devices

Since the appearance of the first quantum generator, an innumerable quantity of a great variety of devices have been created. The manufacture of some of them has been assimilated by industry; others are at the development stage or in practical use in scientific laboratories. At present, their most significant application is in demonstration of quantum electronics possibilities, and this is adding to the huge list of such devices, which now number in the thousands. For orientation in this "ocean" of devices, it is necessary to offer some classification. We can take the following parameters as a basis: the operating range, the aggregate state of the operating body, and the method used to obtain inversion.

In terms of frequency ranges we can distinguish devices in the radiofrequency band; masers in the infrared band; optical quantum generators (OQGs), optical quantum amplifiers (OQAs), and lasers in the optical band; and devices in the extreme ultraviolet laser (EUVL) and soft x-ray laser (XRL) regions (87.4–3.56 nm).

In quantum devices, all types of aggregate states are used: solids, liquids, gases, and plasma. Among solid-state optical quantum generators, semiconductor lasers are especially distinguished.

To obtain inversion, additional radiation, electronic shock, collision of atoms and molecules, chemical energy, and even nuclear reactions are used.

7.2 Gas Quantum Generators and Amplifiers

Among quantum devices using gas, quantum molecular beam generators occupy a special place because they opened the way to the era of quantum electronics. Molecular and atomic beams have been used to measure the fundamental characteristics of particles. The first experiments were performed in 1911. In 1937, Isidor

© Springer Nature Switzerland AG 2020
V. V. Shtykov, S. M. Smolskiy, *Introduction to Quantum Electronics
and Nonlinear Optics*, https://doi.org/10.1007/978-3-030-37614-7_7

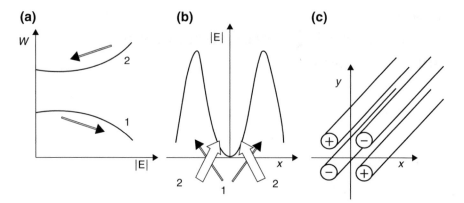

Fig. 7.1 Principle of particle sorting in a molecular beam. (**a**) Function of the energy level versus the modulus of electric field intensity. (**b**) Function of the modulus of electric field intensity versus the coordinate providing focusing of the upper energy level particles. (**c**) Geometry of the quadrupole capacitor. Particles on the upper level tend to be in a weaker field area. 1 corresponds to particles on the lower level, while 2 - to particles on the higher level

Rabi[1] discovered the resonance method for measuring the magnetic moments of atoms and molecules, which subsequently formed the basis of radio spectroscopy.

7.2.1 Quantum Generators Based on Molecular Beams

In the first molecular generator, ammonia was used. The energy diagram was examined in Sect. 2.5 (see Fig. 2.5). Then other molecules were used. Generation was obtained in the millimeter and centimeter bands at a power of $10^{-11} - 10^{-8}$ W.

Population inversion in a molecular quantum generator is achieved by a sorting system whose action is based on the peculiarities of particle motion in a nonhomogeneous electric field.

Figure 7.1a shows the function of the energy level versus the value of the static electric field. This phenomenon is known as the Stark effect. Since any system aspires to occupy a position with minimal energy, particles on the upper level will tend to move in the direction where the field intensity is lower, while particles situated on the lower level will move in the direction where the intensity is greater. Therefore, an electric field that is nonhomogeneous in the space (Fig. 7.1b) will act

---

[1]Isidor Rabi (1898–1988) was an American physicist who offered a measurement method for magnetic moments and in 1939 measured the magnetic moments of the proton and the neutron. In 1944 he received the Nobel Prize in Physics. He was also involved in the development of the cavity magnetron, which is used in microwave radar and microwave ovens. He was awarded the Niels Bohr International Gold Medal, was a member and former president of the American Physical Society, and was a member of the US National Academy of Sciences and the American Academy of Arts and Sciences. In 1985, Columbia University named its physics department after him.

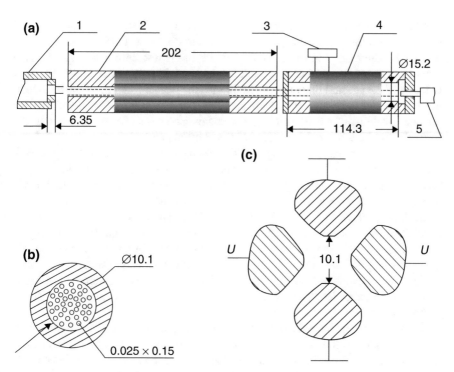

Fig. 7.2 Construction of an quantum generator using ammonia molecules. (**a**) General view: *1* molecular beam source, *2* quadrupole capacitor (focusing system), *3* wave guide input, *4* resonator, *5* adjustment. (**b**) Element of the molecular beam former. (**c**) Cross-section of the sorting system. The directed molecular beam is formed using diaphragms, whose role is played by long channels with a small diameter. Population inversion is formed in the sorting system. A voltage of 6–30 kV is applied to rods in the quadrupole capacitor. The inverse beam appears in the resonator

on particles on the upper level similarly to a convex lens but will act on particles on the lower level similarly to a concave lens. To create a field of the required configuration we can usually use a quadrupole capacitor, which consists of four parallel rods of a special shape, connected in pairs to a high-voltage rectifier. The electric field of such a capacitor is extremely nonuniform (Fig. 7.1c).

A molecular generator circuit is shown in Fig. 7.2. A directed beam of ammonia molecules is formed by the source from chaotically moving particles with the help of one or several narrow channels. The source performance is about 10^{18} molecules per second. Only about 6% of them are situated on the required energy levels. Hence, we can expect approximately $6 \cdot 10^{16}$ useful molecules per second. Under natural conditions, we have 0.4% fewer particles at the upper energy level than at the lower one. That is why the sorting system must eliminate about 1% of the particles from the lower level. After that, as a result of natural relaxation processes, only about $10^{14} - 10^{15}$ molecules per second will appear at the resonator.

During the transit time in the resonator, molecules should not collide, i.e., the free path length should be greater than the length of the resonator (about 15 cm). Taking into consideration the heat molecule velocity in the beam, we can draw a conclusion that the gas pressure in the resonator must be about 10^{-5} mmHg. The spectral line shape width at this pressure will be several tens of kilohertz, and its Q factor will be about 10^7. The total number of active particles in the resonator is so small that the output power of the molecular quantum generator is only about 10^{-11} W. At a Q factor of the spectral line of several millions, the main advantage of the molecular beam generator is its high frequency stability. Measurements have shown that the relative instability of a generator using ammonia molecules is of the order of 10^{-11}. Nevertheless, to achieve this value, additional efforts are required.

The beam of molecules falling into the resonator has a natural velocity spread. Therefore, the spectral line will have nonuniform widening owing to the Doppler effect (see Sect. 2.6). At a temperature near 300 K the average velocity in the beam is about 10^3 m/s. Such a velocity leads to a relative frequency shift of about 10^{-5} (the equivalent Q factor is about 10^5 instead of 10^7). To adjust the situation, it was decided to use resonators, which can be formed from wave guides operating on frequencies close to the cut frequency (i.e., $v_{ph} \to \infty$). In resonators with such a mode the longitudinal eigenvalue is equal to zero. Two examples are the TM_{110} mode of the rectangular resonator and the TM_{010} mode of the cylindrical resonator. At a large resonator length the time of flight value is high and the spectral lines are extremely narrow.

Later, researchers moved to application of hydrogen molecules. The frequency of the operating transition, which is connected with an ultrafine structure of hydrogen levels, is about 1420.205 MHz. The transition is a magnetic dipole one; hence, it is essentially weaker than the above-mentioned transitions of NH_3 molecules. The specific difficulties involved in starting a hydrogen maser are connected with this feature. A hydrogen generator allows a frequency instability of about 10^{-14} to be obtained.

7.2.2   Gas Lasers

The first gas laser (GL) was created in 1961. A mixture of gases (neon and helium) was used as an active medium. Inversion was created as a result of resonance energy transfer of the "buffer" gas (He) to the active gas (Ne). The buffer gas was excited by the electron impact in the plasma of the gas discharge at low pressure (≈ 10 mmHg). An energy level diagram of neon and helium is shown in Fig. 7.3. Since then, the same pumping scheme has been used in other gas lasers. In most cases the discharge is excited by a direct current (DC) source, but it is possible to use radiofrequency discharge. Later, gas lasers appeared with other mechanisms of excitation.

A representation of a gas laser construction is shown in Fig. 7.4. The operating gas is in the discharge tube closed at both ends by windows of optical glass. The specific amplification in the gaseous media in most cases does not exceed several percent. That is why, to fulfill the threshold conditions (see Chap. 5) it is necessary to use mirrors with a reflection factor of nearly 99%. A value so close to 1 can be obtained with the help of interference dielectric mirrors. These mirrors have a

Fig. 7.3 Energy diagram of a helium–neon laser. Neon is the operating gas and helium is the buffer gas. The laser operates on a three-level scheme. Pumping is provided by collisions of helium atoms with neon atoms. The high efficiency of the pumping is connected with the resonance character of the energy transfer from the buffer gas to the operating one. Generation in neon is obtained at almost 150 wavelengths from 0.54 μm to 0.133 mm. Three of the most intensive of these are shown

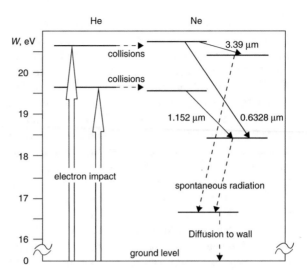

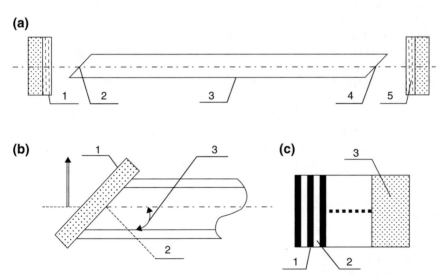

Fig. 7.4 Construction of a gas discharge laser with external mirrors. (**a**) General view: *1* and *5* denote the resonator mirrors, *2* and *4* denote the output windows, and *3* denotes the gas discharge tube. (**b**) Output window of the gas discharge tube: *1* denotes the optical glass, *2* denotes normal, and *3* denotes the Brewster angle. (**c**) Interference mirror of the resonator: *1* and *2* denote the alternate thin film with different permittivity, and *3* denotes the glass substrate. To reduce losses, the output tube windows are oriented under the Brewster angle. To obtain a mirror reflection factor near 99%, mirrors with a multilayer dielectric covering are used

periodic structure with alternating layers of two dielectric sorts. Reflections from the boundaries of the alternating dielectric layers with an optical thickness of a quarter of a wavelength are summed in-phase, which provides the high reflection factor value.

To reduce losses on reflection from the windows closing the gas discharge tube, they are oriented under the Brewster angle to the longitudinal axis. It is known that in this case, for the wave, whose electric field vector lies in the incident plane, the reflection factor is equal to zero. If the self-excitation conditions are fulfilled only for this field, the quantum generator radiation becomes linearly polarized.

Gas lasers can be divided into gas discharge lasers, gas lasers with optical pumping, gas lasers with excitation from a charged particle flow, gas-dynamic molecular lasers (GDLs), and chemical lasers. Depending on the type of operating substance, gas lasers use atomic transitions and ionic lasers and molecular lasers use electronic, vibrational, and rotational transitions. There are also gas lasers with more complicated combined mechanisms of inversion creation.

At the moment, gas lasers account for the largest number of lasing transitions— over 12,000. The output of many lasers may consist of several lines of varying intensities. The output frequencies vary from the ultraviolet range (about 0.1 μm) to the submillimeter range (about 0.3 mm). The gas pressure in a gas laser varies from 10^{-3} mmHg to several atmospheres, but even at such high pressure the medium remains transparent and homogeneous. Therefore, gas lasers allow achievement of high levels of spatial and time radiation coherence. It is precisely this quality that often determines the choice of an optical quantum generator.

In the optical and ultraviolet bands, continuous and pulsed generation are obtained using ions of various orders. Lasers based on rare gas ions are most widely used. A laser based on argon (Ar^+) ions generates in the blue-green area of the spectrum from 0.4545 to 0.5287 μm. Its energy level diagram is shown in Fig. 7.5. The output power of an industrial argon ion laser achieves 1–40 W (in laboratory samples, up to 500 W). A krypton ion laser has worse characteristics.

The current density of the arc discharge in ion lasers is hundreds of amperes per square centimeter. Therefore, the problem of thermal destruction arises, which can

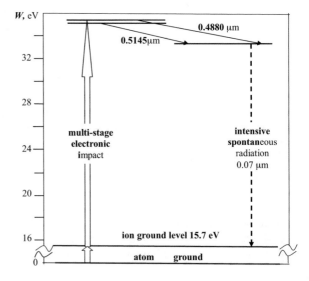

Fig. 7.5 Energy diagram of a laser based on argon atoms. Excitation of ions is provided by multistage electronic impact. At first, ionization occurs; then ion excitation into the upper laser state occurs. The two most intensive generation lines are shown. To date, generation of transitions in almost 30 chemical elements has been achieved. The generation power of industrial ion lasers reaches 1–40 W

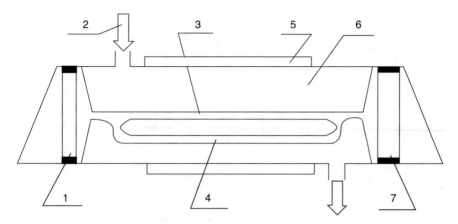

Fig. 7.6 Gas discharge tube of an ion laser: *1* cathode, *2* water input, *3* gas discharge capillary, *4* bypass channel, *5* solenoid, *6* water jacket, *7* anode. To increase efficiency a capillary channel with a diameter of 1–3 mm is used. The problem of thermal destruction is solved by tube cooling with water (4 l/min). The discharge tube is made from ceramics. In the arc discharge the current density is so high that gas transit from the anode to the cathode is observed. To equalize the pressure in the capillary a bypass channel is used. The tube is placed in a solenoid, creating a static longitudinal magnetic field of about 0.1 T. As a result, the output power increases

be solved by intensive cooling with flowing water and by application of a ceramic discharge tube (made of beryllium oxide or graphite). The discharge tube is often placed in a solenoid, creating a longitudinal magnetic field of about 0.1 T, which compresses the plasma cord. This decreases the wall's temperature and leads to power and efficiency growth. A general view of the gas discharge tube of an ion laser is shown in Fig. 7.6.

Excitation transmission from the buffer gas (He or Ne) to the operating one is used in many ion lasers. In widely used gas lasers of this type, a mixture of helium and cadmium is used. The most intensive generation is obtained in spectral lines at 0.4416 and 0.3250 μm. A medium-sized laser has power of 10–50 mW in the line at 0.4416 μm and several milliwatts in the line at 0.3250 μm.

The probability of excitement of the electron states of molecules is comparable to that at the atomic level. As we examined in Chap. 2, vibration of molecules transforms the electron levels into energy bands. Therefore, at excitation by the electronic impact, it is difficult to obtain essential amplification in the operating transition. Nevertheless, generation using electron transition of N_2, H_2, D_2, HD, CO, and NO molecules has been achieved. Lasers based on nitrogen molecules, which generate radiation with a wavelength of 0.3371 μm in the form of pulses with a duration of several nanoseconds and with peak power of 1 MW, are widely used.

Gas lasers based on the vibrational transitions of molecules (see Sect. 1.6) are the most powerful and effective. They generate in the middle part of the infrared band and are called molecular lasers. Carbon dioxide lasers are widely used. Population inversion is formed in the gas discharge of an N_2–CO_2 mixture. The main mechanism of excitation is energy transfer from nitrogen molecules to carbon dioxide

Fig. 7.7 Energy diagram of a quantum generator based on a nitrogen and carbonic acid gas mixture. Excitation of the CO_2 upper level occurs as a result of energy transfer from the buffer gas N_2. The disintegration speed of the lower level can be essentially increased by addition of gases such as He, H, and especially H_2O vapor. The only longitudinal resonator mode falls into the amplification line (~50 MHz). Generation is observed in several transitions. The oscillation frequencies in adjacent transitions differ by approximately 54 GHz

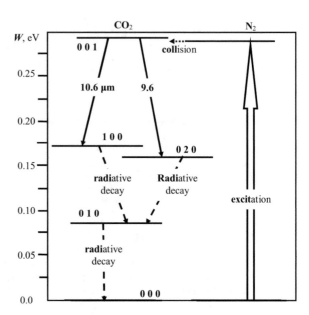

molecules through collisions. A simplified energy diagram of the N_2–CO_2 mixture is shown in Fig. 7.7. Generation is obtained in several spectral lines near wavelengths of 10.6 and 9.6 µm. The generation frequencies correspond to transitions between rotational sublevels of a given pair of vibrational levels with a change of the quantum rotation number J by 1 (see Sect. 1.6). The maximum sublevel population corresponds to $J = 20$–30. Lines with $\Delta J = -1$ (the so-called P branch) of the transition 001–100 are advantageous. The number of molecule gases from which generation is provided has already approached 100.

In lasers with longitudinal discharge, helium is added to the N_2–CO_2 mixture. The helium facilitates a gas temperature decrease (it destroys the population at the lower energy level). In a tube with a length of about 1 m, one can obtain an output power of tens of watts at an efficiency of greater than 10%. The origin of a high output power value was discussed in Sect. 5.5. The relaxation time T_1 in the CO_2 is defined by transition processes between rotational sublevels of the specified vibrational level. The distance between sublevels is small and the relaxation processes occur very fast ($10^{-6} - 10^{-7}$ s). At the same time, the relaxation of the vibrational levels themselves occurs slowly ($10^{-3} - 10^{-4}$ s), which is required to obtain inversion.

A sketch of a molecular generator based on an N_2–CO_2–He mixture is shown in Fig. 7.8. In a general way, its construction looks like the construction of a helium–neon laser. However, there are some differences.

Firstly, with cooling of the operating gas the output power increases to units of kilowatts. Therefore, cooling with flowing water is used and/or continuous bleeding of the gas mixture is used.

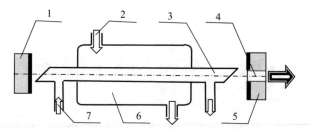

Fig. 7.8 Molecular quantum generator based on an N_2–CO_2–He mixture: *1* mirror, *2* water input, *3* gas discharge tube, *4* coupling hole, *5* output mirror, *6* water jacket, *7* gas input. The peculiarities of this laser are the application of water cooling and continuous bleeding of the operating mixture. The tube length ranges from approximately 50 cm to several meters. The operating diameter is several centimeters. The windows are usually made from a KCl monocrystal. The reflective coverings of the mirrors are made using alumina or gold evaporation on a nontransparent substrate. Radiation is extracted through a hole in the one of the mirrors

Secondly, in the middle part of the infrared range, there are not very many transparent materials. We need to use crystals such as NaCl, KCl, AgCl, KBr, and some others. Some suitable materials dissolve in water and have low rigidity, while others are quite expensive. For output windows we can usually use KCl. The available crystal materials have properties that make them unsuitable for mirror manufacture. Therefore, we must use nontransparent materials as substrates. In this case the radiation outlet is provided through a hole in one of the mirrors. In powerful lasers, mirrors are produced by gold evaporation on a steel substrate.

Continuous-mode molecular quantum generators operate at low pressure (100–1000 Pa). Such pressure creates a sufficient density of active particles and accepted values for the relaxation times of the operating levels. To obtain higher values of pulse power, it is necessary to increase the operating gas density and to provide a fast energy interchange with the pumping source. This problem can be solved in lasers with transverse discharge, which operate at atmospheric pressure. Such quantum generators are called transversely excited atmospheric pressure (TEA) lasers. The output energy of these generators is 1–10 J at a pulse duration of 100 ns, which corresponds to a pulse power of 10–100 MW.

Figure 7.9 shows a sketch of a possible construction of a TEA laser based on an N_2–CO_2–He mixture. The main problem consists in the fact that electrons must maintain the discharge, on the one hand, but, on the other hand, excite the operating medium. Moreover, transition into the arc discharge mode is denied. To solve the problems that arise, the processes of free-electron generation and operating substance excitement by electron impact are divided in the space. In the construction shown in Fig. 7.9, electrons are generated at the expense of the pulse discharge between a series of metal conductors and metal plates. After that, the electrons are accelerated by an electric field and, moving from a cathode to an anode, excite the large volume of the operating mixture. Other variants use an electron gun.

The greatest power (tens of kilowatts) is obtained in the so-called gas-dynamic molecular laser. In this, population inversion is caused by adiabatic cooling of a

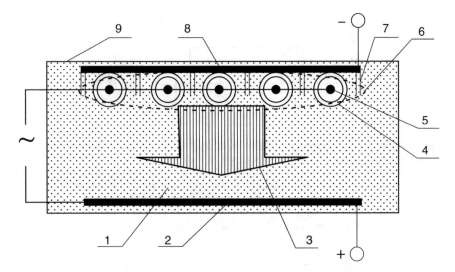

Fig. 7.9 Laser with transverse discharge: *1* N$_2$–CO$_2$–He mixture, *2* anode, *3* electron beam, *4* Pyrex glass tube, *5* metallic wire, *6* area of electron generation, *7* duralumin blade, *8* cathode, *9* case. A pulse or radiofrequency discharge is excited between the cathode blade and metallic wires. The formed electron layer moves toward the anode, exciting the operating mixture. The wires are enclosed in glass tubes to avoid premature discharge and transition to the arc discharge

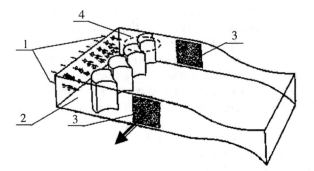

Fig. 7.10 Gas-dynamic molecular laser: *1* injection of air and CO, *2* burning chamber, *3* mirror, *4* nozzles. The original mixture is obtained by burning CO in an air atmosphere. In the burning chamber the operating mixture is under pressure of 5–100 atmospheres at a temperature of 1500–3000 K. It expands during flowing in the supersonic nozzle and cools fast. The nozzles are formed by the rods' surfaces. The difference in level relaxation speeds leads to population inversion of the operating transition and to continuous oscillations in the resonator, formed by a pair of mirrors. The output power in the continuous mode is several tens of kilowatts

heated gas, which is ejected into the resonator through a nozzle. With gas flowing in the supersonic nozzle, the mixture is cooled fast and the level relaxation mechanisms start to act. Inversion is possible in the transients (i.e., until the Boltzmann distribution is stabilized) if the relaxation (cooling) time of the upper level is greater than that of the lower one. The structure of a gas-dynamic molecular laser is shown in Fig. 7.10.

Initially, the operating mixture N_2–CO_2 was obtained by burning CO in an air atmosphere. The pressure in the burning chamber was 5–100 atmospheres and the temperature was 1500–3000 K. The burning products were then transferred to a CO_2 inlet in the supersonic area of the nozzle. As a result, the efficiency of the generator was increased owing to a reduction in the carbon dioxide temperature to 300 K and an increase in the buffer gas temperature to 5000 K.

There is no reliable information on the output power of gas-dynamic molecular lasers in the literature. It has been reported that the radiated energy per mass unit of burned mixture is 20–100 kJ/kg. The output power is probably about 100 kW in the continuous mode.[2]

7.3 Solid-State Quantum Devices

In 1960, Theodore Maiman created the first optical quantum generator, in which a ruby was used as the operating substance. However, this material had already been used in the ultrahigh-frequency (UHF) range in a quantum paramagnetic amplifier (QPA) in 1956. In quantum paramagnetic amplifiers, electron paramagnetic resonance (see Sect. 3.4) is used.

7.3.1 Quantum Paramagnetic Amplifiers

In quantum paramagnetic amplifiers, a particle-sorting application is impossible. Therefore, the two-level variant allows a pulsed operation mode only. All cycles involved in maser operation can be divided into three main stages: excitation, radiation (amplification or generation), and recovery of the equilibrium state. The duration of the stages is the same as the longitudinal relaxation time T_1.

In practice, for creation of excited states in paramagnetic systems, the following are used:

- Pulse excitation
- Adiabatic rapid passage (ARP)

The first method is based on radio pulse application, the duration of which is much less than T_1 and T_2. In this case we can omit items that describes relaxation processes from the Bloch equation. How this can be the case seems surprising, but in this approximation a nonlinear equation (4.14) provides a strict analytical solution. It

---

[2]This is equivalent to the energy of a bullet with a mass of 10 g and traveling at a speed of 1 km/s being only 5 kJ.

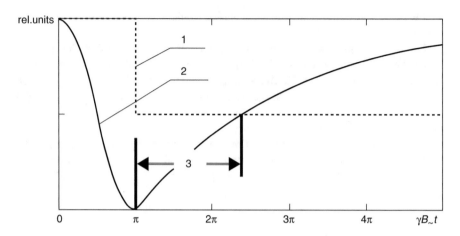

Fig. 7.11 Pulse pumping method in a two-level system. During the action of the radio pulse the longitudinal magnetization M_z changes according to the $\cos(\gamma B_- t)$ law. A pulse of duration $T_{pl} = \pi / \gamma B_-$ (*1*) transfers vector **M** into a state opposite to the direction of the static magnetic field. Until M_z (*2*) is negative, the population difference n_{12} (*2*) is also negative, and it means that the spin system is in an inversion condition (*3*)

turns out that under the action of a field with circular polarization, $\mathbf{B} = B(\cos (\omega_H t)\, \mathbf{1}_x + \sin (\omega_H t)\, \mathbf{1}_y)$, the longitudinal component M_Z of magnetization vector **M** changes in time according to the law

$$M_Z = M_Z^e \cos (\gamma B t).$$

If the cosine argument is equal to 180 degrees, the magnetization vector achieves a minimal negative value. This means that the molecular system proceeds from equilibrium to the inverse state. The pulse fulfilling the population inversion of the two-level system is called the π pulse.

As a matter of fact, relaxation processes trying to keep the population distribution at the equilibrium value contradict inversion formation. It is precisely for this reason that the pumping field amplitude should be chosen on the condition that the duration of the π pulse is less than relaxation times T_1 and T_2. An illustration of the pulse inversion process is presented in Fig. 7.11.

The process of adiabatic rapid passage consists in the fact that a system of paramagnetic particles is placed in a high-frequency field whose frequency changes according to the linear law. The frequency deviation should exceed the line width of the appropriate transition by several times and the pulse duration should be essentially less than the relaxation time. In this sense, the process is rapid. Nevertheless, the frequency variation for the period interval is small in comparison with the frequency itself. That is why the transition process is adiabatic.

Within the scope of this book, it is not possible for us to describe in detail the mathematical substantiation of adiabatic rapid passage. Therefore, we shall provide just a brief description of this process.

The equilibrium condition of the spin system characterized by the definite precession angle of the magnetic moment corresponds to each value of the current frequency. The adiabatic property provides the possibility to consequently pass all of these states following the detuning changes. The precession of the magnetization vector begins in the upper half-sphere. During frequency variation, the precession axis moves into the lower half-sphere. As a result, the magnetic moment turns out to be oriented opposite to the permanent magnetic field, i.e., inversion appears.

From an engineering point of view, it is easier to provide adiabatic rapid passage by means of frequency scanning of the paramagnetic transition with variation of the permanent magnetic field, i.e., $\omega_H = 2\pi\gamma B_0$.

For operation of a maser according to a two-level scheme, the crystal should be cooled to a liquid helium temperature. Cooling is necessary for two reasons. Firstly, the longitudinal relaxation time increases to practically accepted values $(10^{-2} - 10^{-3} \text{ s})$. Secondly, the equilibrium population difference (i.e., the inverse population difference) increases.

The principal uselessness of two-level paramagnetic masers for operation in the continuous mode means these devices have little practical significance, although they have been developed.

The major opportunities are based on the principle of creation of an inverse population by using additional energy levels. This idea was discussed in Sect. 5.1, and it consists in saturation of the auxiliary transition between a pair of levels different from the operating one. Let us recall the facts of the matter.

An energy diagram of a quantum system with three energy levels is presented in Fig. 7.12a. The pumping system has the frequency $\omega_{31} = \frac{W_3 - W_1}{h}$, and its saturated transition is $3 \leftrightarrow 1$. As a result, the sign of paramagnetic susceptibility on frequencies ω_{32} and ω_{21} changes and amplification arises on one of the frequencies. The value of the achievable inverse population slightly increases at operation according to the four-level scheme, one variant of which is shown in Fig. 7.12b. It turns out that paramagnetic crystals are the most suitable class of substances for implementation of this pumping principle.

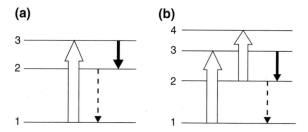

Fig. 7.12 (a) Three-level and (b) four-level pumping methods. In the three-level scheme, the pumping source increases the population of the upper level. Transitions $2 \rightarrow 1$ empty level 2. As a result, inversion arises in the operating transition $3 \leftrightarrow 2$. In the four-level scheme, two pumping sources are used with frequencies f_{31} and f_{42}. The four-level method is more effective

Fig. 7.13 Four-level pumping method using a single source. The energy levels of the chromium ion in the ruby at magnetization at an angle of 54° 44′ are located symmetrically with respect to the initial level; therefore, $f_{31} = f_{42}$. For a magnetization field of 0.42 T the pumping frequency is $f_{pump} = 24.2$ GHz and the operating frequency is $f = 9.4$ GHz

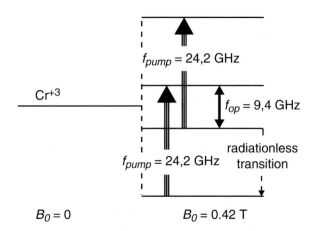

Ruby ($Al_2O_3 : Cr^{3+}$) and rutile ($TiO_2 : Cr^{3+}$ and $TiO_2 : Fe^{3+}$) crystals are most often used in paramagnetic masers. It is known (see Sect. 3.4) that the position of spin levels in paramagnetic crystals depends upon not only the value but also upon the orientation of the external static magnetic field with respect to the crystal axes. This gives additional freedom in the choice of both operating levels and the pumping structure. For example, an angle of 54.7° between the ruby optical axis and the magnetic field allows provision of pumping according to the four-level scheme shown in Fig. 7.13. Its peculiarity consists in the fact that because of the symmetric level position, we can use a single pumping source in the four-level scheme (the push–pull pumping method).

The traveling-wave amplifier is the most evident quantum paramagnetic amplifier scheme. Figure 7.14a shows a simplified structural diagram of such an amplifier. The signal propagates in a transmission line of the "meander" type. Such a line plays the role of slowing the system. Slowing increases the gain because the amplification coefficient

$$\alpha_p = \frac{dP/dz}{P} = \frac{P_{PMC}}{P} = \frac{P_{PMC}}{v_{gr} W_{acc}}.$$

Here, P is the power transmitted by the wave, P_{PMC} is the power delivered by the active medium per unit of length of the paramagnetic crystal, v_{gr} is the group velocity, and W_{acc} is the density energy accumulated by the wave per unit of length. Thus, to increase amplification it is necessary to decrease the group velocity of the wave (i.e., increase the time of interaction of the signal with the active medium).

The pumping wave propagates along the paramagnetic crystal. The wave polarization should be such that a right angle is arranged between the vectors of the pumping wave and the signal wave with respect to the permanent magnetic field.

One possible structure of a traveling-wave maser is shown in Fig. 7.14b. A pumping wave with a frequency of 14 GHz propagates in a rectangular metallic wave guide, inside which the meander strip line of the signal is located. A ferrite

Fig. 7.14 (a) Structural diagram and construction of a quantum paramagnetic traveling-wave amplifier: *1* pumping, *2* input signal, *3* signal transmission line, *4* paramagnetic crystal, *5* output signal. The signal propagates in a line with serpentine geometry. Such a structure slows the wave and increases the specific amplification. Other types of slowing structures can be used in amplifiers as well. (**b**) Sketch of a paramagnetic amplifier: *1* paramagnetic crystal, *2* slowing structure, *3* input line, *4* pumping wave guide, *5* output line, *6* ferrite. The pumping wave propagates in a rectangular metallic wave guide, inside which there is a strip signal line. The ferrite plate provides unidirectionality of the transmission line

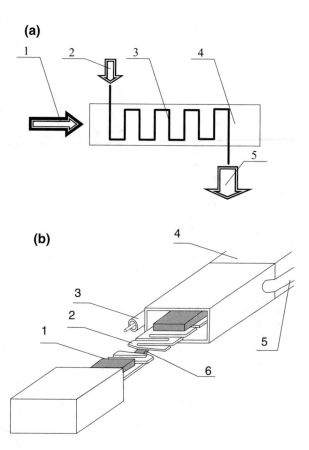

plate provides unidirectionality of the transmission line. The amplifier operates on a frequency of 3 GHz with a pass band of 25 MHz and has a gain of about 30 dB.

Another variety of quantum paramagnetic amplifier is a resonator maser. Figure 7.15 shows a scheme of its construction. The change to application of a resonator allows a decrease in the size of the device. In the case of operation according to a two-level scheme, there is no need for a separate path for pumping, as the pumping frequency coincides with the operating one. Instead, a device for magnetic field scanning, which is necessary for a traveling-wave amplifier, is added (see Table 7.1).

The maser resonator should be selected so that at minimal pumping power the value of the amplification bandwidth (the product of the gain and the bandwidth) is optimized to the fullest possible extent. These are the definite requirements regarding resonator properties in terms of the frequencies of amplification and pumping. These frequencies are determined by the paramagnetic spectrum of the crystal. They are so different from each other that application of one resonator mode for fulfillment of both functions is impossible. Either pumping and amplification are provided in two different modes of the cavity or resonators of a combined type are used.

Fig. 7.15 (**a**) Structural diagram of a resonator quantum paramagnetic amplifier: *1* input, *2* wave guide, *3* absorber, *4* circulator, *5* output, *6* resonator, *7* paramagnetic crystal. (**b**) Construction of an amplifier: *1* resonator, *2* strip line, *3* wave guide, *4* paramagnetic crystal, *5* piston for tuning. The wave reflection from the one-port resonator is used in the amplifier. To separate the incident (input) and reflected (output) waves, the circulator is used. The signal propagates in the shielded strip transmission line, which is connected with the strip resonator through the coupling hole. The walls of the rectangular wave guide in which the pumping wave propagates are the line's shielding. The closed segment of the wave guide forms the resonator tuned to the pumping frequency

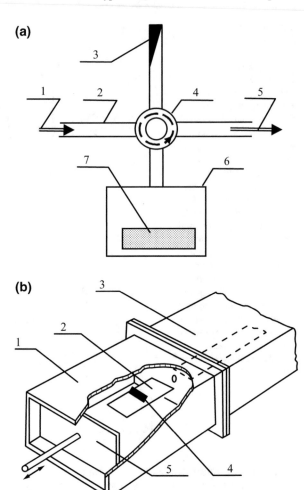

Table 7.1 Parameters of paramagnetic traveling-wave amplifiers

Active substance	Operating frequency (GHz)	Pumping frequency (GHz)	Gain (dB)	Band (MHz)	Tuning range (MHz)	Magnetic field (T)	Operating temperature (K)
Al_2O_3: Cr^{3+}	2.39	13.0	36.0	13	10	0.25	18
Al_2O_3: Cr^{3+}	3.0	14.0	21.5	17		0.29	1.37
Al_2O_3: Cr^{3+}	8.0	25.6	18.0	25	400	–	4.2
TiO_2: Cr^{3+}	23–41	43–82	10.2	–		–	1.5

Table 7.2 Parameters of resonator paramagnetic amplifiers

Active substance	Operating frequency (GHz)	Pumping frequency (GHz)	Amplification area (MHz)	Magnetic field (T)	Pumping power (mW)	Operating temperature (K)
Al_2O_3: Cr^{3+}	0.38–0.45	11.8	0.56	0.007	–	1.7
Al_2O_3: Cr^{3+}	9.4–9.61	22.85	100–200	0.39	30–60	1.35
K_3Co $(CN)_6$: Cr^{3+}	2.8	9.4	1.8	0.2	10	1.25
TiO_2: Cr^{3+}	8.2–10.6	35	25	0.33	10	4.2
TiO_2: Fr^{3+}	70	120	100	0.46	–	4.2

To implement a quantum paramagnetic amplifier on high frequencies, a high enough value of the intensity of the static magnetic field is required. A radical solution is application of superconducting magnets. A superconducting magnet is placed into a cryostat in which the other elements of the maser that require cooling are also placed simultaneously (Table 7.2).

The practical value of the quantum paramagnetic amplifier is due to an extremely low level of inherent noise, which is connected with the heat radiation of the construction elements and with spontaneous radiation of active quanta only. The low operating temperature facilitates the reduction of the construction's thermal noise. The probability of spontaneous transitions is rather low at a low operating frequency. In the UHF range the spontaneous radiation intensity is extremely low. It is acceptable to use the effective noise temperature as a numerical characteristic of the quantum paramagnetic amplifier's noise properties. If the gain $K_p \gg 1$, the noise in a quantum paramagnetic amplifier is determined mainly by the noise of the spin system. Its power spectral density of spin noise is

$$S_{sn} = \frac{\hbar\omega}{1 - \exp\left(-\hbar\omega/k_B T_{spin}\right)},$$

where $T_{spin} = \frac{\hbar\omega}{k_B \ln(n_1/n_2)}$ is the effective temperature of the spin system. If $\hbar\omega \gg k_B|$ $T_{spin}|$, the effective noise temperature achieves its minimal value $T_n^{id} = \hbar\omega/k_B$. On an operating wavelength of 3 cm, T_n^{id} is equal to only 0.5 K. The noise temperature of a real quantum paramagnetic amplifier depends upon the construction of the input line and varies from 5 to 15 K.

The needs of radio astronomy, radio spectroscopy, and radar technology for low-noise receivers have stimulated the search for new operating substances and engineering solutions. At present, the list of active masers is large enough. It covers a wavelength range from decimeter to millimeter waves.

7.3.2   Solid-State Lasers

Quantum generators based on fluorescent dielectric crystals and glass, which are activated by ions of rare-earth elements or by ions in the iron group, are usually called solid-state lasers (SSLs). The energy levels of activators are used for creation of the inverse population.

A ruby—a material whose utility in quantum electronics has been repeatedly demonstrated—was used by Theodore Maiman in the first operating laser, which was created in 1960. Besides the ruby, the list of operating substances includes various crystals and glass activated by rare-earth elements.

On the basis of the pumping principle, solid-state lasers come the closest to three-level masers. A powerful light from an auxiliary source induces transitions from the ground level to some higher level of the quantum system. There is a third level, at which excited particles are quickly deposited, between these levels. When the population of this level adequately exceeds the population at the lower level, the medium becomes amplifying. The transition from the pumping level to the upper operating level happens without radiation. The energy is extracted via that path as heat. We can often use a four-level scheme.

A ruby laser operates according to a three-level scheme (Fig. 7.16), in which level 1 is the ground state, level 2 has two bands of absorption, and level 3 is a doublet level. The generation wavelength of the ruby laser is near 0.69 μm. The typical concentration of Cr^{3+} ions is about 0.05% by weight.

For solid-state laser pumping, as a rule, noncoherent light sources are used: xenon and mercury gas discharge flash lamps, incandescent lamps, etc. These sources

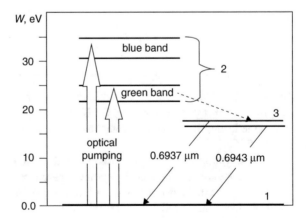

Fig. 7.16 Energy diagram of a ruby laser. Two absorption bands play the role of the upper energy level. Transitions to the band in the green part of the spectrum ($\lambda = 0.5$–0.6 μm) agree well with the radiation of xenon lamps. Radiationless transitions of excited particles effectively occupy the upper operating levels ($T_{32} = 5 \cdot 10^{-8}$ s). Generation is obtained on two wavelengths: 0.6943 and 0.6927 μm. At room temperature the energy in the pulse can reach 300 J. 1, 2, 3 are numbers of the energy levels

radiate in a wide frequency band; therefore, the laser efficiency depends to a great extent upon the absorption line width of the operating substance.

The active element (AE) of such lasers usually takes the form of a circular cylinder or a rod with a rectangular cross-section. Sometimes, active elements with more complicated configurations are used. In the most widely used construction, a cylindrical active element and a gas discharge pumping lamp are placed in the chamber–lighter, concentrating the radiation of the pumping lamp into the active element (Fig. 7.17). Multiple pumping radiation reflections from the internal surface

Fig. 7.17 Main elements of a ruby laser. For pumping we can use spiral (**a**) or rod (**b**) lamps. The reflector increases the effectiveness of the lamp's radiation. For spiral lamps a diffuser reflector is sometimes used. A rod lamp and a ruby rod are placed in the focuses of an elliptic reflector. To increase the pumping intensity we can use several rod lamps (**c**). _1_ active element, _2_ reflector, _3_ pumping lamp

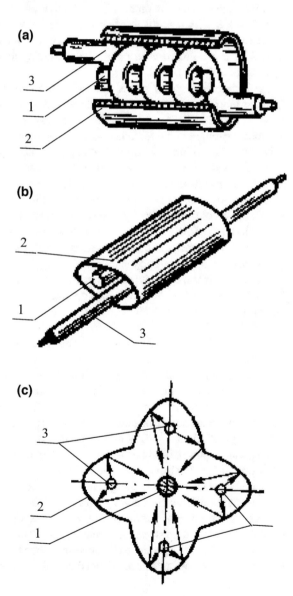

of the chamber provide more complete absorption in the active element. In powerful ruby lasers, circular rods with a diameter of approximately 2 cm and a length of 20–30 cm are used. We can apply lighters in which one pumping lamp operates several active elements or, conversely, one active element is pumped by several lamps. Because of the amazing advantages of semiconductor lasers, recently they have been widely used for solid-state laser pumping.

The range of operating wavelengths of solid-state lasers extends from the ultra-violet spectrum to the region of medium-infrared radiation. Solid-state lasers operate in pulsed and continuous modes. The radiation power in the continuous mode can reach 1–3 kW with specific extraction of up to 1 kW/cm^3 for efficiency of about 3% in existing generators. Average power of 1 kW at a pulse repetition frequency of up to 100 Hz can be realized in the pulsed–periodic mode of free running at a pulse duration of $10^{-3} - 10^{-6}$ s.

To increase the pulse power we can successfully use the mode of resonator Q factor modulation. This issue was discussed in Chap. 5. The usual values of the pulse duration are $10^{-8} - 10^{-7}$ s. Their peak power is about 10^9 W and is limited by the optical strength of the resonator's passive elements, with a value of about $5 \cdot 10^2$ MW/cm^2. The volumetric optical strength of laser materials is usually higher. Q factor modulation is achieved in both a passive manner and an active one. Sometimes, mechanical modulators—for instance, a rotating prism—are used.

Since many proper resonator frequencies fall in the amplification loop, it is rather simple to provide the mode of phase synchronization and to obtain extrashort pulses with a duration of $10^{-12} - 10^{-13}$ s, which is limited by the amplification line width. In addition to Q factor modulation, mode synchronization is provided in both an active manner and a passive manner.

Among activators, neodymium ions (Nd^{3+}) are most widely used in solid-state lasers. Lasers based on rare-earth ions operate according to a four-level scheme. An energy diagram of neodymium ions in a lattice of calcium tungstate (CaWO$_4$) is shown in Fig. 7.18.

Lasers based on silicate and phosphate glasses with neodymium generating radiation in the area of 1.05 μm have found wide applications. The main purpose of lasers based on glasses is generation of single pulses of high power. Active elements made from glass have high optical quality and can have a large volume for the given element form. Lasers based on phosphate glass with neodymium generate the most powerful pulses. For an active element volume of 2 m^3, pulses with energy of about $4 \cdot 10^4$ J and a duration of about 10^{-9} s are obtained, which correspond to power of about $4 \cdot 10^{13}$ W. The radiation on the second harmonic (0.53 μm) and on the third harmonic (0.35 μm) of the main transition frequency for the same pulse duration has energy of about $2 \cdot 10^4$ J.

A crystal of yttrium aluminum garnet (YAG) is often used as a matrix with Nd^{3+}. The necessary spectral luminescence properties of this crystal are successfully combined with its high strength, hardness, and significant heat conductance. Lasers based on YAG: Nd^{3+} operate in all above-mentioned modes. Record power values in the continuous mode have been obtained with them. The generation wavelength of a

Fig. 7.18 Energy diagram of neodymium ions in a lattice of calcium tungstate (CaWO$_4$). Lasers based on rare-earth element ions operate according to a four-level scheme. The lower level 4 is practically unoccupied; therefore, less pumping power is required for inversion. 1, 2, 3, 4 are numbers of the energy levels

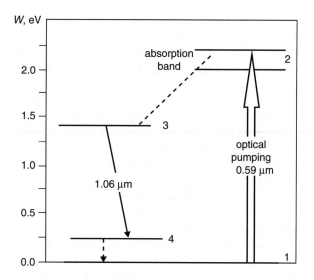

YAG: Nd^{3+} laser in the main transition of neodymium is 1.064 μm. The typical sizes of the active element range from 3 × 50 mm to 10 × 120 mm.

Yttrium aluminate crystals (YAO: Nd^{3+}) and lithium yttrium fluoride crystals (YLF: Nd^{3+}) have also found applications. YAO: Nd^{3+} crystals are preferable to YAG: Nd^{3+} crystals for operation in the mode of Q factor modulation, which is connected with a high value of the relaxation time of the main operating transition and, hence, with the possibility of accumulation of large particle numbers on the upper laser level.

The distinctive properties of YLF: Nd^{3+} crystals are the negative value and small absolute meaning of the temperature coefficient of the refraction index. This feature essentially weakens the manifestations of thermal–optical effects—in particular, the effect of induction of the thermal–optical lens. As a result, the spatial coherence of the laser radiation is increased. The wavelength of laser generation based on a YLF: Nd^{3+} crystal is shifted to the short-wave side ($\lambda \approx 1.05$ μm) in comparison with the wavelength of a YAG: Nd^{3+} laser, which allows the possibility of effective operation of such a laser with an amplifier based on doped glass.

New possibilities for use of trivalent chromium ions as active particles have been demonstrated using an alexandrite (BeAl$_2$O$_4$: Cr^{3+}) crystal. In contrast to a ruby crystal, generation in an alexandrite laser is provided also by electronic–vibrational transitions between the level of the green line of absorption and the main level. With that, this laser provides the possibility of smooth tuning of the wavelength within the limits of 0.73–0.80 μm.

Titanium–sapphire (Al$_2$O$_3$: Ti^{3+}) lasers are tunable lasers, which emit red and near-infrared light in a range from 650 to 1100 nm. These lasers are mainly used in scientific research because of their tunability and their ability to generate ultrashort pulses. The short lifetime of the excited state of Ti^{3+} (about 3 μm) at room temperature makes the lamp pumping of this laser as ineffective. As a rule, pumping is provided either by a continuous argon laser or by pulses of the second harmonic of a neodymium laser. With that, the effectiveness of the laser pumping radiation transformation into titanium ion radiation may exceed 20%.

Joint implementation of Cr^{+3} ions and trivalent ions of rare-earth elements in garnet crystals has led to an essential increase in laser efficiency. High concentrations of rare-earth ions and ions in the iron group allow introduction of the necessary concentrations of both types of particles without degradation of the crystal's optical properties. The specific character of the energy structure of Cr^{3+} ions in a garnet crystal provides complete and fast energy transfer from its electronic–vibrational bands to the upper laser levels of rare-earth elements.

Crystals of gadolinium scandium gallium (GSGG), yttrium scandium gallium (YSGG), and gadolinium scandium aluminum (GSAG) garnets are part of the family of chromium-containing garnets operating in the main transition of neodymium in the area of $\lambda \approx 1.06$ μm. These crystals are used in pulsed and pulsed–periodic operating modes. Efficiency of about 6% is achieved in a laser based on YSGG: $Cr^{3+} - Nd^{3+}$ in the free-running mode pumping 1–3 J. Efficiency of 10% in the free-running mode is achieved with use of a YSGG: $Cr^{3+} - Nd^{3+}$ crystal for pumping about 200 J. In the mode of Q factor modulation for a pulse repetition frequency of up to 50 Hz, efficiency of about 6% is provided at a pulse energy of about 0.4 J, which is limited only by the optical strength of the active element edge. The radiation wavelength of this laser (1.058 μm) corresponds well to the amplification loop of phosphate glass with neodymium. This pair may be successfully used in the system, consisting of two components: the stimulating generator and the amplifier.

The long-wave boundary of effective laser generation with lamp pumping at room temperature achieves 3–3.5 μm. At lower energy gaps the probability of a radiationless transition is essentially greater then the radiation probability. Therefore, it is difficult to select suitable materials for generators on these wavelengths. At lamp pumping the generation of radiation is obtained by Er^{3+} ions with efficiency exceeding 1% in YAG: Er^{3+} and YSGG: $Cr^{3+} - Er^{3+}$ crystals. In the first case the generation wavelength is 2.94 μm; in the second case it is 2.79 μm. A mode of Q factor modulation is realized with a pulse repetition frequency of up to 100 Hz.

The development of semiconductor lasers (SCLs) holds promise for their application in pumping of solid-state lasers. Semiconductor lasers based on arsenide crystals allow radiation to be obtained in the region from 0.75 to 1 μm, which enables the possibility of effectively exciting generation using ions of neodymium (Nd^{3+}), thulium (Tm^{3+}), holmium (Ho^{3+}), erbium (Er^{3+}), and ytterbium (Yb^{3+}). Solid-state lasers with pumping by semiconductor lasers combines the advantages of solid-state and semiconductor lasers.

Pumping by radiation from a semiconductor laser is close to resonance. This, to a great extent, eliminates the problem of thermal distortions in the active element and allows easy achievement of extremely high directivity of the laser beam. Continuous generation is obtained using ions of Ho^{3+} (about 2.1 μm), Tm^{3+} (2.3 μm), and Er^{3+} (2.9 μm), and also different transitions of Nd^{3+} ions. The generation threshold for pumping power in some cases is in milliwatt units. So, for example, the generation

Table 7.3 Materials most widely used in lasers

Active substrate		Operating wavelength (μm)	Activator concentration (%)	Temperature (K)	Operating mode
Base	Activator				
Al_2O_3	C^{3+}	0.6943	0.03	293	Pulsed
	Cr^{3+}	0.6934	0.01	77	Continuous
CaF_2	Dy^{2+}	2.36	0.02–0.2	293/27	Pulsed/continuous
	Nd^{3+}	1.046	–	300	Pulsed
$CaWO_4$	Nd^{3+}	1.0646	0.14	293	Pulsed
Glass	Nd^{3+}	1.06	–	300	Pulsed

threshold for Ho^{3+} ions in a YAG: $Tm^{3+} - Ho^{3+}$ crystal is 4 MW, and the generation threshold for the main transition of Nd^{3+} ions in glass does not exceed 2 mW. Second harmonic generation is obtained with a whole series of crystals containing neodymium. In the main neodymium transition the modes of Q factor modulation and phase synchronization are realized. The total efficiency of a continuous neodymium laser with pumping from a semiconductor laser at a generation wavelength of 1.06 μm reaches 20%.

The applications of solid-body lasers are manifold and include laser technologies for welding, cutting, etc.; electronic devices; medicine; laser radar technology; systems for monitoring atmospheric content; optical information processing; integrated and fiber optics, laser spectroscopy; laser diagnostics of plasma and controllable thermonuclear fusion; laser chemistry and laser isotope separation; nonlinear optics; superfast photography; laser gyroscopes; seismographs; and other accurate physical instruments (Table 7.3).

7.4 Semiconductor Lasers

According to the accepted tradition, lasers based on semiconductor crystals (semiconductor lasers) are in a specific class owing to the peculiarity of their excitation specificity and formation of population inversion in transitions between permitted energy bands of the semiconductor.

The optical properties of semiconductors are determined by the mutual location and population of two upper energy bands. Let us recall that electrons obey the Fermi–Dirac statistics. In nonequilibrium states, the quasi-Fermi levels $W_{Fc} \neq W_{Fv}$ (see Sect. 4.8) and distribution functions do not coincide for different bands. The possibility of such a simple description of nonequilibrium states follows from the extremely fast establishment of quasiequilibrium within the limits of separate bands.

The condition of population inversion in an intrinsic semiconductor is satisfied if at least one of the quasi-Fermi levels is placed inside the appropriate band; in other words, the semiconductor must be degenerate. When this condition is fulfilled, population inversion takes place between levels located near the bottom of the conduction band and the top of the valence band (see Sect. 4.8).

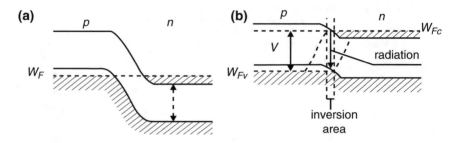

caused by several mechanisms

Fig. 7.19 Energy diagram of the *p–n* junction in the equilibrium state (**a**) and with bias in the forward direction $V \cong W_G$ (**b**). For simplicity the temperature is assumed to be zero. In the laser diode, degenerate semiconductors are used. At zero bias the Fermi level lies inside the bands. With application of bias in the transition area, all states in the conduction band before the quasi-Fermi level W_{Fe} are occupied, and in the valency band the levels above the quasilevel W_{Fv} are free, which is a condition of population inversion existence

A whole series of methods for inverse population achievement in semiconductors is known: optical pumping, excitation of a fast electron beam, avalanche breakdown under the action of an external electric field, and carrier injection through the *p–n* junction. Only the last of these methods is specific to semiconductors. Generation using the *p–n* junction was achieved by Robert Hall and Marshall Nathan in 1962.

Injection lasers based on gallium arsenide (GaAs) were created prior to other semiconductor quantum generators. At this time, GaAs remains the classical laser semiconductor, and charge carrier injection through the *p–n* junction is the most effective and suitable pumping method. Such semiconductor lasers are called laser diodes (SLDs).

Energy diagrams of the *p–n* junction, which explain the process of inversion formation, are shown in Fig. 7.19. The main role is played by an electron injection in the *p* region. Here, near the boundary, the inverse population arises.

Two opposite edges of the laser crystal usually serve as resonator mirrors. Because of the high value of the semiconductor refraction index, the reflection is several tens of percent, so the resonator Q factor is high enough even without reflective covering. The construction of an injection laser is explained in Fig. 7.20. The laser diode is the most miniature type of laser. The length of its active part is measured in fractions of a millimeter. This is sufficient because of the high value of the gain.

It should be noted that the refraction index is slightly higher in the region of the *p–n* junction than in the other volume of the crystal. Experimental data allow estimation of the relative value of the refraction index jump: it is 10^{-3}. The dielectric permeability jump is caused by several mechanisms.

Firstly, reduction of dielectric permeability is connected with the intraband motion of free charge carriers in the *p* and *n* regions, which are adjacent to the depletion layer. According to (4.42),

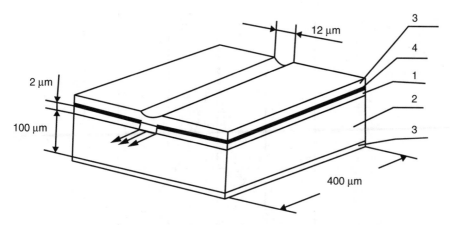

Fig. 7.20 Construction of a semiconductor diode laser: *1* area of *n*-type, *2* area of *p*-type, *3* electrodes, *4* isolating oxide layer. The modern construction is shown, in which an injection area restriction of up to 12 μm is used with the help of an oxide layer. The natural crystalline edges fulfilled the role of mirrors

$$\varepsilon \approx 1 - \frac{\omega_0^2}{\omega^2},$$

where $\omega_0^2 = \frac{n_e e^2}{\varepsilon_0 m_e^*}$ is the square of the plasma frequency (see Sect. 4.4).

Secondly, variation of the electric properties can be caused by the transition from population inversion in the active layer to equilibrium distribution in the environment. In this case, the medium properties should be described by complex dielectric permeability:

$$\dot{\varepsilon}_a = \varepsilon_a - \frac{\sigma}{j\omega}.$$

At the region's boundary the contribution to dielectric permeability of band-to-band transitions changes the sign.

The third possible mechanism consists in the dependence of the position of the optical absorption edge upon the degree of impurity doping.

In simple diode structures the conditions vary between cases, and it is impossible to indicate exactly which mechanism plays the decisive role.

Experimental results confirming the existence of the dielectric wave guide in the *p–n* junctions are shown in Fig. 7.21, where the intensity (measured in arbitrary units) is indicated on the vertical axis and the distance from the *p–n* junction (measured in microns) is shown on the horizontal axis.

Such a profile facilitates radiation field localization near the plane of the *p–n* junction. This is very important because beyond the active layer, the absorption on the operating wavelength is extremely high because of the band-to-band transitions and the influence of free electrons in the bands themselves (see Chap. 4). Moreover,

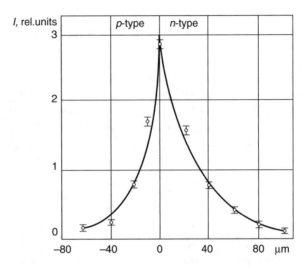

Fig. 7.21 Distribution of radiation intensity in the cross-section of an injection laser. The exponential character of the field decrease beyond the active area is well observed. The experimental point from the left matches well with the function 2.8 exp (x/15.9), and from the right, 2.8 exp (x/30.5). The value $x = 0$ corresponds to the middle of the p–n junction. The significant part of the wave power is transmitted in the area that has losses. This is one of the reasons for the low effectiveness of the first injection lasers, which required huge densities of electric current for excitation

the planar wave guide decreases diffraction losses in the open resonator. The Fresnel number ($a^2/\lambda L$) for the size of the semiconductor laser ($a \approx 4$ µm, $L \approx 400$ µm, $\lambda \approx 1$ µm) does not exceed the value 0.1, and the diffraction loss value should be extremely high. Initially, this caused anxiety that the threshold current would be unacceptably high. However, these anxieties were groundless; the diffraction losses were reduced by the natural planar wave guide.

Initially, the current was passed through the whole plane of the p–n junction. As was soon clear, the result of this was that the laser beam broke down into threads. The waves of the separate threads were not synchronized. Therefore, the first semiconductor laser had bad spatial–time coherence. Moreover, the output laser characteristics could not be repeated from one pulse to the next one. The greater part of the current passing through the junction was wasted, and this led to high values of the threshold current and to diode heating. Transfer to planar electrodes did not yield the desired reduction in the current density. In the first semiconductor generators, the threshold current density was of the order of 1 kA/cm^2. The high threshold current density and difficulties with heat transfer were two circumstances that did not allow provision of continuous generation using diodes with a simple structure at room temperature. Such diodes in the pulsed mode developed power measured in tens of watts. Originally, continuous radiation with the power of several watts was obtained only in lasers that were cooled by liquefied gases.

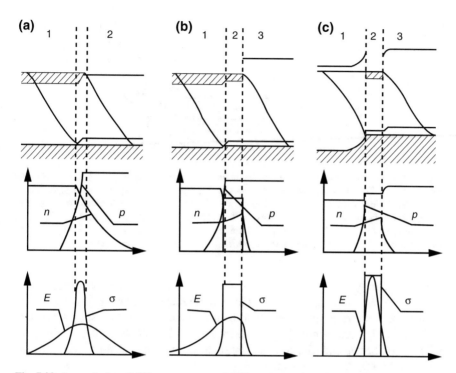

Fig. 7.22 Laser diodes of different structures. (**a**) Homostructure: *1* nGaAs, *2* pGaAs. (**b**) Single heterostructure: *1* nGaAs, *2* pGaAs, *3* pAl$_x$Ga$_{1-x}$As. (**c**) Doubled heterostructure: *1* nAl$_x$Ga$_{1-x}$As, *2* pGaAs, *3* pAl$_x$Ga$_{1-x}$As. In the *upper horizontal line* the energy diagrams are shown, followed by the distribution of carriers (*n* and *p*) and the functions of the field intensity (*E*) and the conductivity of the active area (*σ*) versus the coordinate. In the homogeneous structure the junction thickness is determined by diffusion and recombination processes, and is a value of the order of 1 μm, while in the heterojunctions the value is measured in fractions of microns. The field concentration in the space and areas of population inversion can be seen

The problem was solved with the transfer to semiconductor heterostructures. Semiconductors with different chemical structures are now used in diodes with heterojunctions. To create a qualified junction it is important that the periods of the crystal lattices coincide, with accuracy of about 0.1%. More often, the solid solution Al$_x$Ga$_{1-x}$As is used. The forbidden band width of gallium arsenide is 1.5 eV, and that of AlAs is 2.2 eV. By changing the chemical composition of an alloy, one can control the semiconductor parameters. Heterojunction structures are shown in Fig. 7.22. The applied technology allows us to obtain very sharp junctions with a thickness of only 4–5 atom layers.

Continuous generation at room temperature was achieved at the end of the 1960s on diodes with a single and doubled heterostructure. Lasers based on a three-layer (double) heterostructure (DHLs), with an active layer made from a narrow-band

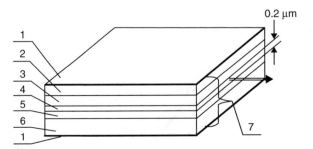

Fig. 7.23 Laser diode with a double heterostructure: *1* metallic contacts ({100} face), *2* p^+GaAs (contact layer), *3* pAl$_x$Ga$_{1-x}$As (emitter), *4* pGaAs (active layer), *5* nAl$_x$Ga$_{1-x}$As (emitter), *6* n^+GaAs (substrate), *7* cleaved {110} face. Layers of strongly doped semiconductors, which are adjacent to metallic contacts, are necessary for creation of the ohmic contact, and they are absent in Fig. 7.22. The active layer has a thickness of about 0.2 μm and forms the integrated planar wave guide. The total thickness of the active area is about 1 μm. The output power can reach tens of milliwatts

semiconductor placed between two wide-band ones, have the best parameters. Figure 7.22 shows energy diagrams and distributions of the charge carrier concentration, the modulus of the electric field E, and conductivity σ for a simple structure and a double heterostructure. We can see from Fig. 7.22 that localization in the space of nonequilibrium and electron–hole plasma and the optical field occurs in the heterostructure.

The refraction index of a narrow-band semiconductor is higher (GaAs: 3.59) than that of a wide-band one (AlAs: 2.97). As a result, the active area becomes the planar wave guide of a surface wave of high quality.

Nonequilibrium carriers can be localized in a smaller area than the light field. The thickness of the narrow-band active layer can be reduced to the size of the de Broglie wavelength of an electron with kinetic energy close to the height of the potential barrier at the boundaries (approximately 6–8 nm).

Two-sided optical and electronic limitations lead to coincidence of the population inversion region and the electromagnetic field region, which allows generation at a small pumping current. The crystal edges are usually used as mirrors in lasers on heterojunctions. The structure of such a laser diode is shown in Fig. 7.23.

Spectral radiation characteristics can be essentially improved if we use external reflectors in the form of periodic structures. The relative reflection band is of the order of $1/N$ for a mirror consisting of N irregularities.

To form such a mirror on the surface of the wave guide layer, a diffraction lattice with period Λ (Fig. 7.24) multiplied by the integer number of half-waves of the wave guide mode can be created. We can distinguish lasers with distributed feedback (DFB—when the light wave interacts with a lattice in the amplification area) and with distributed Bregg reflection (DBR—when the lattice is deposited on the passive part of the wave guide structure). Lasers with diffraction lattices are characterized by high degrees of coherence, which are commensurable with the bands of gas lasers, and high temperature stability. A periodic structure is used in the laser with

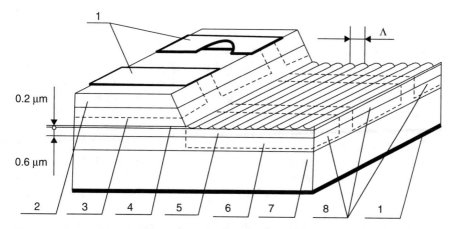

Fig. 7.24 Monolithic pair of lasers with a reflective lattice: *1* contacts, *2* $pGaAs$, *3* $pAl_{0.45}Ga_{0.55}As$, *4* $pGaAs$, *5* $nAl_{0.1}Ga_{0.9}As$, *6* $nAl_{0.3}Ga_{0.7}As$, *7* $n^{+}GaAs$, *8* implanted areas. The structure is similar to that presented in Fig. 7.23. Localization of the injection current in two flows is achieved using impurity implantation. The periodic structure provides the necessary reflection factor on a wavelength equal to 2Λ. Active areas form the three-dimensional integrated wave guides. Similar lasers are included in the structure of phased radiator arrays with power of 5–50 W. Phasing is provided by distributed coupling of three-dimensional dielectric wave guides

distributed feedback for radiation extractions. This improves the radiation directivity pattern and increases its power. A distributed Bregg reflection structure can be formed in a unified technological process with other elements of integrated optics based on semiconductor wave guide heterostructures.

A monolithic pair of laser diodes is shown in Fig. 7.24. The total number of laser diodes can reach the 50. Their radiation turns out to be synchronized at the expense of coupling of the dielectric wave guides. The output power of the generator series can achieve units or even tens of watts.

Heterostructure lasers were first created in the USSR (in 1968), and then in the USA (in 1969) with GaAs and AlAs. They emit from the yellow-green region of the visible radiation in wavelengths of tens of micrometers (1980). Solid solutions of Ga_xIn_{1-x} and As_yP_{1-y}, which are isoperiodic with a substrate of GaP_x and As_{1-x}, allow creation of a very-short-wave laser diode (at $T = 300$ K). These solid solutions, which are isoperiodic with an InP substrate, allow a low-threshold laser diode in the range of 1–1.6 µm (the most promising range for fiber-optic communication lines) to be obtained. Solid solutions of In_xGa_{1-x} and As_ySb_{1-y}, which are isoperiodic with a substrate of GaSb and AlSb, are promising for wavelengths from 2 to 4 µm. The far-infrared region (more than 5 µm) can be assimilated with the help of the solid solutions PbS_xSe_{1-x} and $Pb_xS_{1-x}Te$.

The significance of these investigations is difficult to overestimate. Development of lasers using heterojunctions opened the way to wide practical application of quantum generators. As a mark of the esteem in which the work of the Russian scientist Jores Alfiorov[3] is held by the international scientific community, he was awarded the 2000 Nobel Prize in Physics.

Bombardment of a semiconductor crystal by an electron flow represents one more method of laser pumping in the visible and ultraviolet bands. Electrons in a beam falling normally on the plane crystal surface penetrate it to some depth, wasting energy on creation of electron–hole pairs. The penetration depth is only several microns. It depends upon the electron energy, and to ensure that the active layer is not too thin, the beam is accelerated to an energy of more than 20 keV. Each electron in the beam gives birth to about 10^4 electron–hole pairs in the semiconductor.

Electrons transmitted on the conduction band take away not only the energy but also the impulse from the beam electrons. For this reason they fall at first on the levels outlying from the bottom of the band. This limits the laser efficiency by a value of 40%.

The electron excess in the conduction band of the semiconductor that is necessary for generation can be created also by light absorption. Because of the short lifetime of the excited state, this purpose can be achieved only with application of a powerful pumping source. A laser is usually used for pumping. The depth of light penetration into the crystal and then the active layer thickness does not exceed tens of microns. For understandable reasons, this method of pumping cannot compete with electron excitement or with carrier injection through the p–n junction.

7.5   Liquid Lasers

In 1961, Sergei Rautian and Igor Sobelman introduced optical quantum generators based on organic liquids.

Liquid lasers (LLs) surpass solid-state ones in terms of their specific power and energy because at practically the same concentration of active particles they allow effective cooling of the active substance by bleeding through the resonator and a heat exchanger.

Originally, compositions of trivalent ions of rare-earth elements with organic molecules were used in liquid lasers. Such compositions are called chelates. However, the huge absorption of pump radiation (up to 10^4 m^{-1}!) meant that inversion could be created only in a 1 mm layer. Therefore, at the present time, two types of liquid lasers have practical significance.

---

[3]Jores Alfiorov (1930–2019) was a Soviet physicist and academician. His main publications focused on investigation of heterojunctions and creation of devices based on them. He discovered the "superinjection" phenomenon in heterostructures and devised the principles of heterostructure application in semiconductor electronics. In 1993–1994, for the first time, creation of a heterolaser based on structures with quantum points ("artificial atoms") was realized. In 1995, Alfiorov and colleagues demonstrated the first injection heterolaser using quantum points, operating in the continuous mode at room temperature. In 2000 he received the Nobel Prize in Physics.

   The first type has no major differences from solid-state lasers based on doped glass. In their properties, such lasers occupy an intermediate position between solid-state neodymium lasers based on glass and lasers based on crystals.

   Investigation of chelates showed that radiation absorption is connected with hydrogen and deuterium ions. This eliminated the application of organic solvents. Moreover, the solvent should be transparent at a working wavelength. Therefore, nonorganic liquid luminophores with addition of rare-earth element ions (mainly Nd^{3+}) are used in liquid lasers. The luminophore represents a mixture of oxychloride (for instance, $POCl_3$) with an acid (for example, $POCl_3$).

   In a liquid $POCl_3$–$SnCl_4$: Nd^{3+} luminophore the neodymium ion is surrounded by eight oxygen atoms and has several wide absorption bands (at 0.58, 0.74, 0.8, and 0.9 µm), which are used for pumping according to a four-level scheme. In a general way, the energy diagram does not differ from the diagram shown in Fig. 7.18. Generation is obtained at a wavelength of 1.051 µm. Pumping is provided with the help of xenon lamps. The lower operating level is elevated above the main level so far that it is practically unoccupied. For this reason, the generation threshold can be easily overcome and the laser has rather high efficiency (up to 5%). The generation energy in the pulsed mode is greater than 1 kJ. The output power in the continuous mode is about 1 kW.

   A schematic image of a liquid laser is shown in Fig. 7.25. The laser cell usually has a length of about 15 cm and an internal diameter of ≈6 mm. Mirrors are either produced directly on the window edges or located outside.

   One of the problems is connected with thermal liquid expansion. In the pulsed mode, the shock heat is able to destroy the cell. Compensation volumes at both tube ends partially solve this problem. In the continuous mode, optical irregularity of the

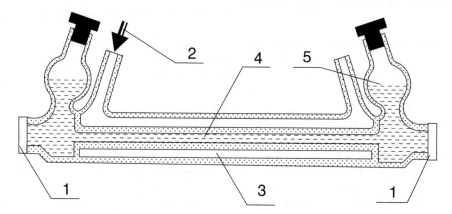

Fig. 7.25 Scheme of a liquid laser: *1* mirror, *2* water input, *3* water jacket, *4* active liquid, *5* compensation volumes. The length of the laser element is about 15 cm. The internal tube diameter is about 6 mm. To compensate for heat expansion of the operation liquid, compensation volumes are located on the left and right. The cell is surrounded by a case in which the coolant circulates. In lasers that operate continuously the operating liquid is steadily bled in a closed cycle through an external radiator

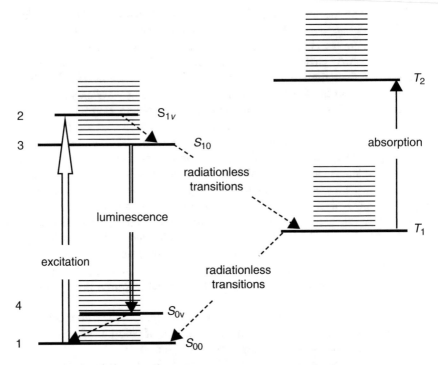

Fig. 7.26 Energy diagram of organic molecules: S singlet levels, T triplet levels. The short lines are vibrational sublevels. At room temperature, levels S_{0v} are practically empty. Radiation is intensively absorbed at the transitions $S_{00} \rightarrow S_{1v}$. During time $10^{-11} - 10^{-13}$ s, molecules are transmitted to states S_{10} and return to their initial state owing to luminescence and radiationless transitions. Between levels S_{10} and S_{1v} there is a population inversion

medium is a consequence of operating liquid heating. Therefore, additional water cooling and continuous bleeding of the operating liquid are used.

The key application areas for this type of liquid laser are laser technologies, medicine, pumping of other lasers, etc.

In a laser of the second type, solutions of organic colorants are used. Usually, water, alcohol, and benzol components are used as solvents.

The operating transitions are connected with electronic–vibrational bands. A typical structure of the energy levels of an organic colorant is shown in Fig. 7.26. The energy bands consist of vibrational and rotational sublevels. The distance between sublevels is about 0.1 eV.

A laser based on a colorant operates according to a four-level scheme (see Fig. 7.26). The pumping source transfers molecules to one of the sublevels of the upper band. Generation arises between the electron level of this band and one sublevel of the lower band.

Some difficulties arise as a result of energy absorption with participation of triplet states (shown in the right part of Fig. 7.26). The influence of these states can be

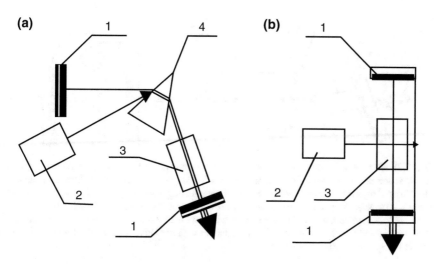

Fig. 7.27 Schematic image of a laser based on a colorant with longitudinal pumping (**a**) or with transverse pumping (**b**): *1* mirrors, *2* lasers, *3* dish with colorant, *4* prism. Longitudinal pumping provides greater uniformity but requires additional optical elements—for instance, a prism. A prism may also play the role of a frequency-selective element

eliminated either by using intensive pulse pumping or by adding an impurity to destroy the triplet states.

Pumping is provided by pulsed gas discharge lamps or by other lasers. Lasers based on colorants with pumping by lamps generate pulses with a duration of about a microsecond. The pulse lamp and a dish are placed inside the lighter. The frequency of the pulse repetition can reach 100 Hz at an average power measured in hundreds of watts.

Laser pumping allows realization of the continuous generation mode. For pumping, powerful ion lasers based on Ar and Kr are used. In the pulsed mode, the pumping is provided by a laser based on nitrogen. In this case, a laser based on a colorant generates pulses with a duration of 1–10 ns and peak power ranging from units to tens of kilowatts. Application of another laser for pumping allows achievement of greater power. Pumping is provided according to transverse or longitudinal schemes (see Fig. 7.27).

To make the radiation spectrum narrower, an additional spectral-selective element is introduced into the resonator. A prism (see Fig. 7.27a), a diffraction lattice, and/or a Fabry–Pérot resonator (Fig. 7.28) are used as such elements. These measures allow us to obtain a relative width of the generation line of the order of 10^{-7}. Lasers based on colorants allow continuous tuning of the operating wavelength from 0.3 to 1.2 μm. Accurate tuning is performed by rotating the selective element, while rough tuning is performed by changing the colorant. For tuning within the mentioned range, a set of approximately 30 colorants is required.

A special class of lasers based on colorants is formed by generators with distributed feedback. We mentioned such lasers earlier. However, in a liquid

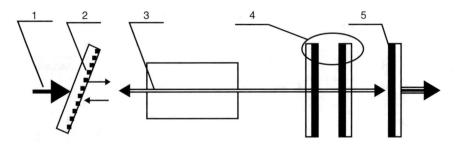

Fig. 7.28 Scheme of a laser based on a colorant with elements of frequency selection: *1* pumping, *2* diffraction lattice, *3* dish with colorant, *4* Fabry–Pérot resonator, *5* mirror. A diffraction lattice is used instead of using mirrors. For the specific orientation angle with respect to the longitudinal axis, there is a wavelength (frequency) on which the direction of the first order of diffraction coincides with the longitudinal axis. Additional selection is provided by the Fabry–Pérot passive resonator. This structure allows us to obtain coherent radiation with a relative bandwidth of about 10^{-7} and smooth frequency tuning of about 10%

medium the spatial–periodic structure is formed directly in the operating state owing to interference of the pumping beams. The interference pattern gives birth to periodic variation of the refraction index. By changing the angle between the pumping wave beams, it is easy to provide smooth tuning of the generation frequency.

Pulsed lasers based on colorants can generate pulses with a duration of 10^{-14} s using mode synchronization (see Sect. 5.10).

Lasers based on colorants are used for superaccurate spectroscopic investigations. This scientific direction is known as laser spectroscopy. Exact frequency tuning is required also for resonance control of chemical reactions. Such processes form the basis of laser photochemistry.[4]

7.6 Quantum Generators Based on Free Electrons

The free-electron laser (FEL) is a device that converts the kinetic energy of free (unbound) electrons into electromagnetic radiation. It is closely related to a general class of devices called traveling-wave tube amplifiers and oscillators and, in its present embodiment, it employs a spatially alternating static magnetic field. It was first proposed in 1951 by Hans Motz and collaborators. A microwave version, known as the ubitron, was operated in 1960 by Ronald Phillips.

---

[4]Laser chemistry deals with chemical transformations under the action of laser radiation. In laser photochemistry, resonance phenomena are used. For instance, under radiation from an ArF excimer laser ($\lambda = 193$ nm), decomposition of methylamine produces CH_3H_2, HC, and CH_3C, which, under normal conditions, are absent from the final reaction product. Laser thermochemistry is connected only with reagent heating. The heating itself has a resonance character, which increases its effectiveness. Thus, for example, we can noticeably increase the ammonia outlet at its synthesis from nitrogen and water steam.

There was a revival of interest in devices capable of operating in the visible ultraviolet and infrared regions of the spectrum. Free-electron lasers were invented by John Madey at Stanford University in 1976.

In generators based on free electrons, the flow of free relativistic electrons is used as an active medium beam. The electrons perform cross vibrations with respect to the direction of forward movement under the action of the external electric or magnetic field. As a result of the Doppler effect, the radiation frequency exceeds the frequency of these vibrations many times:

$$\omega = \frac{\omega_0}{1 - \frac{v}{c}\cos(\varphi)}.$$

Here, ω_0 is the vibrational frequency of electrons, $v_{\parallel}$ is the electron velocity in the direction of wave propagation, c is the light speed, and φ is the angle between the radiation direction and the particle forward movement direction.

Several approaches are used to excite electron periodic vibrations in the plane perpendicular to the wave propagation direction. In the most short-wave part of the spectrum, the spatial–periodic static magnetic field of a powerful pumping wave is used.

The circuit of a ubitron is shown in Fig. 7.29. The periodic magnetic field is created by an undulator—a spatial structure of permanent magnets. Each electron passed through the undulator radiates a wave train whose duration depends upon the device's length and the longitudinal velocity of the electron beam. The packing frequency is determined by the above-mentioned formula at $\omega_0 = \frac{\pi v}{d}$. Each electron radiates independently; therefore, the beam radiation is noncoherent (spontaneous). Nevertheless, if we introduce an external electromagnetic wave into the undulator

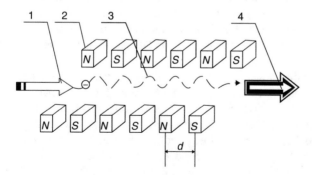

Fig. 7.29 Scheme of a laser based on free electrons with an undulator: *1* electron beam, *2* magnets, *3* electron trajectory, *4* output radiation. A relativistic electron beam with a speed close to light speed moves between two rows of magnets in a spatial–periodic field. The frequency of the particles' oscillations is $\omega_0 = \pi v/d$. An external electromagnetic wave is amplified by induced coherent radiation. If the undulator is placed inside the resonator, one may obtain continuous oscillations. The radiation frequency of this generator extends from the millimeter wavelength range to the ultraviolet range

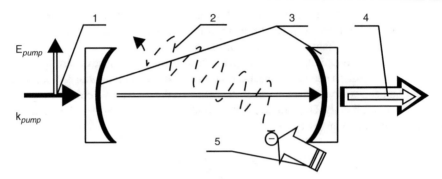

Fig. 7.30 Laser based on free electrons (laser Compton scattering) with pumping by the field of a powerful electromagnetic wave: *1* pumping, *2* trajectory of electrons, *3* Fabry–Pérot resonator, *4* output radiation, *5* electron beam. The powerful pumping wave causes periodic electron movement in the space. The energy part of the longitudinal forward movement is transformed into transverse movement equivalent to the current source, which is periodic in time and moves along the longitudinal axis of the beam. When the threshold is overcome, continuous oscillations arise in the resonator. It is possible to obtain generation from the millimeter range to the ultraviolet range

space, induced coherent radiation arises and the initial wave will be amplified. In the presence of feedback we can obtain continuous oscillations.

Laser Compton scattering (LCS) (Fig. 7.30) is the closest to a traditional laser in terms of its structure. The pumping wave causes periodic motion of electrons in the space. The rest of the process happens as in the ubitron. At fulfillment of the threshold conditions, continuous oscillations arise in the resonator. From the point of view of quantum mechanics, laser Compton scattering processes are connected with the inverse Compton effect—photon dispersion by relativistic electrons.

Quantum generators that use the braking radiation of electrons rotating in a homogeneous magnetic field have been successfully developed. Here, predominance of radiation over absorption becomes possible because of nonequidistance of the Landau levels (see Sect. 4.5). The generation frequency is limited to the submillimeter wavelength range because of limitations on the achievable value of the magnetic field intensity.

In all described devices, in each elementary act, the electron radiates the quantum $\hbar\omega$, whose energy is much less than the initial energy of the particle. Therefore, in the process of interaction with the wave, so many quanta are emitted (up to 10^8) that it is possible to account for the process using a classical description, not just a quantum one. Lasers based on free electrons are related to the well known traveling-wave tubes and klystrons. Forced radiation, according to the classical description, transfers to the self-regulating process of grouping of the electron beam into clots under the action of an inoculating wave, and further amplification occurs as a result of coherent radiation of the electromagnetic wave by these clots.

The frequency range covered by lasers based on free electrons extends from millimeter waves to the ultraviolet range.

A detailed consideration of lasers based on free electrons is beyond the scope of this book. It would also be inappropriate to include it here, because the processes that occur in lasers based on free electrons can be accounted for by a classical description, which is included in other disciplines involving electronic profiles. In including the material that is provided in this book, the authors have simply tried to show how unexpectedly close these things are to each other although they might seem to be very far apart at the first glance. The reason lies in the uniformity of the surrounding world and in the limitations of our knowledge about it.

7.7 Atomic Standards of Frequency and Time

We have already mentioned the extremely high stability of molecular generator oscillations. This property is caused by the high frequency stability of the quantum transition. Devices for obtaining electromagnetic oscillations with a time-stable frequency in which stimulated quantum transitions are used are called quantum frequency standards. The relative root-mean-square deviation as well as the relative error the recovered real frequency value can achieve 10^{-14} (-14 - the exponent!) in these devices[5]. In 1964 the International Committee on Measures and Weights accepted the atomic frequency standard of cesium atoms as the standard. The second, equal to 9192631770 periods of the resonance transition of the ^{133}Cs isotope, is accepted as the unit of time.

For most applications, high-stability frequency oscillations, which lie in the radiofrequency range, are required, while oscillations with a frequency of 1 Hz are necessary for the time standard. Therefore, quantum frequency standards should be added by an electronic circuit for conversion of the oscillation frequency. Thus, the quantum device plays the role of a frequency reference.

Active and passive quantum references are distinguished. An active reference contains a quantum generator. In a passive reference a spectral line is used for automatic control of the frequency of the auxiliary oscillator. Structural diagrams of two types of frequency standards are shown in Fig. 7.31.

In an active reference a hydrogen generator is used (see Sect. 7.2) or a generator based on rubidium vapors with optical pumping is used. In the UHF range the natural stability of a quantum generator is used (see Sects. 5.4 and 5.8). If a laser is used as the generator, we need to apply special measures for stabilization of its frequency. For this a cell containing a gas at low pressure is placed into the resonator. Laser radiation saturates the transition of the auxiliary gas. As a result, a narrow dip (see Sect. 2.12) appears in the nonuniformly widened absorption line, which is used as a reference in a phase-locked loop system.[6] A structural diagram of an optical frequency standard with an active reference is shown in Fig. 7.32. The best results are

---

[5]This is equal to $\approx 10^{-9}$ s per day or ≈ 0.35 µs per year.

[6]For frequency stabilization we can use saturation of the operating gas, which is manifested in reduction of the output power with tuning on the center of the amplification line (the Lamb dip). However, at operating pressure this dip is so wide that this phenomenon cannot be used for development of frequency standards.

Fig. 7.31 Structures of active (**a**) and passive (**b**) frequency standards. In an active standard a quantum generator is used, whose frequency is transmitted in the required range. In a passive standard the atomic reference is used as the frequency discriminator of the electronic circuit

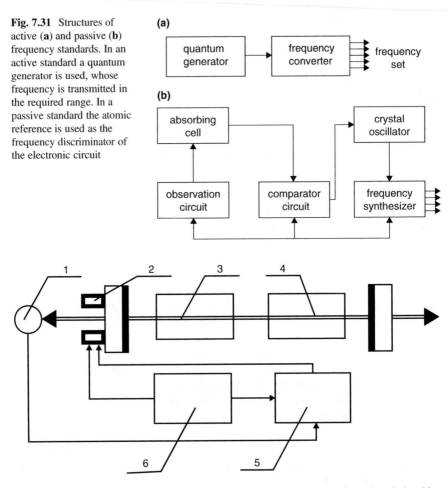

Fig. 7.32 Optical frequency standard: *1* optical detector, *2* piezo elements for tuning, *3* absorbing cell, *4* active medium, *5* synchronous detector, *6* low-frequency oscillator. A cell with a substance that has a nonuniformly widened absorption line is placed inside a resonator. The saturation effect leads to a narrow dip in the absorption line (see Sect. 2.12). An automatic control system tracks the center of the dip

obtained with application of a helium–neon laser ($\lambda = 3.39$ μm) and a CH_4 cell. In the optical range at a wavelength of 0.6328 μm, a cell based on iodine $^{127}I_2$.vapors is used. However, optical frequency standards are essentially more complicated than quantum frequency standards using the radiofrequency range.

In the references the quantum transition operates as the frequency discriminator. In the UHF range, magnetic dipole transitions between levels of an ultrathin structure are used. The ultrathin level splitting results from interaction of the magnetic moments of electrons and the nucleus. The distance between sublevels of the

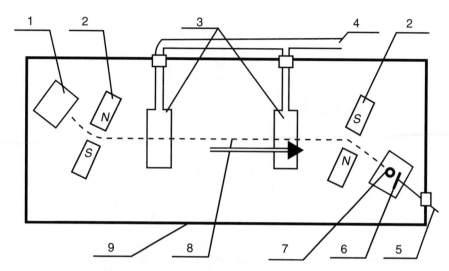

Fig. 7.33 Passive cesium frequency standard: *1* cesium oven, *2* sorting magnetic system, *3* pair of resonators, *4* wave guide input, *5* output current, *6* ion collector, *7* ionizer, *8* atomic beam, *9* vacuum chamber. The atom source is a cavity filled with liquid cesium at a temperature of about 100 °C. Atoms pass through a narrow channel to an atomic-molecular tube. Atom sorting is provided by two magnets, and focusing is provided by the uniform field in the center. A glowing tungsten wire ionizes the atoms. The collector current is proportional to the particle flow density

ultrathin structure corresponds to the radiofrequency range. For example, in the hydrogen atom there are transitions on a frequency of about 1420.205[7] MHz.

Hydrogen, thallium, and alkaline metals are suitable for frequency references. The best results are obtained with application of cesium, with a frequency of about 9.19263177 GHz, and rubidium (^{87}Rb), with a frequency of about 6.834682614 GHz. The device shown in Fig. 7.33 detects variations in the atom's magnetic moments at its excitation by an electromagnetic field. Alternate fields are created by two cavities. At accurate phase adjustment the joint action of two in-phase resonators looks like the effect observed with application of one long resonator. However, in a configuration with two resonators we can reduce the effective width of the spectral line, since at the field frequency deviation from the resonance value an interference phenomenon arises. Interference in the space causes fast ripples in the transmission frequency characteristic of the device.

A beam of cesium atoms goes out from the thermostat at a temperature of about 100 °C through a nozzle, which gives the beam a band shape with a thickness of about 0.5 mm. The cesium flow is approximately 10^{-11} g/s. Atoms pass through a nonuniform magnetic field formed between the magnetic poles; atoms with the appropriate magnetic moment are deflected in the direction of the device's axis. After that the particles pass the first resonator in which a high-frequency magnetic field changes the populations of the energy levels. Having traveled 1 m, the atoms

---

[7]The exact value is 1,420,405,751,768 Hz.

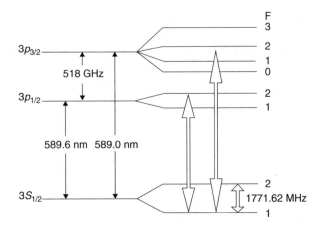

Fig. 7.34 Energy diagram of sodium. The ultrathin structure is shown. The sublevels of the ultrathin structure are designated by the quantum number F, which corresponds to the total magnetic moment of the atoms' electrons and nuclei. Optical transitions occur between level $3S_{1/2}F = 1$ and levels $3P_{1/2}F = 2$ and $3P_{3/2}F = 2$. The transition frequencies between sublevels fall in the radiofrequency range and are used as a frequency reference

appear in the second resonator and then again appear in the nonuniform magnetic field. Atoms that perform the stimulated transition are now deflected in the opposite direction. They are focused on the detector, which contains a heated tungsten wire (ionizer). Cesium atoms touching the wire give back an electron and, in the form of ions, are picked up by the collector. In the circuit between the ionizer and the collector, a current flows, which is proportional to the intensity of the atomic beam and hence provides a measure of the number of atoms performing the stimulated quantum transition. This means that the current depends upon the field frequency detuning with respect to the center of the absorption line.

Typical samples of cesium standards are equipped with a frequency synthesizer device to ensure that the standard frequency at the output is related by a factor of 10—for instance, 100 kHz, 1 MHz. Test results show that the atomic and molecular resonances are reproduced in a frequency with accuracy of 10^{-14}. Secondary cesium standards manufactured industrially are as accurate as the primary standard.

In atomic standards with optical pumping, magnetic dipole transitions between sublevels of the ultrathin structure are also used. For optical frequency standards, atoms of alkaline elements are most suitable. As an example, Fig. 7.34 shows the energy diagram of natrium.[8] This quantum reference with optical pumping acts on the basis of a three-level scheme. Optical radiation causes transitions from level $3S_{1/2}F = 1$ to levels $3S_{1/2}F = 2$ and $3P_{3/2}F = 1$. As a result of the saturation effect, the population of the lower level decreases (see Sect. 2.12). The population of level $3S_{1/2}F = 2$ remains practically unchanged because particles leaving the upper levels are distributed over a large number of lower states. Radiation with a frequency of

---

[8]The quantum number F describes the interaction of electron and nuclear magnetic moments.

Fig. 7.35 Structural diagrams of a reference based on rubidium vapor (**a**) and one of the variants of a frequency standard (**b**): *1* light sources, *2* optical filter, *3* ultrahigh-frequency (UHF) resonator, *4* optical detector, *5* output signal, *6* UHF oscillator, *7* operating cell. A gas discharge tube containing vapor of the isotope ^{87}Rb is used as a light source. A filter with vapor of the isotope ^{85}Rb passes only frequencies of transitions between level $3S_{1/2}F = 13$ and levels $3P_{1/2}F = 2$ and $3P_{3/2}F = 2$. An operating cell containing ^{87}Rb vapor is placed in the UHF resonator. A photodetector detects the absorption variation under the action of the UHF field

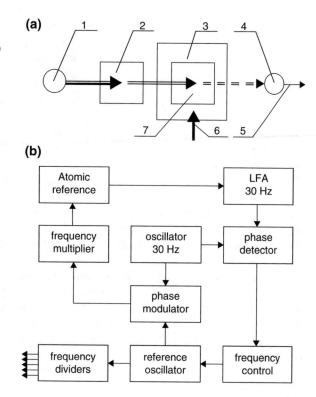

about 1771.62 MHz changes the population of the lower level. This is a result of saturation of the transition $3S_{1/2}F = 1 \leftrightarrow 3S_{1/2}F = 2$. Changing of populations is demonstrated in absorption variations in optical transitions. According to Eq. (2.41) the variation in the level population depends upon the radiation frequency offset with respect to the transition frequency. Measuring the absorption in the optical range, we can construct a frequency discriminator in the UHF range for inclusion in the structure of an automatic frequency control system.

The real reference for the frequency standard is based on rubidium vapor. Its structural diagram is shown in Fig. 7.35. Pumping radiation from a gas discharge lamp using ^{87}Rb vapor, before arriving in the operating area, passes through a filter of ^{85}Rb vapor. This filter rejects unwanted radiation, which causes transitions from level $3P_{3/2}F = 2$. To increase the electromagnetic field intensity, the operating cell is placed in a UHF resonator. The passing radiation is registered by a photoreceiver. To form an error signal in the electronic circuit, we can use the phase modulation of UHF oscillation (see Fig. 7.35b).

Table 7.4 shows the characteristics of atomic frequency standards.

Atomic frequency standards form the basis of International Atomic Time (TAI), a high-precision atomic coordinate time standard based on the notional passage of proper time on the Earth's surface. It is the basis for Coordinated Universal Time (UTC), which is used for civil timekeeping all over the Earth's surface, and for

Table 7.4 Characteristics of atomic frequency standards

Characteristic	Type of standard		
	Hydrogen generator	Cesium standard	Rubidium standard
Frequency accuracy	$1 \cdot 10^{-12}$	$1 \cdot 10^{-13}$	$1 \cdot 10^{-10}$
Relative error of reproduction of the real frequency value	$5 \cdot 10^{-14}$	$5 \cdot 10^{-14}$	$1 \cdot 10^{-12}$
Root-mean-square deviation of the relative frequency per second	$5 \cdot 10^{-13}$	$1 \cdot 10^{-12}$	$1 \cdot 10^{-12}$
Root-mean-square deviation of the relative frequency per 24 hours	$5 \cdot 10^{-15}$	$5 \cdot 10^{-15}$	$1 \cdot 10^{-13}$

Terrestrial Time, which is used for astronomical calculations. Since June 30, 2012, when the last leap second was added, TAI has been exactly 35 s ahead of UTC. This difference results from the initial difference of 10 s at the start of 1972, plus 25 leap seconds in UTC since 1972.

Time coordinates on TAI scales are conventionally specified using traditional means of specifying days, carried over from nonuniform time standards based on the rotation of the Earth. Specifically, both Julian dates and the Gregorian calendar are used. TAI in this form was synchronized with Universal Time at the beginning of 1958, but the two have drifted apart ever since, because of the changing motion of the Earth.

Appendixes

Appendix 1: Fundamental Constants

Table 1 lists some fundamental constants

Appendix 2: Relations Between Physical Quantities' Values in Different Unit Systems

Several sources have been used to create Table 2. Table 3 lists some useful relations.[1]

Appendix 3: Designation of Planes and Directions in a Crystal

Crystal anisotropy requires definition of a definite system of plane and direction designations in a crystal.

The system of Miller indices is accepted for plane designation. It is based on the coordinate system, whose axes coincide with the edges of the elementary cubic lattice (x: a; y: b; z: c). The plane position is specified by three numbers A, B, C, corresponding to the coordinates of the points of intersection with the axes (Fig. 1). Indices l, m, n result from reduction of the ratios of inverse quantities $\frac{a}{A}$, $\frac{b}{B}$, $\frac{c}{C}$ to the least mutual denominator, i.e., $l : m : n = \frac{a}{A} : \frac{b}{B} : \frac{c}{C}$.

---

[1]Rounded numerical values are presented.

© Springer Nature Switzerland AG 2020
V. V. Shtykov, S. M. Smolskiy, *Introduction to Quantum Electronics and Nonlinear Optics*, https://doi.org/10.1007/978-3-030-37614-7

Table 1 Fundamental constants in quantum electronics

Name	Designation or formula	Numerical value	Multiplier	Dimension
Light speed	c	2.99792458	10^8	m/s
Electron charge	e	1.60217653	10^{-19}	C
Electron mass	m_e	9.109534	10^{-31}	kg
Unified atomic mass unit		1.660539	10^{-27}	kg
Proton mass	M_p	1.6726485	10^{-27}	kg
Neutron mass	M_n	1.6749543	10^{-27}	kg
Planck constant	h	6.626176	10^{-34}	J $\cdot$ s
	$\hbar = h/2\pi$	1.0545887		
Bohr radius	$a_0 = \frac{4\pi\varepsilon_0 \hbar^2}{e^2 m_e}$	5.2917706	10^{-11}	m
Bohr magneton	$\mu_B = \frac{e\hbar}{2m_e}$	9.274078	10^{-24}	J $\cdot$ T
Nuclear magneton	$\mu_n = \frac{e\hbar}{2M_p}$	5.050824	10^{-27}	J $\cdot$ T
Ratio charge/mass of an electron	e/m_e	1.7588047	10^{11}	C/kg
Gyromagnetic relation for an electron	$\gamma = e/m_e$	1.7588047	10^{11}	1/(s T)
	$\gamma/2\pi$	2.799225	10^{10}	Hz/T
Gyromagnetic relation for a proton in water	γ_p	2.6751301	10^8	1/(s T)
	$\gamma_p/2\pi$	4.257602	10^7	Hz/T
Boltzmann constant	k_B	1.380662	10^{-23}	J/K

Plane indices are enclosed in parentheses. Figure 2 shows some main planes of a cubic crystal ($a = b = c$).

Direction indices in the crystal represent a set of least numbers l, m, n, whose ratio to each other is equal to the ratio of vector projection, which is parallel to the specified direction. Direction indices are enclosed in square brackets. The perpendicular plane corresponds to each direction. In cubic crystals, numbers designating the direction coincide with the indices of this plane. Negative values in the index are designated by an upper line. In crystals of lower symmetry, this regulation is not fulfilled.

Some directions in a cubic crystal are presented in Fig. 3.

Appendix 4: Point Groups of Symmetry

In crystallography, all groups of symmetry can be divided into six large classes or systems. A full list of all 32 point groups of symmetry is given in Table 4, as are crystal examples of different point groups.

In the first column the system name is shown. In the second column the abbreviated international designation of the group is presented (the generative symmetry elements are given). The third column lists crystal examples concerning the specified point group.

Table 2 Units of quantities in quantum electronics

Name of quantity	Unit system			
	SI	ESU	EMU	CGS
Main units				
Length	1 m	10^2 cm	10^2 cm	10^2 cm
Mass	1 kg	10^3 g	10^3 g	10^3 g
Time	1 s	1 s	1 s	1 s
Current	1 A	$3 \cdot 10^9$	10^{-1}	$3 \cdot 10^9$
Mechanical units				
Speed	1 m/s	10^2 cm/s	10^2 cm/s	10^2 cm/s
Acceleration	1 m/s^2	10^2 cm/s^2	10^2 cm/s^2	10^2 cm/s^2
Energy and work	1 J	10^7 erg	10^7 erg	10^7 erg
Power	1 W	10^7 erg/s	10^7 erg/s	10^7 erg/s
Force	1 N	10^5 dyn	10^5 dyn	10^5 dyn
Pressure	1 Pa	0.1 Ba	0.1 Ba	0.1 Ba
Electric units				
Charge quantity	1 C	$3 \cdot 10^9$	10^{-1}	$3 \cdot 10^9$
Difference of potentials	1 V	1/300	10^8	1/300
Electric capacity	1 F	$9 \cdot 10^{11}$	10^{-9}	$9 \cdot 10^{11}$
Electric resistance	1 Ω	$1/9 \cdot 10^{-11}$	10^{-9}	$1/9 \cdot 10^{-11}$
Dielectric permeability	1 F/m	$36\pi \cdot 10^9$	$4\pi \cdot 10^{-11}$	$36\pi \cdot 10^9$
Electric field intensity	1 V/m	$1/3 \cdot 10^{-4}$	10^6	$1/3 \cdot 10^{-4}$
Electric inductance	1 C/m^2	$12\pi \cdot 10^5$	$4\pi \cdot 10^{-4}$	$12\pi \cdot 10^5$
Magnetic units				
Magnetic field intensity	1 A/m	$12\pi \cdot 10^7$	$4\pi \cdot 10^{-3}$ Oe	$4\pi \cdot 10^{-3}$ Oe
Magnetic induction	1 T	$1/3 \cdot 10^{-6}$	10^4 G	10^4 G
Magnetic flow	1 Wb	1/300	10^8 Mx	10^8 μs
Magnetic moment	1 A/m^2	$3 \cdot 10^{13}$	10^3	10^3
Induction	1 H	$1/9 \cdot 10^{-11}$	10^9 cm	10^9 cm
Magnetic permeability	1 H/m	$1/36\pi \cdot 10^{-13}$	$1/4\pi \cdot 10^7$	$1/4\pi \cdot 10^7$

Ba barye [equal to 1 dyne per square centimeter], *CGS* centimeter–gram–second, *EMU* electromagnetic unit, *ESU* electrostatic unit, *Mx* Maxwell [the unit of flow], *SI* International System of Units

Table 3 Some useful relations in different systems of units

Initial unit	Numerical value
1 year	$3.1557 \cdot 10^7$ s
1 kg-force	$9.81 \cdot 10^5$ N
1 atmosphere	$1.0133 \cdot 10^5$ Pa
1 mm Hg	$1.333 \cdot 10^2$ Pa
1 eV	$1.60 \cdot 10^{-19}$ J
1 eV	1.24 μm
1 eV	$2.42 \cdot 10^{14}$ Hz
1 eV	$5.93 \cdot 10^5$ m/c[1] (the electron speed is presented)
1 eV	$11.6 \cdot 10^3$ K
1 km mass	9.81 J
1 calorie	$3.60 \cdot 10^{-2}$ J
1 horsepower	735.5 W

Fig. 1 The plane position is specified by three numbers A, B, C corresponding to the coordinates of the points of intersection with the axes

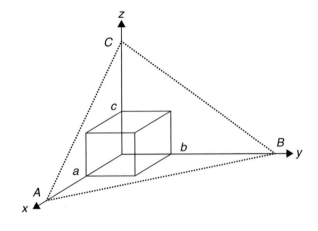

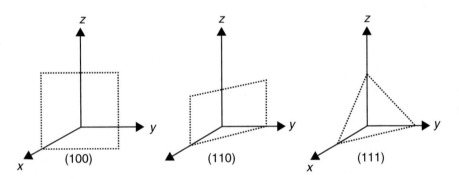

Fig. 2 Some main planes of the cubic crystal ($a = b = c$)

Fig. 3 Some directions in the cubic crystal

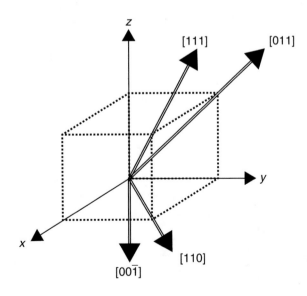

Table 4 The full list of all 32 point groups of symmetry

System	Designation of point group	Crystal examples
Triclinic	1	HgK, $(NH_3)_2PdI_2$
	$\bar{1}$	Al_2SiO_5; $CuSO_3 \times 5H_2O$
Monoclinic	2	$LiSO_4 \times H_2O$
		$NaKC_4H_4O_6 \times 4H_2O$ ($-18°C < t < 24°C$)
	m	As_2Te; $SrSO_4$
	2/m	As_4S_4; $C_{10}H_8$
Orthorhombic	222	$NaKC_4H_4O_6 \times 4H_2O$ ($t > 24°C; t < -18°C$)
	mm2	$BaNaNb_5O_{15}$; $BaTiO_3$ ($t < 0°C$)
	mmm	$PbZnO_3$; S
Tetragonal	4	$Cr_2KNb_5O_{15}$
	422	ZnP_2; CdP_2; $CdAs_2$
	4/m	$CaWO_4$
	4mm	$BaTiO_3$ [$t = (5 - 20)°C$]; $HgF \times H_2O$
	4/mmm	$ZrSiO_4$; Sb; TiO_2; SnO_2
	$\bar{4}$	$C(CH_2OH)$
	$\bar{4}2m$	$NH_4H_2PO_4$; KH_2PO_4
Trigonal	3	TeO_2; BiI_3
	32	HgS; SiO_2; ($t < 576°C$); $K_2S_2O_6$
	3m	$LiNbO_3$; $BiFeO_3$; Ag_3AsS_4
	$\bar{3}$	$Cu_6Si_6O_{12} \times 6H_2O$; B_2SiO_4; $NaTiO_3$
	$\bar{3}m$	Bi_2Te_3; $CaCO_3$; Al_2O_3
Hexagonal	6	$LiKSO_4$; $LiIO_3$
	$\bar{6}$	$Ag_2(PO_4)H$; GaSe
	$\bar{6}m2$	$BaTiGe_3O_9$; $CsNO_3$
	622	SiO_2 ($575°C < t < 870°C$)
	6/m	$CaF(PO_4)_3$; $Ba(ClO_4)_2 \times 3H_2O$
	6mm	CdS; CdSe; ZnO
	6/mmm	CrSb; MnTe; Zn; Cd
Cubic	23	$NaClO_3$; $Na_3SbO \times 9H_2O$
	$m\bar{3}$	FeS_2; Mn_2O_3; $CrK(SO_4)_2 \times 12H_2O$
	$\bar{4}3m$	GaAs; InSb; ZnS
	432	$CsPF_6$; KPF_6; NH_4Cl
	$m\bar{3}m$	PbS; PbTe; CaF_2; NaF; Si; Ni

Appendix 5: Tensors of Magnetic Susceptibility

Table 5 presents linear susceptibility tensors for all crystal classes. Symmetric tensors of the second rank can be reduced to the main axes. Such a transformation ensures that all class tensors of crystals are diagonal.

Table 5 The view of linear susceptibility tensors for all crystal classes

System	Point group	Optical properties of crystals	Tensor of linear susceptibility	Crystal examples
Triclinic	1 $\bar{1}$	Biaxial	$\begin{vmatrix} \dot{\chi}_{xx} & \dot{\chi}_{xy} & \dot{\chi}_{xz} \\ \dot{\chi}_{yx} & \dot{\chi}_{yy} & \dot{\chi}_{yz} \\ \dot{\chi}_{zx} & \dot{\chi}_{zy} & \dot{\chi}_{zz} \end{vmatrix}$	Mica—point group $\bar{1}$ $n_x = 1552$ $n_y = 1582$ $n_z = 1558$
Monoclinic	2 m 2/m		$\begin{vmatrix} \dot{\chi}_{xx} & 0 & \dot{\chi}_{xz} \\ 0 & \dot{\chi}_{yy} & 0 \\ \dot{\chi}_{zx} & 0 & \dot{\chi}_{zz} \end{vmatrix}$	
Orthorhombic	222 2mm mmm		$\begin{vmatrix} \dot{\chi}_{xx} & 0 & 0 \\ 0 & \dot{\chi}_{yy} & 0 \\ 0 & 0 & \dot{\chi}_{zz} \end{vmatrix}$	
Tetragonal	4 $\bar{4}$ 4/m 422 4mm $\bar{4}$2m 4/ mmm	Single axis	$\begin{vmatrix} \dot{\chi}_{xx} & 0 & 0 \\ 0 & \dot{\chi}_{xx} & 0 \\ 0 & 0 & \dot{\chi}_{zz} \end{vmatrix}$	*Negative* Rutile (TiO_2)—point group 4/mmm $n_o = 2616$ $n_e = 2903$ Quartz (SiO_2)—point group 32 $n_o = 1544$ $n_e = 1553$
Trigonal	3 $\bar{3}$ 32 $\bar{3}$m $\bar{3}$m		$\begin{vmatrix} \dot{\chi}_{xx} & 0 & 0 \\ 0 & \dot{\chi}_{xx} & 0 \\ 0 & 0 & \dot{\chi}_{zz} \end{vmatrix}$	*Positive* Beryl/emerald ($Al_2O_3 \cdot 6SiO_2 \cdot 3BeO$) $n_o = 1598$ $n_e = 1590$
Hexagonal	6 $\bar{6}$ 6/m 622 6mm $\bar{6}$m2 6/ mmm		$\begin{vmatrix} \dot{\chi}_{xx} & 0 & 0 \\ 0 & \dot{\chi}_{xx} & 0 \\ 0 & 0 & \dot{\chi}_{zz} \end{vmatrix}$	Lithium niobate ($LiNbO_3$)—point group 3m $n_o = 2300$ $n_e = 2208$ $BaTiO_3$—point group 4mm $n_o = 2416$ $n_e = 2364$
Cubic	$\bar{4}$3m 432 m$\bar{3}$ 23 m$\bar{3}$m	Isotropic	$\begin{vmatrix} \dot{\chi}_{xx} & 0 & 0 \\ 0 & \dot{\chi}_{xx} & 0 \\ 0 & 0 & \dot{\chi}_{xx} \end{vmatrix}$	Silicon (Si)—point group m3m $n_x = 3.56$ Gallium arsenide (GaAs)—point group $\bar{4}$3m $n_x = 3.40$

All materials have a frequency dispersion. The given values of refraction indices give only a representation of the quantity order. More detailed information can be found in the reference literature.

Appendix 6: Properties of Some Semiconductor Crystals

In Table 6 the most important characteristics of some semiconductors are given. Parameter values are obtained as a result of data processing from the reference literature.

The second-rank tensor of the inverse effective mass (see Sect. 1.9) can always be reduced to the main axes. In this representation the matrix of its components becomes diagonal. Therefore, in the general case for each type of particle, there are three values of the effective mass. The isoenergy surfaces of germanium and silicon represent ellipsoids, whose rotational axis is oriented along the direction [111]. Therefore, two values of mass coincide, $m_1 = m_2 = m_\perp$, and the third, $m_3 = m_\parallel$. The isoenergy surfaces of other semiconductors presented in Table 6 are almost spherical. For them, some averaged values of the effective mass are given.

Appendix 7: Tensors of Quadratic Susceptibility

Quadratic nonlinearity is described by

$$\dot{P}_{quad}(\omega_3) = \chi(\omega_1, \omega_2, \omega_3)\dot{E}(\omega_1)\dot{E}(\omega_2).$$

The third-rank tensor describing quadratic nonlinearity is nonzero only in crystals that have no inversion center. The tensor forms for all point groups in which it is nonzero are presented in Table 7.

Numerical values of tensor elements correspond to second harmonic generation. Nevertheless, they give accepted accuracy values for nonlinear susceptibility and for other processes.

Table 6 The most important characteristics of some semiconductors

Semiconductor	Width of forbidden band (eV)		Mobility at $T = 300$ K (m²/ (V · s))		Effective mass ($m^* = m/m_e$)		ε
	300 K	0 K	Electrons	Holes	Electrons	Holes	
Ge	0.67	0.75	0.395	0.340	$m^*_\parallel = 1.6$ $m^*_\perp = 0.082$	0.3; 0.16	16
Si	1.1	1.15	0.190	0.0425	$m^*_\parallel = 0.97$ $m^*_\perp = 0.19$	0.5; 0.16	12
InSb	0.16	0.26	7.800	0.075	0.013	0.6	17
InAs	0.33	0.46	3.300	0.046	0.04	0.4	14.5
InP	1.29	1.34	0.460	0.0150	0.07	0.4	14
GaSb	0.67	0.80	0.400	0.140	0.042	0.5	15
GaAs	1.39	1.58	0.850	0.040	0.072	0.5	12.5
GaP	2.24	2.40	0.011	0.0075	–	–	10

Table 7 The tensor form for all point groups, in which it is non-zero

System	Point group	Tensor of quadratic susceptibility	Crystal example $d_{ij}\ 10^{-22}\ \frac{F}{V}$
Triclinic	1	$\begin{vmatrix} d_{11} & d_{12} & d_{13} & d_{14} & d_{15} & d_{16} \\ d_{21} & d_{22} & d_{23} & d_{24} & d_{25} & d_{26} \\ d_{31} & d_{32} & d_{33} & d_{34} & d_{35} & d_{36} \end{vmatrix}$	
Monoclinic	m	$\begin{vmatrix} d_{11} & d_{12} & d_{13} & 0 & 0 & d_{16} \\ d_{21} & d_{22} & d_{23} & 0 & 0 & d_{26} \\ 0 & 0 & 0 & d_{34} & d_{35} & 0 \end{vmatrix}$	
	2	$\begin{vmatrix} 0 & 0 & 0 & d_{14} & d_{15} & 0 \\ 0 & 0 & 0 & d_{24} & d_{25} & 0 \\ d_{31} & d_{32} & d_{33} & 0 & 0 & d_{36} \end{vmatrix}$	
Orthorhombic	mm2	$\begin{vmatrix} 0 & 0 & 0 & 0 & d_{15} & 0 \\ 0 & 0 & 0 & d_{24} & 0 & 0 \\ d_{31} & d_{32} & d_{33} & 0 & 0 & 0 \end{vmatrix}$	
	222	$\begin{vmatrix} 0 & 0 & 0 & d_{14} & 0 & 0 \\ 0 & 0 & 0 & 0 & d_{25} & 0 \\ 0 & 0 & 0 & 0 & 0 & d_{36} \end{vmatrix}$	$\alpha\text{-HIO}_3$ $\lambda = 1.06\ \mu m$ $d_{14} = 0.6$
Tetragonal	4mm	$\begin{vmatrix} 0 & 0 & 0 & 0 & d_{15} & 0 \\ 0 & 0 & 0 & d_{15} & 0 & 0 \\ d_{31} & d_{31} & d_{33} & 0 & 0 & 0 \end{vmatrix}$	BaTiO_3 $\lambda = 1.06\ \mu m$ $d_{33} = -0.78,$ $d_{31} = -2.0,$ $d_{15} = -2.0$
	$\bar{4}$	$\begin{vmatrix} 0 & 0 & 0 & d_{14} & d_{15} & 0 \\ 0 & 0 & 0 & -d_{15} & d_{14} & 0 \\ d_{31} & -d_{31} & 0 & 0 & 0 & d_{36} \end{vmatrix}$	CdGaS $\lambda = 1.06\ \mu m$ $d_{36} = 3.5$
	$\bar{4}2m$	$\begin{vmatrix} 0 & 0 & 0 & d_{14} & 0 & 0 \\ 0 & 0 & 0 & 0 & d_{14} & 0 \\ 0 & 0 & 0 & 0 & 0 & d_{36} \end{vmatrix}$	CdGeAs_2 $\lambda = 10.6\ \mu m$ $d_{36} = 31.0$
	4	$\begin{vmatrix} 0 & 0 & 0 & d_{14} & d_{15} & 0 \\ 0 & 0 & 0 & d_{15} & -d_{14} & 0 \\ d_{31} & d_{31} & d_{33} & 0 & 0 & 0 \end{vmatrix}$	KNbO_3 $\lambda = 1.06\ \mu m$ $|d_{33}| = 2.3,$ $|d_{31}| = 1.3,$ $|d_{15}| = 1.5$
	422	$\begin{vmatrix} 0 & 0 & 0 & d_{14} & 0 & 0 \\ 0 & 0 & 0 & 0 & -d_{14} & 0 \\ 0 & 0 & 0 & 0 & 0 & 0 \end{vmatrix}$	TeO_2 $\lambda = 1.06\ \mu m$ $d_{14} = 0.5$
Trigonal	3	$\begin{vmatrix} d_{11} & d_{12} & d_{13} & d_{14} & d_{15} & d_{16} \\ d_{21} & d_{21} & d_{21} & d_{21} & d_{21} & d_{21} \\ d_{31} & d_{31} & d_{31} & d_{31} & d_{31} & d_{31} \end{vmatrix}$	
	32	$\begin{vmatrix} d_{11} & d_{12} & d_{13} & d_{14} & d_{15} & d_{16} \\ d_{21} & d_{21} & d_{21} & d_{21} & d_{21} & d_{21} \\ d_{31} & d_{31} & d_{31} & d_{31} & d_{31} & d_{31} \end{vmatrix}$	
	3m	$\begin{vmatrix} d_{11} & d_{12} & d_{13} & d_{14} & d_{15} & d_{16} \\ d_{21} & d_{21} & d_{21} & d_{21} & d_{21} & d_{21} \\ d_{31} & d_{31} & d_{31} & d_{31} & d_{31} & d_{31} \end{vmatrix}$	LiNbO_3 $\lambda = 1.06\ \mu m$ $d_{33} = -3,$ $d_{31} = -2.5,$ $d_{22} = 0.36$

(continued)

Table 7 (continued)

System	Point group	Tensor of quadratic susceptibility						Crystal example $d_{ij}\ 10^{-22}\ \frac{F}{V}$
Hexagonal	$\bar{6}m2$	d_{11}	d_{12}	d_{13}	d_{14}	d_{15}	d_{16}	
		d_{21}	d_{21}	d_{21}	d_{21}	d_{21}	d_{21}	
		d_{31}	d_{31}	d_{31}	d_{31}	d_{31}	d_{31}	
	$\bar{6}$	d_{11}	d_{12}	d_{13}	d_{14}	d_{15}	d_{16}	
		d_{21}	d_{21}	d_{21}	d_{21}	d_{21}	d_{21}	
		d_{31}	d_{31}	d_{31}	d_{31}	d_{31}	d_{31}	
	6mm	0	0	0	0	d_{15}	0	CdSe $\lambda = 10.6\ \mu m$ $d_{33} = 5.0$, $d_{31} = -2.5$, $\lvert d_{15}\rvert = 2.6$
		0	0	0	d_{15}	0	0	
		d_{31}	d_{31}	d_{33}	0	0	0	
	6	0	0	0	d_{14}	d_{15}	0	LiIO$_3$ $\lambda = 1.064\ \mu m$ $d_{33} = -0.6$, $d_{31} = -0.6$, $\lvert d_{14}\rvert = 0.27$
		0	0	0	d_{15}	$-d_{14}$	0	
		d_{31}	d_{31}	d_{33}	0	0	0	
	622	0	0	0	d_{14}	0	0	
		0	0	0	0	$-d_{14}$	0	
		0	0	0	0	0	0	
Cubic	$\bar{4}3m$	0	0	0	d_{14}	0	0	GaAs $\lambda = 10.6\ \mu m$ $d_{14} = 12.0$
		0	0	0	0	d_{14}	0	
		0	0	0	0	0	d_{14}	
	23	0	0	0	d_{14}	0	0	
		0	0	0	0	d_{14}	0	
		0	0	0	0	0	d_{14}	

A view of the tensor of linear electro-optical effect **r** can be obtained by transposition of reduced matrices **d**. To get an impression of the value of tensor **r** elements, we can use the equation $r \approx \frac{2d}{\varepsilon_0 n^4}$, for which substantiation is given in Sect. 6.12. Detailed tables can be found in the literature.

Appendix 8: Cubic Nonlinearity in an Isotropic Medium

Cubic nonlinearity is described as

$$\dot{\mathbf{P}}_{cub}(\omega_4) = \chi(\omega_1, \omega_2, \omega_3, \omega_4)\dot{\mathbf{E}}(\omega_1)\dot{\mathbf{E}}(\omega_2)\dot{\mathbf{E}}(\omega_3).$$

The four-rank tensor describing cubic nonlinearity is nonzero in crystals in all symmetry groups. Its elements are symmetric over the three last indices. A view of tensors with this property can be found in the literature for all symmetry groups.

To describe cubic phenomena in an isotropic medium, one number is enough. Its values for some isotropic substances (liquids and glasses) are shown in Table 8. The data were obtained in experiments on four-wave mixing at a wavelength of 0.694 μm. However, they give quite acceptable results in terms of accuracy for other cubic transformations in the optical range.

The Kerr effect is connected with cubic nonlinearity. The four-rank tensor describing this effect is symmetric on the index pair's transposition and the indices in each pair. The form of the tensor can be found in the literature.

In isotropic media the variation of the refraction index is usually written in the form

$$n_e - n_o = K\lambda E^2,$$

where K is the Kerr constant. Its values are presented in Table 9.

Table 8 Values for some isotropic substances (liquids and glasses)

Substance	n	$\chi\ 10^{-34}\ \frac{Fm}{V^2}$
Carbon bisulfide (CS$_2$)	1.612	441.0
Carbon(IV) chloride (CCl$_4$)	1.454	6.2
Fused quartz (SiO$_2$)	1.455	1.5
Yttrium aluminum garnet (YAG)	1.829	7.41
Benzol (C$_6$H$_6$)	1.493	68.9
Glass (LSO)	1.505	2.26
Glass (ED-4)	1.557	2.8
Glass (SF-7)	1.631	9.8
Glass (LaSF-7)	1.910	12.4

Table 9 Kerr constant values

Substance	λ (μm)	n	$K\ 10^{-15}\ \frac{m}{V^2}$
Benzine	0.546	1.503	4.90
	0.633	1.496	4.14
Carbon bisulfide	0.546	1.633	38.8
	0.633	1.619	31.8
	0.694	1.612	28.3
	1.000	1.596	18.4
	1.600	1.582	11.1
Carbon(IV) chloride	0.633	1.456	0.74
	0.546	1.460	0.8.6
Water	0.589		51.0
Nitrotoluene	0.589		1370
Nitrobenzine	0.589		2440

Index

© Springer Nature Switzerland AG 2020
V. V. Shtykov, S. M. Smolskiy, *Introduction to Quantum Electronics
and Nonlinear Optics*, https://doi.org/10.1007/978-3-030-37614-7

Printed in the United States
by Baker & Taylor Publisher Services